Textbook of Nutrition in Health and Disease

Kaveri Chakrabarty • A. S. Chakrabarty

Textbook of Nutrition in Health and Disease

 Springer

Kaveri Chakrabarty
Department of Zoology
Hansraj College, University of Delhi
New Delhi, India

Department of Physiology and Biophysics
School of Medicine, Case Western Reserve
University
Cleveland, OH, USA

A.S. Chakrabarty
Department of Physiology
Maulana Azad Medical College
New Delhi, India

Department of Physiology
Jawaharlal Institute of Post-graduate,
Medical Education and Research
Puducherry, India

Department of Physiology
Vardhman Mahavir Medical College
New Delhi, India

ISBN 978-981-15-0964-3   ISBN 978-981-15-0962-9   (eBook)
https://doi.org/10.1007/978-981-15-0962-9

This Springer imprint is published by the registered company Springer Nature Singapore Pte Ltd.
The registered company address is: 152 Beach Road, #21-01/04 Gateway East, Singapore 189721, Singapore

Dedicated to my Mother
Late Dr Krishna Chakrabarty (MD, AIIMS, New Delhi)
Former Professor of Physiology
Maulana Azad Medical College
New Delhi

Kaveri Chakrabarty

Preface

Human nutrition is a multidisciplinary science, which involves physiology, bio-chemistry, pathology, immunology, food science, zoology, medicine, and other fields. Knowledge of human nutrition is essential for maintenance of health. The book is concise yet a comprehensive account of human nutrition, food, and nutrition-related health problems. Addiction-related health problems, lifestyle-related disorders, social health problems, and poor maintenance of food hygiene and food safety are highlighted because of adverse effects on health. Age-related decline of cognitive functions, nutritional therapies, and additional measures for successful aging is highlighted. Mental disorders are common throughout the world. The role of nutritional therapies for mental disorders is included. Nutritional interventions for cancer and various diseases, lifestyle-related diseases, and addiction-related health problems are discussed. Preventive strategies for food-borne and water-borne diseases along with management and nutritional therapy of diarrheal diseases are reviewed. The applied aspect of human nutrition is emphasized in each chapter. Many illustrations, figures, and tables are designed to amplify the text. The readers can supplement with other recommended books mentioned in the references. We look forward to suggestions and criticisms from the readers for further improvement of this book. We also solicit any correction from the readers.

New Delhi, India   Kaveri Chakrabarty
New Delhi, India   A. S. Chakrabarty

Acknowledgements

I am short of suitable words to express my deep sense of respect and heartfelt gratitude to my father Dr. A. S. Chakrabarty, Former Director Professor and Head, Department of Physiology, Maulana Azad Medical College, New Delhi, Former Professor and Head, Department of Physiology, Jawaharlal Institute of Postgraduate Medical Education and Research, Puducherry, Former Professor of Physiology, Vardhman Mahavir Medical College and Safdarjung Hospital, New Delhi, for his invaluable guidance and constant supervision that provided me the strength to complete this book. I feel privileged to have gained from his immense knowledge of physiology, biochemistry, and medicine. I am thankful to my brother Dr. Abhijit Chakrabarty who has motivated me for preparation of the book. Lastly, I would like to express my deepest heartfelt respect to my mother Late Dr. Krishna Chakrabarty for her encouragement in every sphere of my life. Her divine blessings enabled me to write this book. My prayers are due to her. I would like to thank Dr. Madhurima Kahali (Editor, Springer Nature) and Sivachandran Ravanan (Project Coordinator, Springer Nature) for their invaluable suggestions in shaping of this book.

Kaveri Chakrabarty

Contents

About the Authors

Kaveri Chakrabarty was awarded a PhD (Physiology) from the Faculty of Medical Sciences, University of Delhi. She is currently an Assistant Professor at the Department of Zoology, Hansraj College, University of Delhi. She is a member of the American Physiological Society (APS) and life member of the Association of Physiologists and Pharmacologists of India. She co-authored the book entitled *Fundamentals of Respiratory Physiology*, and the book chapter "Gender and Human Nutritional Safety in Gender and Space: Multidisciplinary Insights. She has published her work in journals like the *Journal of Applied Physiology* and *Federation of American Society for Experimental Biology*. She was former Research Associate at the Department of Physiology and Biophysics, School of Medicine, Case Western Reserve University, Cleveland, Ohio, USA. She was invited as a guest speaker for the Women in Science Program, Inspiring Women in STEM Speaker Series at the Department of Biological Sciences, Winston Salem State University, North Carolina, USA in 2017. She has presented papers in various international and national conferences, and has recently presented a paper at the international conference Experimental Biology, in Chicago, USA. She was awarded a national scholarship under the CAS scheme during her MSc (University of Delhi).

A. S. Chakrabarty was the former Director Professor and Head of the Department of Physiology, Maulana Azad Medical College, New Delhi, and former Professor and Head of the Department of Physiology, Jawaharlal Institute of Postgraduate Medical Education and Research, Puducherry. He was also a Professor at the Department of Physiology, Vardhman Mahavir Medical College (VMMC) and Safdarjung Hospital, New Delhi. He has been awarded innumerable honors during his long career, prominent being the Hari Om Alembic Research Award from the Medical Council of India and the Medical Education Prize from the Association of Physiologists and Pharmacologists of India. He has many publications in various international and national journals. He has co-authored a book entitled *Fundamentals of Respiratory Physiology*. He was felicitated by Maulana Azad Medical College Old Students Association (MAMCOS) in MIDCON-2010. Recently he was felicitated by Maulana Azad Medical College, and VMMC and Safdarjung Hospital. He has been invited to deliver lectures and Orations for Indian Science Congress and many other scientific organizations.

Abbreviations

ACE	Angiotensin-converting enzyme
Acetyl-CoA	Acetyl coenzyme A
ACP	Acyl carrier protein
ACTH	Adrenocorticotropic hormone; same as corticotropin
AD	Anxiety disorder
ADH	Antidiuretic hormone
ADHD	Attention deficit hyperactivity disorder
ADP	Adenosine diphosphate
AIDS	Acquired immune deficiency syndrome
ALA	Aminolevulinic acid
ALAD	Aminolevulinic acid dehydratase
AMP	Adenosine monophosphate
AN	Anorexia nervosa
ANP	Atrial natriuretic peptide
AP	Allergic protocolitis
APO	Apoprotein or Apolipoprotein
ARFID	Avoidant-restrictive food intake disorder
ATP	Adenosine triphosphate
AZT	Azidothymidine
BD	Bipolar disorder
BDNF	Brain-derived neurotrophic factor
BED	Binge eating disorder
BMI	Body mass index
BMR	Basal metabolic rate
BN	Bulimia nervosa
BNP	Brain natriuretic peptide
C peptide	Connecting peptide
C terminal	–COOH end of a peptide or protein
cAMP	$3'5'$-cyclic adenosine monophosphate; cyclic AMP
CAPN10	Calpain 10
CCK	Cholecystokinin
CD	Crohn's disease

cGMP	Cyclic 3′5′guanosine monophosphate
CIH	Chronic intermittent hypoxia
CIHD	Coronary ischemic heart diseases
CO	Carbon monoxide
CoA	Coenzyme A
COPD	Chronic obstructive pulmonary disease
CSU	Chronic spontaneous urticaria
DA	Dopamine
DAG	Diacylglycerol
DASH	Dietary approaches to stop hypertension
DHA	Docosahexaenoic acid
DIPF	Diisopropylphosphofluoridate
DMT	N,N-dimethyltryptamine
DNA	Deoxyribonucleic acid
Dopa	Dihydroxyphenylalanine
DPG	2,3-Diphosphoglycerate
DSM	Diagnostic statistical manual of mental disorders
ECF	Extracellular fluid
ED	Eating disorder
EDNOS	Eating disorder not otherwise specified
ELISA	Enzyme-linked immunosorbent assay
EPA	Eicosapentaenoic acid
ETC	Electron transport chain
FAD	Flavin adenine dinucleotide (oxidized)
$FADH_2$	Flavin adenine dinucleotide (reduced)
FFA	Free fatty acids
Figlu	Formiminoglutamic acid
FMN	Flavin mononucleotide
FPE	Food protein-induced enteropathy
FPIES	Food protein-induced enterocolitis syndrome
FSH	Follicle-stimulating hormone
G6PD	Glucose-6-phosphate dehydrogenase
GABA	Gamma-aminobutyric acid
GAG_S	Glycosaminoglycans
GALT	Galactose-1-phosphate uridyl transferase
GGT	Gamma glutamyl transpeptidase
GI	Glycemic index
Gla	Υ-Carboxyglutamate
GLP-1	Glucagon-like polypeptide-1
HAART	Highly active antiretroviral therapy
HAV	Hepatitis A virus
HbA_{1C}	Hemoglobin A_{1C}
HDL	High density lipoprotein
HEV	Hepatitis E virus
HIV	Human immunodeficiency virus

HLA	Human leukocyte antigen
HMG-CoA	3-hydroxy-3-methylglutaryl-coenzyme A
HNF	Hepatocyte nuclear factor
IBD	Inflammatory bowel disease
IBS	Irritable bowel syndrome
ICF	Intracellular fluid
ICMR	Indian Council of Medical Research
ICSD	International Classification of Sleep Disorders
IGF	Insulin-like growth factor
IDDM	Insulin-dependent diabetes mellitus
IL	Interleukin
IRDS	Infant respiratory distress syndrome
kcal	Kilocalorie
K_m	Michaelis constant
LDH	Lactate dehydrogenase
LDL	Low-density lipoprotein
LH	Luteinizing hormone
LPL	Lipoprotein lipase
LSD	Lysergic acid diethylamide
MAO	Monoamine oxidase
MCH	Mean corpuscular hemoglobin
MCHC	Mean corpuscular hemoglobin concentration
MC4R	Melanocortin-4-receptor
MCV	Mean corpuscular volume
MDD	Major depressive disorder
MDMA	3,4 methylenedioxymethamphetamine
MOD	Maturity onset diabetes
mRNA	Messenger RNA
MSH	Melanocyte-stimulating hormone
mtDNA	Mitochondrial DNA
MTHFR	Methylene tetrahydrofolate reductase gene
MUFA	Monounsaturated fatty acid
NAC	N-acetylcysteine
NAD^+	Nicotinamide adenine dinucleotide (oxidized)
NADH	Nicotinamide adenine dinucleotide (reduced)
$NADP^+$	Nicotinamide adenine dinucleotide phosphate (oxidized)
NADPH	Nicotinamide adenine dinucleotide phosphate (reduced)
NDM-1	New Delhi metallo-beta-lactamase-1
NE	Norepinephrine
NES	Night eating syndrome
NIDDM	Non-insulin-dependent diabetes mellitus
NIN	National Institute of Nutrition
NMDA	N-methyl-D-aspartate
NNRTI	Non-nucleoside reverse transcriptase inhibitors
NO	Nitric oxide

NPY	Neuropeptide Y
NREM	Non-rapid eye movement
NRTI	Nucleoside analogues reverse transcriptase inhibitor
OCD	Obsessive compulsive disorder
OGTT	Oral glucose tolerance test
ON	Orthorexia nervosa
OSFED	Other specified feeding or eating disorders
OVLT	Organum vasculosum of the lamina terminalis
PABA	Para-aminobenzoic acid
PAH	Polycyclic aromatic hydrocarbon
PAI	Plasminogen activator inhibitor
PAL	Physical activity level
PCP	Phencyclidine
PCV	Packed cell volume
PEM	Protein energy malnutrition
PGs	Prostaglandins
PFC	Prefrontal cortex
PKU	Phenylketonuria
PTH	Parathyroid hormone
PUFA	Polyunsaturated fatty acid
PYY	Peptide YY
RAAS	Renin-angiotensin-aldosterone secretion
RBP4	Retinol-binding protein 4
RDA	Recommended dietary allowance
REM	Rapid eye movement
RNA	Rinonucleic acid
RO	Reverse osmosis
ROS	Reactive oxygen species
RQ	Respiratory quotient
SAM	S-adenosylmethionine
SCFA	Short-chain fatty acid
SCID	Severe combined immunodeficiency disease
SDA	Specific dynamic action
SFO	Subfornical organ
SGLT	Sodium-glucose linked transporter
SRED	Sleep-related eating disorder
SSRI	Selective serotonin reuptake inhibitor
T_3	Triiodothyronine
T_4	Tetraiodothyronine; thyroxine
TCA	Tricarboxylic acid cycle
TF	Transferrin
Tmg	Transport maximum of glucose
TNF	Tumor necrosis factor
TPP	Thiamine pyrophosphate (Thiamine diphosphate)
TSH	Thyroid-stimulating hormone or Thyrotropin

UDP	Uridine diphosphate
UL	Tolerable upper intake level
UC	Ulcerative colitis
UV	Ultraviolet
VLDL	Very low density lipoprotein
V_{max}	Maximal velocity
WHO	World health organization
Zn/Cu SOD	Zinc/copper superoxide dismutase

List of Figures

List of Tables

An Integrated View of Human Nutrition and Health

1

Abstract

Optimal nutrition containing essential nutrients is one of the major determinants of health. It prevents diseases, slows the progression of degenerative diseases, and promotes optimal growth. The quantities of nutrients must be sufficient, since the body is nourished by nutrients. The recommended dietary allowance (RDA) ensures nutritional requirement of the individual and maintains energy homeostasis. Active daily living (both indoor and outdoor activities) can be carried out if there is adequate nutrition. Lack of nutrients can disturb the functions of cells, tissues, organs, and the body as a whole. Overnutrition is to be avoided as it may lead to obesity and toxicity. Homeostatic imbalance results in disorders or diseases (deviation from normal health). Nutritional interventions for inborn errors of metabolism in the infants will ameliorate the serious consequences. Food intake depends on the interaction between feeding and satiety centers of the hypothalamus. Various appetite-stimulating and appetite-inhibiting hormones will modulate food intake.

Keywords

Essential nutrients · RDA · Energy homeostasis · Feeding center · Satiety center

1.1 Concept of Nutrition

1.1.1 Essential Nutrients of the Diet

Essential nutrients of the diet consist of energy sources (carbohydrates, fats, and proteins), essential amino acids, essential fatty acids, vitamins, minerals, water, choline and carnitine. Essential nutrients except arginine cannot be synthesized or can be synthesized in small amount. Arginine can be synthesized by urea cycle.

A certain amount of carbohydrate can be synthesized de novo from glycerol and protein.

1.1.2   Component of Food Nutrients (Table 1.1)

Protein, fat, and carbohydrate are called macronutrients as they are needed in large amounts, and the intake of these three nutrients is larger than that of the other dietary nutrients whereas vitamins and minerals are called micronutrients as these nutrients are needed in lesser amounts. Minerals can be further subdivided into macrominerals and microminerals. If the daily requirement of minerals is more than 100 mg, they are called macrominerals. If the daily requirement of minerals is less than 100 mg, they are called microminerals or trace elements. Apart from macronutrients and micronutrients, water, choline, and carnitine are considered as essential nutrients. Water is essential for the survival of all living organisms and takes part in various metabolic processes. The amount of body water depends upon the balance between water intake and water loss. Normal water intake is about 2.4 L/day (includes drinking water/liquid, water from the solid/semisolid food, and water from the oxidation of food) and normal water loss at normal body temperature, which is about 2.4 L/day (through urine, feces, sweat, skin, and lungs). Excessive loss of water due to gastroenteritis (vomiting and diarrhea) can cause marked lowering of blood pressure, resulting in hypovolemic shock followed by coma and eventually death. On the other hand, high intake of water, especially after administration of antidiuretic hormone (ADH) can cause water intoxication. Swelling of the brain cells can lead to convulsion, coma, and eventually death. In the infants and lactating mothers, the amount of water intake should be adequate. The infants require more water due to high ratio of surface area to volume. Adequate intake of water for the infants should not be less than 0.7 L/day. The lactating mothers require additional water intake for milk production. Choline is a lipotropic factor and is synthesized in the body in small amounts, using methyl group donated by methionine (an essential amino acid). Deficiency of choline may result in fatty infiltration of the liver. Carnitine is synthesized from lysine and methionine and is essential for the oxidation of fatty acids. Carnitine takes part in the transport of fatty acids across the mitochondrial membrane. Due to defective synthesis of carnitine, the body is unable to utilize fatty acids as fuel.

1.1.3   Nutritional Requirements and RDA

Nutritional requirements are the minimum amount of nutrients that are essential for physio-biochemical functions of the body. RDA is the average daily amount of a nutrient required by the body. The RDA reflects the standard of good nutrition and ensures nutritional requirement of the individual. The RDA maintains energy homeostasis. The RDA of nutrients varies with different age groups, gender, body weight, and during pregnancy and lactation. The tolerable upper intake level (UL) is the maximum average daily intake of nutrients of an individual. UL may not cause health hazard. Above UL, the individual may have adverse health effects.

Table 1.1 Macro- and micronutrients

Macronutrients	Micronutrients			
Carbohydrates	Vitamins		Minerals	
Proteins	Fat Soluble Vitamins	Water Soluble Vitamins	Macrominerals (> 100 mg)	Microminerals/ Trace elements (< 100mg)
Fats	Vitamin A	Vitamin B$_1$	Calcium	Iron
	Vitamin D	Vitamin B$_2$	Phosphorus	Copper
	Vitamin E	Vitamin B$_3$	Sodium	Iodine
	Vitamin K	Vitamin B$_5$	Potassium	Zinc
		Vitamin B$_6$	Chloride	Fluoride
		Biotin	Magnesium (infants need magnesium <100 mg/day)	Manganese
		Vitamin B$_{12}$		Selenium
		Folic acid		Molybdenum
		Vitamin C		Cobalt
				Sulfur
				Chromium

Convalescence (gradual recovery of strength of the body after a disease) and growth period require additional nutrients with high-protein diet. Various diets lacking in one nutrient or the other can significantly affect the RDA, for example,

1. Diets poor in vitamin C will impair the absorption of iron.
2. Protein synthesis will be impaired, even if there is a lack of a single essential amino acid in the diet.
3. Prolonged consumption of raw eggs can cause biotin deficiency.
4. Chronic alcoholics will suffer from deficiencies of vitamins B$_1$ and B$_6$.

Nutritional requirements of different age groups (infants, preschool children, schoolchildren, adolescents, adults, and elderly people), and during pregnancy and lactation are described below.

1.1.3.1 Infants (0–1 Year)

The term nutrition is derived from the Latin word "Nutrire," which means to breast-feed and nurse. Mother's milk should be preferred rather than cow's milk for the first 6 months. Mother's milk is sterile and contains IgA, lactoferrin, and lactoperoxidase. Lactoferrin is a glycoprotein, having antimicrobial activity. Human colostrum contains maximum amount of lactoferrin followed by mother's milk and then cow's milk. Lactoperoxidase catalyzes reactions, resulting in oxidized products that are antimicrobial. Low pH of mother's milk inhibits the growth of pathogens. Thus, mother's milk protects the infants from infection. Mother's milk also protects against allergens and reduces the chances of allergic reactions in infants. Mortality rate of infants is reduced. Mother's milk contains lactalbumin, which is easily digestible compared to casein present in cow's milk. The ratio of lactalbumin and casein in mother's milk is 2:1 and in cow's milk is 1:3. Carbohydrate and highly emulsified fat are better digested as amylase and lipase are present in mother's milk, and prevent abdominal pain in the infants. Mother's milk contains high amount of docosahexaenoic acid (DHA), which improves cognitive functions as well as development of the brain in the infants and children (Krol and Grossmann 2018). Longer breast-feeding duration is associated with better cognitive and motor development in children aged 2–3 years (Bernard et al. 2013). Breast milk provides 8.3% linoleic acid, whereas cow's milk contains 1.6%. It also, provides high amount of vitamin A, vitamin C, nicotinic acid, thiamin, and pyridoxine. Breast-feeding is an important birth control device. Suckling reflex through the hypothalamus stimulates the secretion of prolactin from the anterior pituitary. Prolactin stimulates the formation of milk and simultaneously inhibits the actions of follicle-stimulating hormone (FSH) and luteinizing hormone (LH) on the ovaries. As a result, ovulation is inhibited. Breast-feeding stimulates the maternal behavior. Lactating mother must avoid caffeine, alcohol, and as far as possible antibiotics, since they are secreted into the mother's milk and could be harmful to the infant.

Weaning (Table 1.2) is the transition period of introducing the liquid and semisolid diet to the infant other than the breast milk from 6 months to 1 year. Weaning foods should contain adequate amount of macronutrients and micronutrients, for the growth and development of the infants. The newborn infants need a prophylactic dose of vitamin K to prevent hemorrhagic disease. Flours of germinating cereals should be preferred as they contain high amount of amylase.

1.1.3.2 Preschool Children (1–6 Years)

Adequate protein intake is needed, since kwashiorkor generally occurs in the children after 1 year of age if their diet consists mainly of carbohydrate. High-energy intake is necessary for growth and activity. Oral dose of vitamin A should be given below the age of 5 years to prevent blindness. Additional amount of calcium, phosphorus and vitamin D is needed as rickets occurs mostly between the age of 1 and 3 years. Excess fat, salt, and sugar should be avoided from the beginning to decrease

Table 1.2 Weaning foods

7 months	1–2 spoons of fruit juices (containing vitamin C) mixed with less sugar and 1–2 spoons of cooked green leafy vegetable soup (containing iron). Excess sugar is to be avoided in order to prevent overweight
8 months	2 spoons of well-cooked cereal in water or milk or 2 spoons of powdered germinated cereals dissolved in water or milk should be fed. Flours of germinated cereals are rich in alpha-amylase and will increase digestibility. 2 spoons of half-boiled egg yolk. 2 spoons of boiled and mashed potato with butter or mashed banana with milk
9 months	Small amounts of well-cooked vegetables and rice with pulses
10–12 months	Start with small quantities and then increase the amounts of soft-boiled egg, cooked vegetables, cereals, pulses, and milk. 2 spoons of germinated cereals and multivitamin syrup should be given

the desire to eat these nutrients. The habit of eating a balanced diet should be initiated from the age of 5–6 years. Soluble dietary fibers and plenty of fluid should be started to prevent constipation and diverticular disease (a pouch or sac formed at the weak point of the intestinal tract). Deficiencies of nutrients during pregnancy may cause cognitive decline during childhood (including the child's interaction with the environment) (Prado and Dewey 2014).

1.1.3.3 Schoolchildren (6–12 Years)

Continuous growth of the body demands an increased calorie intake. All the nutrients according to the RDA should be taken. Nutrients in excess of the RDA are not recommended. Special care should be considered for bone formation. Balanced diet is to be followed. Dietary fibers, plenty of fluids, and fruits are to be consumed. Regular exercise must be encouraged. Excess salt, sugar, saturated, and trans fat should be avoided. Fluoridated toothpaste is recommended to prevent dental caries. Flavored milk (Huxtable et al. 2018) is a palatable choice of schoolchildren. Flavored milk contains nutrients like plain milk and meets the requirements of proteins, carbohydrates, essential fatty acids, calcium, potassium, phosphorus, folate, ascorbic acid, and vitamins (B_1, B_2, B_3, B_6, B_{12}). Desire to drink flavored milk and having an excessive amount of sugar and chocolate can lead to childhood/adult obesity due to high-energy intake. The disadvantage is that flavored milk, containing added sugar and chocolates, has been associated with obesity and lower nutrient density of children's diet (Fayet-Moore 2016). It may lead to dental caries/dental erosion later in childhood. Obesity due to high intake of flavored milk is common in developed countries. However, healthy eating and active sports motivated by "Maternal child health services" and parents can prevent obesity later in the childhood.

1.1.3.4 Adolescents (8–13 Years in Girls and 9–14 Years in Boys)

Rapid growth of the body in adolescence needs adequate nutrients. Protein requirement is more in adolescents (about 1.5 g/kg/day) compared with adults (about 0.8 g/kg/day) for the maturation of reproductive organs and for the repair of old tissues. Calcium and vitamin D are needed for building bone density. Dietary fibers, plenty of fluids, and fruits should be consumed.

1.1.3.5 Adults (21–59 Years)

Nutritional needs, as described under the RDA, are essential to slow age-related progressive decline of physiological and cognitive functions. Adult females require more iron to compensate the loss of blood during menstruation. Daily exercise and active daily living will minimize the progression of lifestyle-related diseases or disorders like type 2 diabetes, hypertension, obesity, and atherosclerosis. Minimize saturated and trans fat (<10% of total energy), cholesterol-rich diet (<250 mg/day), and salt intake (<4 g/day). Smoking must be avoided as it increases the synthesis of cholesterol and causes constriction of coronary artery, leading to myocardial infarction. Extra calcium and vitamin D are needed to prevent osteomalacia. Fruits, vegetables, and dietary fibers should be regularly consumed.

1.1.3.6 Elderly People (60 Years and Above)

Low protein intake due to anorexia causes depletion of protein store with aging. Elderly people may suffer from edema and lower resistance to infection if protein intake is not adequate. Positive nitrogen balance should be maintained. Extra vitamin D and calcium are needed to prevent osteoporosis and fractures. Antioxidants are needed to decrease oxidative stress. Regular walking or non-strenuous exercise is essential to slow the progress of aging. Exercise improves muscle mass, strength, and balance. Maintenance of balance and equilibrium is difficult in old age, probably due to cerebellar dysfunction. Saturated fat, trans fat, and cholesterol-rich diet must be restricted to prevent atherosclerosis. Soluble dietary fibers with plenty of fluid should be taken to prevent constipation and diverticular disease. Apart from the RDA, certain nutrients may be required in increased amounts due to defective absorption, for example, vitamin B_{12} and iron are required to prevent anemia, and zinc is required for wound healing. Tobacco smoking must be avoided, since it produces ROS. It increases metabolic rate and aggravates insomnia. Alcohol consumption must be avoided to prevent loss of cognitive function. It has been reported that senile cataract is due to low intake of antioxiodants (vitamin C, vitamin E, and carotenoids). The carotenoids lutein and zeaxanthin may prevent degeneration of macula lutea present in the center of the retina and thus may prevent loss of vision in old age. Dark green vegetables contain lutein, whereas orange, capsicum, and yellow maize contain zeaxanthin (Mann and Truswell 2007). Stimulants like coffee and tea should be avoided before night sleep to prevent insomnia.

1.1.3.7 Pregnant Women

Additional requirement of nutrients (more than all age groups) must be considered in pregnant women for the benefit of fetus. The nutritional requirements of the pregnant women for "feeding two" are more than those of nonpregnant women (Edwards and Bouchier 1991). Adequate nutrition during pregnancy accounts for the fetal brain development as it stimulates neuron proliferation and myelination. Malnutrition in the mother will impair brain development of the infant (Prado and Dewey 2014). Sufficient energy is needed for the development of the fetus and placenta and for the secretion of breast milk. Extra 250 calories/day is essential during last 3 months. Excessive amount of vitamin A should be avoided, since it may be teratogenic. Adequate folate intake is required to prevent abnormal fetal development. Zinc supplementation during pregnancy leads to a significant decrease in preterm births without affecting infant birth weight. Adequate amount of essential fatty acids is essential for the fetal brain development. It was reported that DHA might act as a precursor of the anti-inflammatory lipid mediator resolving D, suggesting its anti-inflammatory role in pregnancy (Rogers et al. 2013). Sufficient amount of vitamin C is needed for the absorption of iron. Adequate vitamin D and calcium are needed for building bone density. Carbohydrate intake should not be less than 100 g/day as it will lead to ketosis. High-caloric intake due to increased consumption of fat and carbohydrate with adequate protein in pregnancy may result in neonatal adiposity (Pereira-da-Silva et al. 2014). If the pregnant women are underweight or suffer from anemia and vitamin deficiencies, many infants are born with low birth weight. Zinc, iron, iodine, folate, and vitamin A deficiencies cause perinatal complications and increased risk of morbidity and mortality of the pregnant women. The pregnant women must eat a well-balanced diet and should not suffer from malnutrition (Danielewicz et al. 2017). Food hygiene is essential. The pregnant woman may suffer from nausea and vomiting during the first trimester. Small frequent meals as desired should be given. Inadequate intake of iodine may cause congenital hypothyroidism of the infants. Regular non-strenuous exercise should be advised. Soluble dietary fibers with plenty of fluids should be taken to prevent constipation. Alcohol consumption during pregnancy must be avoided, since it may cause abnormal fetal development and prenatal death. Maternal smoking may cause miscarriage and fetal death. Pregnant women are susceptible to ketosis due to insulin resistance, increased lipolysis, and free fatty acids (FFA). They suffer from hyperlipidemia and ketoacidosis after short period of fasting (even after 1 day fasting) and vomiting. Metabolic acidosis should be treated with fluid resuscitation (lactate, 5% dextrose, and insulin therapy) (Hui and Shuying 2018).

1.1.3.8 Lactating Mother

Nutritional requirements in the lactating mothers are higher than all age groups and pregnant women. Adequate fluid intake and nutrients are required for the production of milk. Undernourished mother produces less quantity of milk. Smoking, alcohol, and few contraindicated medications should be avoided during lactation.

1.2 Inborn Errors of Metabolism

1.2.1 Galactosemia (Edelstein and Sharlin 2009)

The infants are not able to synthesize glucose from galactose in the liver due to congenital deficiency of the enzyme galactose-1-phosphate uridyl transferase (GALT). This is caused by mutation in the GALT gene. Ingested galactose accumulates in the blood, leading to the delay of growth with decreased height and development. Vomiting, diarrhea, enlargement of the liver, jaundice, and failure to gain body weight may occur. Galactosemia can be diagnosed by measuring the enzyme GALT. Large amount of galactose will be excreted in the urine after galactose intake. Galactose-free diets improve growth and development.

1.2.2   Phenylketonuria (Elia et al. 2013)

Mutation in the gene for phenylalanine hydroxylase prevents the conversion of phenylalanine to tyrosine in the liver. Accumulation of a large amount of phenylalanine and its keto acid derivatives causes mental retardation, brain damage, and epilepsy in the newborn infants. Dietary intervention, i.e., reducing the amount of phenylalanine in the diet can ameliorate the serious consequences of PKU. Screening of the newborn infants by testing a blood sample for phenylalanine enables the condition to be detected and must be compulsory. Certain amount of phenylalanine is essential for protein synthesis. Nitrogen balance becomes negative after complete removal of phenylalanine from the diet, which may lead to malnutrition.

1.2.3   Fructose Intolerance (Fructosemia) (Edelstein and Sharlin 2009)

Fructose is converted to fructose-1-phosphate by fructokinase. Fructose-1-phosphate is converted to dihydroxyacetone phosphate by fructose-1-phosphate aldolase. Congenital deficiency of fructose-1-phosphate aldolase leads to the accumulation of fructose-1-phosphate, which inhibits glycogenolysis and glucose synthesis and thus causes severe hypoglycemia. Hypoglycemia causes agitation, delirium, impaired judgement, and progressive liver damage. Apart from hypoglycemia, consumption of fructose or sucrose causes severe vomiting.

Dietary intervention, i.e., removal of fructose, sorbitol, and sucrose from the diet results in alleviation of the serious symptoms of fructose intolerance.

1.3   Concept of Energy Metabolism

1.3.1   Energy Derived from Food (Ganong 2003)

The metabolism of food converts about 40% of the energy of food to adenosine triphosphate (ATP) and 60% of energy is dissipated as heat. Heat cannot be utilized for the energy and increases body temperature. The energy content of food is measured from the heat released by the total combustion of food in a calorimeter and is calculated in kilocalorie (kcal), i.e., equal to 1000 calories per gram. A calorie (equal to 4.185 J) is the amount of heat that raises the temperature of 1 g of water from 14.5 to 15.5 °C. Energy derived from the macronutrients ("Energy nutrients") in kcal/g is 9 for fats, 4 for carbohydrates, and 4 for proteins. The food provides energy, which is essential to power all the body functions. Normal health depends on the supply of optimal energy. Energy supply from the food is utilized for the storage of energy, internal heat production, and external work.

1.3.2   Metabolic Rate

Energy supply is reduced by energy expenditure of basal metabolic rate (BMR). The BMR is defined as the minimum energy expenditure necessary to carry on the basic physiologic functions of the body and the vital life processes of the body (heart rate, respiratory rate, etc.) and to maintain metabolic functions of the tissues when a person is at rest and is awake. The BMR is thus called "metabolic cost of living." The measurement of BMR is taken at least 12 h after a meal in a room at a comfortable thermoneutral temperature (about 25 °C). The subject must be at complete physical and mental rest (standardized conditions). One liter of oxygen consumption used to oxidize food releases 4.82 kcal. Oxygen consumption (milliliter per unit time) is measured by a spirometer (oxygen-filled chamber with a device to absorb carbon dioxide) and is corrected to standard temperature and pressure. Metabolic rate = liters of oxygen consumption per unit time × 4.82 kcal. An adequate diet must have an energy value sufficient to provide the requirement of basal metabolism. The BMR of men (about 60 kg) is about 1800 kcal/day and for women (about 50 kg) is about 1400 kcal/day. Active persons [agricultural workers, swimming daily for an hour, jogging for about 1 h, rowing (15 strokes/min), blacksmith, etc.] require more calories (80–100%) above the BMR. The metabolic rate (total energy expenditure per unit time) depends on various factors (Table 1.3) and diseases/disorders (Table 1.4). Conditions altering the metabolic rate are listed in Table 1.5.

Ingested food increases the metabolic rate due to specific dynamic action (SDA) of the food. The SDA is also called "thermic effect of feeding" (Widmaier et al. 2006). The SDA of a food is the energy expenditure due to digestion and absorption. The SDA contributes to about 10% of total energy expenditure. The SDA of the food is exerted mainly by protein food. An amount of protein sufficient to provide 100 kcal increases the metabolic rate of about 30 kcal. A similar amount of

Table 1.3 Factors affecting the metabolic rate

Factors	Metabolic rate
1. Age	(↓ with increasing age and ↑ in young children)
2. Sex	(↑ in male compared with female)
3. Sleep	↓
4. Height or weight or body surface	↑
5. Daytime (because of food intake and physical activities)	↑
6. Exercise	↑
7. Recent ingestion of food (because of SDA)	↑
8. Fasting and starvation	↓
9. High environment temperature	↑
10. Stress	↑
11. Growth	↑
12. Pregnancy and lactation	↑
13. Tobacco smoking	↑
14. Caffeine	↑
15. High circulating levels of calorigenic hormones	↑

↑ indicates increase; ↓ indicates decrease

Table 1.4 Diseases/disorders affecting the metabolic rate

Diseases/disorders	Metabolic rate
Hyperthyroidism (e.g., Graves' disease)	↑
Hypothyroidism (Myxedema)	↓
Adrenal insufficiency (Addison's disease)	↓
Fever	↑
Protein malnutrition (kwashiorkor)	↓
Energy malnutrition (marasmus)	↓
Over-nutrition (obesity)	↓

↑ indicates facilitation; ↓ indicates inhibition

Table 1.5 Factors affecting the metabolic rate

Factors	Metabolic rate
Exercise	Increases the metabolic rate due to increased secretion of glucagon and due to increased sympathetic activity. Exercise increases the metabolic rate by about 10 times or even more. Trained athletes can increase the metabolic rate even up to 20 times
Prolonged fasting and starvation	Decreases the metabolic rate by about 30–40%. Decline is due to (1) decline of plasma thyroxine and triiodothyronine, (2) increased FFA and ketones due to lipolysis, resulting in inhibition of glucagon secretion, and (3) decreased sympathetic discharge
Stress	Increases the metabolic rate due to the activation of sympatho-adrenocortical system
Fever	Increases the metabolic rate (1 °C rise in body temperature increases BMR by about 10%). This is mainly due to cutaneous vasoconstriction and shivering (involuntary muscular contraction) during the rising phase of fever (see section on fever)
Severe cold exposure	May increase the metabolic rate. This is due to (1) increased hunger, (2) cutaneous vasoconstriction, (3) shivering, (4) increased secretion of epinephrine and norepinephrine

carbohydrate increases it by 6 kcal and a similar amount of fat increases it by 4 kcal. Total energy expenditure = BMR + SDA + Physical activity. Physical activity level (PAL) can be determined from total energy expenditure and BMR. PAL = total energy expenditure/BMR. PAL reflects physical activity and is much higher in athletes compared to people with sedentary life (Longo et al. 2011).

1.3.3 Energy Balance

Energy balance depends on the balance between energy intake and energy expenditure. Energy balance is negative when the energy expenditure is more than the energy content of ingested food. Because of negative energy balance, endogenous stores of fat, glycogen, and protein are catabolized, resulting in loss of body weight. Energy balance is positive when the energy expenditure is less than the energy content of ingested food. Because of positive energy balance, energy is stored that results in gain of body weight. However, under physiologic condition, food intake is regulated to the point where energy intake equals the energy expenditure (energy homeostasis). Body weight depends on the balance between energy intake and energy expenditure. Food intake increases after illness or starvation until the individual regains the lost weight. Food intake decreases after overfeeding for days until the individual loses body weight to the control level.

1.3.4 RDA of Energy Intake

RDA of energy intake depends on age, gender, body weight, and physical activity (NIN 2011). The RDA of energy intake indicates the average dietary intake that is essential to maintain energy homeostasis. The RDA for energy intake (kcal/day) for schoolchildren (boys 10–12 years, weight 34.3 kg) is about 2190 whereas for men (sedentary work, weight 60 kg) is about 2320. Energy intake increases proportionately with moderate and heavy activities. Energy intake of adult women is lesser compared to adult men due to high-fat content of the body. Obesity in women begins at puberty. Pregnant women require an additional energy for the development of fetus, placenta, and maternal tissues. Indian Council of Medical Research (ICMR) recommended an additional intake of 300 kcal/day during second and third trimesters. In view of sufficient milk production, ICMR recommended an additional energy of 550 kcal/day during the first 6 months of lactation. Because of the loss of lean body mass and poor physical activity, energy intake of elderly people (>60 years) is less than the adult (about 1970 kcal/day for males).

1.3.5 Respiratory Quotient

The RQ is the ratio of the volume of CO_2 production to the volume of O_2 consumption per unit time.

1.3.5.1 RQ for Carbohydrate

$$\underset{\text{(Glucose)}}{C_6H_{12}O_6} + 6O_2 \rightarrow 6CO_2 + 6H_2O, \quad RQ = 6/6 = 1.0$$

1.3.5.2 RQ for Fat

$$\underset{\text{(Palmitic acid)}}{CH_3(CH_2)_{14}COOH} + 23O_2 \rightarrow 16CO_2 + 16H_2O, \quad RQ = 16/23 = 0.7$$
$$\text{(moreO}_2\text{ is required for the formation ofH}_2\text{O)}$$

1.3.5.3 RQ for Protein

As protein is not simply oxidized to CO_2 and H_2O, determination of RQ of protein is a complex process. Generally the RQ value of protein is around 0.82. The amount of oxidation of carbohydrate, fat, and protein can be determined from the RQ and the urinary nitrogen excretion (metabolism of about 6.3 g of protein produces 1.0 g of urinary nitrogen).

The RQ is reduced under the following conditions:

1. Diabetes mellitus due to increased utilization of fat, which is secondary to decreased utilization of carbohydrate.
2. High-fat diet.
3. Increased oxidation of ketone bodies.

1.3.6 Energy Sources of the Cells

Energy is released when a cell breaks down into macronutrients and is transferred to adenosine triphosphate (ATP). ATP consists of a molecule of adenosine and three phosphate groups, having high-energy phosphate bonds. The stored energy of ATP is released due to hydrolysis. ATP is converted to adenosine diphosphate (ADP), inorganic phosphate, and H^+ with the liberation of energy. Loss of another phosphate forms adenosine 5′ monophosphate (AMP) with the release of energy. Energy is utilized for various functions of the body such as physical activity and muscular contraction, active transport of molecules across the membrane, synthesis of organic molecules, regulation of body temperature, and other physiological functions, which require energy. Other high-energy phosphate compounds are

1. Phosphoenolpyruvate, which is converted to pyruvate with the formation of ATP.
2. 1,3-Bisphosphoglycerate forms ATP from ADP.
3. Phosphocreatine is a high-energy phosphate compound used during muscular contraction.

Formation of ATP associated with oxidation by the flavoprotein-cytochrome system (by the reaction between hydrogen and oxygen to form water) is called oxidative phosphorylation. As the electrons are passed down the electron transport chain to molecules of oxygen, energy is stored by the formation of ATP from ADP and Pi. Uncoupling of oxidative phosphorylation releases energy without the formation of ATP (e.g., thyroid hormones). Mitochondria of brown adipose tissue in the infants contain thermogenin (uncoupling protein), which causes thermogenesis due to uncoupling oxidative phosphorylation. Energy of oxidation is not used for the formation of ATP and causes oxidation of fatty acids of adipocytes, dissipating energy as heat. Brown fat is present around the scapula and acts as an "electric blanket" for the newborn and infants.

1.3.7 Calorigenic Hormones

1. Thyroid hormones (thyroxine and triiodothyronine) increase the metabolic rate at a very high level. The hormones increase the oxygen consumption and heat production of metabolically active organs, except the adult brain, anterior pituitary, uterus, spleen, and testes. This calorigenic action is mainly mediated by uncoupling oxidative phosphorylation. The hormones increase the activity of Na^+-K^+ ATPase, which hydrolyzes ATP to ADP with the release of energy.

2. Epinephrine and norepinephrine increase the metabolic rate due to cutaneous vasoconstriction, which decreases heat loss and increases heat production. The calorigenic action is also due to lipolysis.
3. Ghrelin secreted by the stomach and neuropeptide Y secreted by arcuate nucleus of the hypothalamus increase the metabolic rate by stimulating food intake.
4. Growth hormone, glucocorticoids, and glucagon increase lipolysis by activating hormone-sensitive lipase and thus can increase the metabolic rate. The calorigenic action of glucagon is also due to increased deamination of amino acids in the liver. Growth hormone increases the metabolic rate in the infants and children due to metabolism of new tissue formed.
5. Thermogenin increases thermogenesis in the infants due to uncoupling oxidative phosphorylation.

1.4   Concept of Health

The cells of our body are surrounded by the extracellular fluid (ECF). Cells take up oxygen and nutrients from ECF. The interstitial fluid of ECF is outside the blood vessels, bathing the cells, and is the actual internal environment of the cells. Constancy of the interstitial fluid is maintained in order to have normal functions of the cells by various physio-biochemical regulatory mechanisms. Maintenance of the constant internal environment is known as homeostasis (coined by W.B. Cannon). Homeostatic imbalance results in disorders or diseases (deviation from normal health). The excess or deficiency of amino acids, glucose, fatty acids, vitamins, minerals, and water of the internal environment will lead to nutritional disorders. Various factors, namely physical and mental state, poverty, chronic infection, lack of exercise, inadequate intake of nutrients, genetics, hereditary, educational status, occupational hazards, and food habits can affect health. Poor digestion and absorption of the nutrients will deteriorate health. Hippocrates described "positive health," which depends on the primary human constitution (may be considered as genetics), diet, and exercise. He thought that proper diet and exercise were essential for health and well-being. The World Health Organization (WHO) formulated its definition of health in 1948, which states that "Good health is a state of complete physical, mental and social well being and not merely the absence of disease or infirmity."

1.4.1   Physical

Normal appearance of the skin, face, eye, leg, muscle mass without excess fat deposition, and normal posture indicates good health. Cracking, erosion, and ulceration of the skin are present in kwashiorkor. Vitamin A deficiency causes rough skin. Niacin deficiency causes dermatitis. Cheilosis, i.e., fissures at the corners of the mouth is caused by deficiency of vitamin B_2. Xerophthalmia, Bitot's spots, and keratomalacia of the eyes occur because of vitamin A deficiency. Bowing of legs occurs due to rickets in children and deformed legs in adults due to osteomalacia. Emaciation is predominant in marasmus. Excess fat deposition is found in obese individuals. A pale mucus membrane of the eyelids indicates anemia. Gum bleeding from tender swollen gums is suggestive of vitamin C deficiency.

1.4.2   Mental

Normal mental state of human beings is essential for the maintenance of health. An individual with an abnormal state of mind cannot take care of his/her health. Poor memory, deterioration of personal care, hallucination, decline of intellectual functions (due to niacin deficiency, chronic alcoholism/drug abuse) will affect health. Paranoid personality disorder (suspiciousness, hostility, delusion of sexual infidelity, etc.) deteriorates health.

1.4.3   Social

Socially grouped individuals can take better care of their health compared with isolated individuals (recluse). An individual living in a good family will have cheerful attitude of life. Prevalence of alcoholism, smoking, and drug abuse is less in joint families. Marital tension and unemployment deteriorate health.

1.4.4   Genetics

Expression of deoxyribonucleic acid (DNA), transcription, and translation determine the genetic make-up of an individual. Nutrients influence the synthesis of protein and enzymes by influencing gene expression within the cells. Glucose increases the transcription of glucokinase synthesis in liver. Similarly, vitamin K-dependent carboxylase increases the transcription of glutamic acid residues in the liver to form Υ-carboxyglutamate of prothrombin.

1.4.5   Hereditary

A vast number of diseases are hereditarily acquired, for example, type 2 diabetes mellitus.

1.4.6   Economic Status

Poverty is the major cause of malnutrition and high infant/child mortality rate. It affects normal pregnancy. Elderly people with low economic status are more susceptible to age-related diseases.

1.4.7   Educational Status

Educated people are more aware and concerned about health. Regular exercise and optimal diet promotes health. Smoking, alcohol, and illicit drugs are avoided due to awareness of their detrimental effects.

1.4.8 Type of Occupation and Employment

Occupational hazards, for example, welders suffer from health hazards (even fibrosis of the lungs) due to inhalation of zinc oxide. Painters due to inhalation of lead may suffer from anemia (due to inhibition of aminolevulinic acid (ALA) dehydrogenase and thus hemoglobin synthesis) and renal and brain damage. Bluish marking of the gums (lead line) is seen in chronic lead poisoning.

1.4.9 Food Habits

Excessive intake of iodine present in seafoods and seaweeds may cause hyperthyroidism. Storage and boiling of food in brass vessels may cause adverse effect on health due to copper toxicity characterized by anorexia and hepatitis. Preparation of food in iron utensils causes iron toxicity characterized by hemosiderosis.

1.4.10 Digital Environment

Both health and knowledge need to be understood and clarified as a part of digital society. Excessive usage of electronic gadgets can have detrimental effects on our health as given below.

1. Digital gadget use at night with intake of stimulants like coffee and tea may lead to insomnia. Minimum 6–8 h night sleep is essential for being healthy. Insomnia may have adverse effect in health.
2. Prolonged use leads to bad posture, headaches, back aches, shoulder strain, muscle strain in the neck (Tech neck), carpal tunnel syndrome (characterized by paresthesia of the thumb index and middle fingers due to pressure on the median nerve as it passes through the wrist), cell phone elbow known as cubital tunnel syndrome (aching, burning, or tingling sensation in ulnar nerve on the forearm and hand).
3. Computer vision syndrome characterized by eyestrain, blurred vision, and redness in eyes. Temporary blindness can occur by continuously viewing mobile phone with one eye while lying down.
4. Hearing loss due to continuous plugging of earphones the entire day.
5. Physical fatigue, stress, anxiety, depression, cognitive dysfunction (lack of focus and concentration) occurs due to constant usage of electronic gadgets.
6. Obesity occurs especially in adolescents due to sedentary lifestyle and lack of outdoor activities as they spend more time on gadgets.
7. Radiation from cell phones may increase the incidence of cancer.

1.5 Control of Food Intake (Chakrabarty and Chakrabarty 1972)

Hunger (an intrinsic instinct of sensation to eat food) and appetite (desire to eat food associated with pleasant sensation) depend on the interaction of feeding center (lateral hypothalamus) and nearby satiety center (ventromedial nucleus of the hypothalamus). Feeding center is active all the time and initiates hunger and appetite. But, increased activity of satiety center after the ingestion of food inhibits the feeding center, leading to cessation of eating. Inhibition of satiety center (e.g., in the morning) leads to unchecked activity of feeding center, leading to hunger and appetite. The activity of satiety center ultimately determines the activity of feeding center (Fig. 1.1).

1.5.1   Interaction of Hypothalamus with Appetite-Stimulating and Appetite-Inhibiting Hormones in Controlling Food Intake

During evolution, the anaerobic organisms, which could survive in the absence of oxygen, were transformed into multicellular organisms with highly developed process of oxidative phosphorylation. The life of higher animals is dependent upon:

1. Adequate supply of oxygen to the tissues, by which cells derive energy in the form of ATP. Constant and abundant supply of ATP, "the energy currency of the cells," depends on constant and abundant supply of oxygen. Every cell of the body is an oxygen sensor, i.e., senses oxygen concentration. A vast number of enzymatic reactions utilize oxygen as a substrate. Oxygen utilization results in the formation of highly toxic reactive oxygen species (ROS). Therefore, there must be continuous supply of oxygen through internal respiration (oxygen consumption and carbon dioxide production due to mitochondrial respiration) via external respiration (exchange of oxygen and carbon dioxide, occurring between alveolar air and pulmonary capillaries).
2. Gastrointestinal system for digestion and absorption of the food, liberating nutrients into the blood.
3. Cardiovascular system to distribute nutrients and oxygen to the cells.
4. Neural and endocrine systems to coordinate and integrate the functions of various systems of the body in response to various nutrients.

Fig. 1.1 Diagrammatic representation of satiety center [ventromedial nucleus of the hypothalamus (VMN)] and feeding center [lateral hypothalamus (LH)]

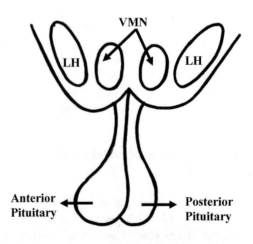

Physio-biochemical and nutritional management studies determine the following:

1. The role of various macronutrients and micronutrients needed for the optimum physiological functioning of every cell of the body.
2. The problems of undernutrition, malnutrition, and overnutrition.
3. "How much to eat and when to stop eating" depends on the interactions of the hypothalamus with appetite-stimulating and appetite-inhibiting hormones, as mentioned below.

Various theories/factors have been proposed for the control of food intake (Fig. 1.2).

1. Glucostatic theory states that the satiety center contains cells known as glucostats (glucoreceptors). Activity of the satiety center is determined by the level of glucose utilization. Inability to utilize glucose, for example, in case of diabetes mellitus will cause inhibition of satiety center and will lead to unchecked activity of feeding center, leading to polyphagia.
2. Food present in the gastrointestinal tract releases the hormone cholecystokinin (CCK), which inhibits feeding by stimulating satiety center and ghrelin, an appetite-stimulating hormone, secreted by the stomach, will have the opposite effect. Neuropeptide Y (NPY) secreted by arcuate nucleus of the hypothalamus

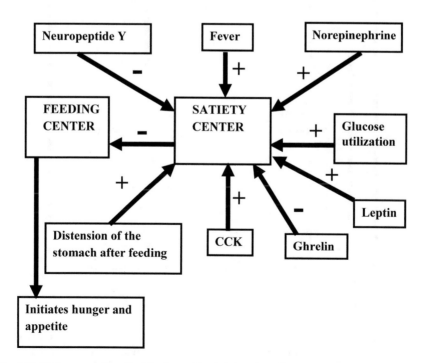

Fig. 1.2 Factors controlling food intake. + stimulation, − inhibition

increases food intake, whereas norepinephrine decreases food intake (amphet-amine discussed under drug abuse causes anorexia by releasing norepinephrine in the hypothalamus). Leptin, an appetite-inhibiting hormone, released from fat depots due to increased fat deposition inhibits food intake by activating leptin receptors of the hypothalamus, possibly in the satiety center. The anorexigenic alpha-melanocyte stimulating hormone (α-MSH) secreted by arcuate nucleus of the hypothalamus inhibits food intake (Nelson and Cox 2008). Peptide YY_{3-36} (PYY_{3-36}) is a peptide hormone secreted by the small intestine and colon. It inhibits NPY release from the arcuate nucleus and reduces food intake. It has been reported that Neuromedin U, a peptide, decreases energy intake and body weight, acting at the level of arcuate nucleus of the hypothalamus. The hormone adiponectin released by the adipocytes reduces energy intake and prevents tri-glyceride deposition in the adipocytes. These interactions among hormones and polypeptides controlling food intake are complex (Fig. 1.3).

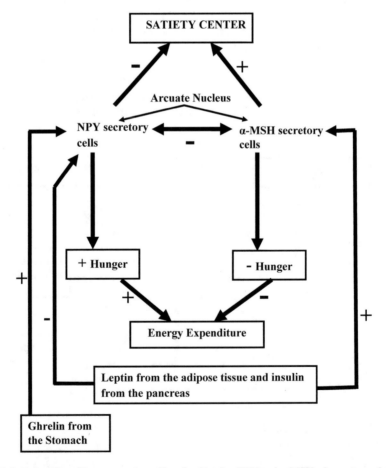

Fig. 1.3 Interaction of hormones controlling food intake. NPY and α-MSH of arcuate nucleus of the hypothalamus are reciprocally innervated. Leptin from the adipose tissue and insulin from the pancreas inhibit NPY secretory cells and stimulate α-MSH secretory cells. Ghrelin stimulates NPY secretory cells. Balance of these interactions will determine food intake and energy expenditure

Table 1.6 Some causes of anorexia

Lack of macro- and micronutrients	Protein (kwashiorkor), thiamin (beriberi), vitamin A, biotin and zinc
Gastrointestinal diseases	Malignancy, ulcerative colitis, celiac disease (mucosal cell defect), Crohn's disease (edematous inflammation of the bowel), and abdominal pain
Liver diseases	Fatty liver followed by cirrhosis
Chronic infectious diseases	Tuberculosis
Miscellaneous causes	AIDS, chronic alcoholism, severe depression, physical and mental stress, major surgery, burns, acute sepsis, and malignancy in an organ of the body

3. Distension of the stomach after eating will try to terminate feeding by stimulating satiety center.
4. Thermostatic theory states that a fall in body temperature increases food intake and a rise in body temperature (fever) decreases food intake, acting at the level of hypothalamus.
5. Various factors causing anorexia are shown in Table 1.6.

1.6 Summary

Apart from macro- and micronutrients, water, choline, and carnitine are considered as essential nutrients. Nutritional and dietary requirements at different age groups and during pregnancy and lactation are discussed. Advantage of breast-feeding for the infants as well as for the mothers is highlighted. Dietary interventions for inborn errors of metabolism in infants in order to ameliorate the serious consequences are mentioned. Energy homeostasis maintains a healthy life. The interactions of feeding and satiety centers of the hypothalamus with hormones and polypeptides in controlling the food intake are discussed.

References

Bernard JY et al (2013) Breastfeeding duration and cognitive development at 2 and 3 years of age in the EDEN mother-child cohort. J Pediatr 163(1):36–42

Chakrabarty K, Chakrabarty AS (1972) Neural systems in the regulation of food intake. Review article bull of the JIPMER (India). Clin Soc 8(4):121–128

Danielewicz H et al (2017) Diet in pregnancy-more than food. Eur J Pediatr 176(12):1573–1579

Edelstein S, Sharlin J (eds) (2009) Life cycle nutrition. Jones Bartlett Publishers LLC, Burlington

Edwards CRW, Bouchier IAD (eds) (1991) Davidson's principles and practice of medicine, 16th edn. Churchill Livingstone, London

Elia M et al (eds) (2013) Clinical nutrition, 2nd edn. Wiley-Blackwell, Hoboken

Fayet-Moore F (2016) Effect of flavoured milk vs plain milk on total milk intake and nutrient provision in children. Nutr Rev 74(1):1–17

Ganong WF (2003) Review of medical physiology, 21st edn. McGraw-Hill, New York

Hui L, Shuying L (2018) Acute starvation ketoacidosis in pregnancy with severe hypertriglyceridemia a case report. Medicine (Baltimore) 97(19):e0609

Huxtable A et al (2018) Parental translation into practice of healthy eating and active play mes-
sages and the impact on childhood obesity: a mixed methods study. Nutrients 10(5):545–560

Krol KM, Grossmann T (2018) Psychological effects of breastfeeding on children and mothers.
Bundesgesundheitsblatt Gesundheitsforschung Gesundheitsschutz 61(8):977–985

Longo DL et al (eds) (2011) Harrison's principles of internal medicine, 18th edn. McGraw-Hill,
New York

Mann J, Truswell AS (eds) (2007) Essentials of human nutrition, 3rd edn. Oxford University Press,
Oxford

Nelson DL, Cox MM (2008) Lehninger principles of biochemistry, 5th edn. W.H Freeman and
Company, New York

NIN (2011) Dietary guidelines for Indians—a manual, 2nd edn. National Institute of Nutrition
(NIN), Hyderabad

Pereira-da-Silva L et al (2014) The adjusted effect of maternal body mass index, energy and macronutrient intakes during pregnancy, and gestational weight gain on body composition of full-
term neonates. Am J Perinatol 31(10):875–882

Prado EL, Dewey KG (2014) Nutrition and brain development in early life. Nutr Rev 72(4):267–228

Rogers LK et al (2013) DHA supplementation: current implications in pregnancy and childhood.
Pharmacol Res 70(1):13–19

Widmaier EP et al (2006) Vander's human physiology the mechanism of body function, 10th edn.
McGraw-Hill, New York

Further Reading

Elia M et al (eds) (2013) Clinical nutrition, 2nd edn. Wiley-Blackwell, Hoboken

Krol KM, Grossmann T (2018) Psychological effects of breastfeeding on children and mothers.
Bundesgesundheitsblatt Gesundheitsforschung Gesundheitsschutz 61(8):977–985

Svalastog AL et al (2017) Concepts and definitions of health and health-related values in the
knowledge landscapes of the digital society. Croat Med J 58(6):431–435

Macronutrients

2

Abstract

Carbohydrates are aldehyde or ketone derivatives of polyhydroxy alcohols. Some carbohydrates contain nitrogen, phosphorus, and sulfur. Carbohydrates consist of simple or free sugars (monosaccharides and disaccharides) and complex carbohydrates (oligosaccharides and polysaccharides). Dextran, inulin, digitalis glycosides, and cellulose are clinically important carbohydrates. Lipids are heterogeneous compounds characterized by relative insolubility in water, but free solubility in organic solvents like ether, chloroform, acetone, etc. Dietary lipids mainly consist of triglycerides (about 90%). The remainder dietary lipids consist of cholesterol, cholesteryl esters, phospholipids, and free fatty acids. Lipids are transported by plasma as lipoproteins. Lipoproteins are important constituents of plasma membrane and mitochondrial membrane. Physio-biochemical important lipids are steroids, prostaglandins, lipid-soluble vitamins, carotenoids, etc. Lipids are the major source of energy fuel. About 80% of energy fuel of the body is derived from lipids. Proteins are polymers of amino acids. Proteins differ from carbohydrates and fats in containing nitrogen, sulfur, and occasionally phosphorus. Amino acids take part in the formation of neurotransmitters, hormones, and digestive enzymes and maintain the contents of enzyme systems of every cell. Essential amino acids cannot be synthesized by humans and must be supplied in diet. Whereas nonessential amino acids are synthesized in vivo by metabolic pathways.

Keywords

Simple carbohydrates · Complex carbohydrates · Triglycerides · Lipoproteins · Essential amino acids · Nonessential amino acids

K. Chakrabarty, A. S. Chakrabarty, *Textbook of Nutrition in Health and Disease*,
https://doi.org/10.1007/978-981-15-0962-9_2

2.1   Carbohydrate Metabolism and Nutrition

2.1.1   Classification, Structure, Properties, and Functions

2.1.1.1 Classification (Tables 2.1, 2.2, 2.3, 2.4, and 2.5)

Table 2.1 Classification (based on molecular size or degree of polymerization)

1. Monosaccharides cannot be hydrolyzed by digestive enzymes further into any simpler carbohydrate and can be classified according to the number of carbon atoms present in the structures		
(a)	5 Carbon pentoses ($C_5H_{10}O_5$)	Ribose
(b)	6 Carbon hexoses ($C_6H_{12}O_6$)	Glucose, galactose, and fructose
2. Disaccharides. On hydrolysis yield two molecules of monosaccharides		
(a)	Maltose	Glucose + glucose
(b)	Sucrose	Glucose + fructose
(c)	Lactose	Glucose + galactose
Note: Trehalose is a disaccharide and consist of two molecules of glucose. Certain bacteria, fungi, and invertebrate animals synthesize it as a source of energy		
3. Oligosaccharides. On hydrolysis yield 2–10 molecules of monosaccharide		
(a) Maltodextrins (α-glucans)		
(b) Raffinose and inulin (non-α-glucans)		
4. Polysaccharides. On hydrolysis, yield >10 molecules of monosaccharide		
(a) Starch	(b) Glycogen	
Note: Simple or free sugars include monosaccharides, disaccharides, and sugar alcohols (e.g., sorbitol). Sorbitol may be used in diabetic patients instead of other simple sugars as sorbitol is partially absorbed. Complex carbohydrates include oligosaccharides and polysaccharides. In addition, complex carbohydrates include amino sugars and uronic acids		

Table 2.2 Monosaccharides

1. Glucose	Produced due to hydrolysis of maltose, sucrose, lactose and starch. It is utilized by the tissues for various metabolic functions (see glycolysis)
2. Fructose (levulose)	It is the sweetest sugar present in honey and nectar of flowers. Hydrolysis of sucrose produces fructose. It is changed to glucose in the liver
3. Galactose	Produced by hydrolysis of lactose. It is synthesized in the mammary gland to produce lactose of milk and is required for the synthesis of glycolipids and GAGs
4. Ribose	It is a component of nucleic acids and coenzymes, for example, NAD, NADP, and flavoproteins. Ribose-5-phosphates are intermediates of pentose phosphate pathway and are converted to glyceraldehyde-3-phosphate and fructose-6-phosphate

Table 2.3 Disaccharides

1. Lactose	Present in milk and milk products. Lactase of the intestinal mucosa hydrolyzes lactose to glucose and galactose. Due to lactose deficiency, ingestion of milk is not tolerated, leading to lactose intolerance. It is a reducing disaccharide
2. Sucrose	Present in beet, pineapple, sorghum, and carrots. It can be extracted from sugar cane. Sucrase of the intestinal mucosa hydrolyzes sucrose to glucose and fructose. It is not a reducing disaccharide
3. Maltose	Present in germinating cereals, malt, barley, and sprouted wheat. Maltase of the intestinal mucosa hydrolyzes maltose into two molecules of glucose. Amylase of either saliva or pancreas digests starch into maltose along with dextrin and maltotriose. It is a reducing disaccharide

Table 2.4 Polysaccharides

1. Starch	It is a polymer of glucose linked by 1–4 glycosidic linkage. It consists of amylose (long unbranched chain) and amylopectin (branched chain). It is present in potatoes, legumes, rice, wheat, and other food grains. Salivary and pancreatic amylase hydrolyze starch to dextrins, maltotriose, and maltose. These products are further digested by isomaltase and maltase of brush border of the intestine into glucose. Starch that is incompletely digested is known as resistant starch
2. Glycogen	It is a branched structure (more than amylopectin). Consists of glucose units combined by glycosidic linkages. The liver contains excessive amount of glycogen. Glycogen is converted to glucose via glucose-1-PO_4 and glucose-6-PO_4

Table 2.5 Some clinically important carbohydrates

1. Dextran	Is a highly branched polymer of glucose units. Because of large molecular weight, it cannot go out of vascular compartment and is used for intravenous transfusion to increase the plasma volume
2. Inulin	Is a long-chain D-fructose unit with a molecular weight of 5200 and is present in oats, onion, garlic, chicory, and other bulbs and tubers. Clinically it is used for measuring glomerular filtration rate as it is completely filtered through the glomerular capillaries and is neither reabsorbed nor secreted by the renal tubules
3. Digitalis glycosides	Important in medicine because of their inotropic action on cardiac muscle. They increase the force of contraction of the cardiac muscle by increasing the intracellular calcium concentration. Yellow oleander contains cardiac glycosides
4. Cellulose	Chief carbohydrate of plants and polymer of glucose. It is made up of glucose units linked by β-1,4 linkage. It cannot be digested by humans, because their intestine lacks hydrolase that can attack the β-linkage. Bacteria of the intestine of herbivorous animals, including ruminants can attack the β-linkage, resulting in the utilization of glucose for energy. It is useful as a dietary fiber in humans

2.1.1.2 Structure (Champe and Harvey 1994)

Isomers

Compounds that have the same chemical formula are called isomers. Glucose and fructose are isomers of each other having the same chemical formula $C_6H_{12}O_6$. They differ in structural formula (keto group in position 2 of fructose and aldehyde group in position 1 of glucose).

Epimers

Different configuration of two monosaccharides around one specific carbon is called epimerism (Fig. 2.1).

Enantiomers

Some dissolved molecules can rotate the plane of vibration of polarized light either to the right or to the left. If the plane of vibration is rotated to the right, it is designated as dextrorotatory or D form. If the plane of vibration is rotated to the left, it is designated as levorotatory or L form. Monosaccharides exist either in D or L form. Enantiomers are mirror images of the pairs of structure. The two members are designated as D-sugar and L-sugar. Sugars in humans are mainly D-sugars (Fig. 2.2).

2.1.1.3 Properties

1. Most of the carbon atoms are linked to a hydrogen atom and a hydroxyl group. Carbohydrates are soluble in water due to hydroxyl groups.
2. Monosaccharides commonly consist of unbranched carbon atoms, which are linked by single bonds. However, one of the carbon atoms in open-chain form is double-bonded to oxygen to form a carbonyl group. Sugars are termed reducing sugars because oxygen on the anomeric carbon (carbonyl group) is not attached to any other structure. A reducing sugar can react with chemical reagents (e.g., Benedict's solution) and the carbonyl group becomes oxidized. Benedict's reagent will be reduced by any sugar with free aldehyde or keto group. Except sucrose, most sugars are reducing sugars (capable of reducing cupric ion). Sugars are strong reducing agents in alkaline medium (Vasudevan et al. 2011).
3. Glucose can be oxidized to form gluconic acid. The enzyme glucose oxidase oxidizes glucose to gluconic acid and H_2O_2.
4. The oxidation of CH_2OH at last carbon to COOH results in the formation of uronic acid. Uronic acids are carboxylated derivatives of glucose and galactose. Glucose is oxidized to glucoronic acid. Glucoronic acid is present in small amount in the diet.
5. Deoxy sugars are formed by the replacement of the hydroxyl group by hydrogen atom. Ribose and deoxyribose are linked by N-glycosidic bonds to purine and pyrimidine bases.
6. Amino sugars are formed by the replacement of hydroxyl group of sugars by amino groups to form amino sugars. Mainly the amino group is derived from amide group of glutamine. Amino sugars are glucosamine (a constituent of

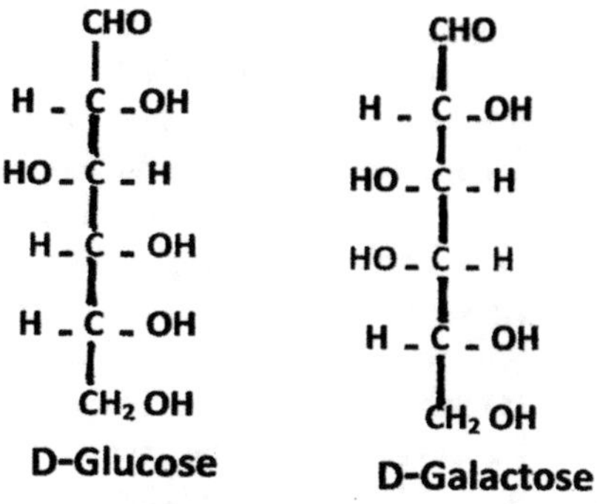

Fig. 2.1 Shows epimerism. Glucose and galactose differ only in the position at carbon 4

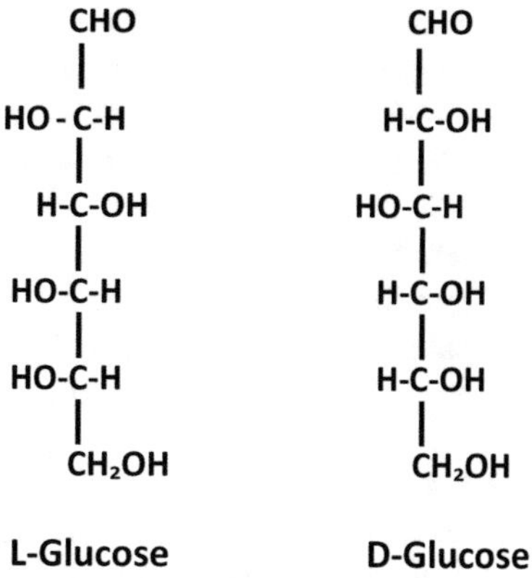

```
        CHO                        CHO
         |                          |
  H _ C _OH                  H _ C _OH
         |                          |
 HO_ C _ H                 HO_ C _ H
         |                          |
  H_ C _ OH                 HO_ C _ H
         |                          |
  H _ C _ OH                 H _ C _ OH
         |                          |
      CH₂ OH                     CH₂ OH
    D-Glucose                 D-Galactose
```

Fig. 2.2 Shows enantiomerism of L-glucose and D-glucose

```
        CHO                        CHO
         |                          |
  HO - C-H                   H-C-OH
         |                          |
   H-C-OH                    HO-C-H
         |                          |
  HO-C-H                     H-C-OH
         |                          |
  HO-C-H                     H-C-OH
         |                          |
    CH₂OH                      CH₂OH

  L-Glucose                  D-Glucose
```

hyaluronic acid) and galactosamine (a constituent of chondroitin). Glycoproteins, some antibiotics (e.g., erythromycin) and glycosaminoglycans (GAGs) contain amino sugar.

7. Glycosides are formed by condensation between the hydroxyl group of ano-meric carbon of a monosaccharide and a second compound, which may be a monosaccharide or an aglycone (Murray et al. 1996). Benedict's reagent cannot be reduced by glycosides. Cardiac glycosides contain steroids as the aglycone compound and include derivatives of digitalis such as ouabain and digitalis glycosides. They inhibit Na^+-K^+ ATPase of the cell membrane. Na^+-K^+ ATPase moves Na^+ out of the cell and takes K^+ into the cell. Due to inhibition of the Na^+-K^+ ATPase by cardiac glycosides, the intercellular Na^+ concentration increases and Ca^{2+} extrusion decreases due to the closure of voltage-gated chan-nel (because of depolarization). As a result, increased intercellular Ca^{2+} concen-tration increases the force of contraction of cardiac muscle (inotropic action). Thus, cardiac gylcosides are useful for the treatment of heart failure.

8. Phosphorylation (introduction of a phosphate) is the initial process of sugar metabolism. Sugar phosphates (e.g., glucose-1-phosphate and glucose-6-phosphate) are essential for the metabolism of sugar.

9. Sugar alcohol, for example, sorbitol present in cherries provides less energy as they are partially absorbed. Cataract of lens may be due to increased sorbitol. Sorbitol is present in human lens. Increased concentration of sorbitol in diabetes mellitus may lead to diabetic cataract.

10. GAGs consist of repeating disaccharide units, containing amino sugars. GAGs have shock-absorbing properties as they bind with water to form gel-like substance in the bone and cartilage. Mucopolysaccharides (long chain of sugar molecules) are essential in building of bones, cartilages, skin, and other tissues. Genetic deficiencies of lysosomal hydrolases involved in the degradation of GAGs cause abnormal accumulation of GAGs, resulting in mucopolysaccharidosis. Higher accumulation of mucopolysaccharides damages the cells. Hurler's syndrome is a rare hereditary disorder (autososmal recessive) caused by deficiency of α-L-iduronidase enzyme. It is characterized by mental deficiency, enlargement of the liver and spleen, facial deformity (thickening of facial features), and deformities of the bones. The patients are not able to walk on their own because of stiffness of joints and knees. Hunter's syndrome is also a rare hereditary disorder due to the deficiency of iduronate sulfatase. This disorder is sex-linked and restricted to male, and female may be carriers. It is characterized by stiff joints, mental retardation, hepatomegaly, and coarse facial features. The patients of Hurler's syndrome and Hunter's syndrome, if not treated with enzyme replacement therapy, may not survive more than 10 years.

2.1.1.4 Functions

1. Glucose is the most important carbohydrate. Glucose is called "blood sugar" as it is the main sugar present in the blood. Dietary carbohydrate is absorbed mainly as glucose into the blood stream and to a certain extent fructose and galactose. Liver cells convert fructose and galactose to glucose. Glucose provides energy to the tissues and is "a universal fuel of the fetus" as brain cannot oxidize FFA. Glucose is easily utilized by the cells. Brain cells and RBCs are practically dependent on glucose as the energy fuel. The brain requires around 100 g of glucose daily for energy, whereas the rest of the organs require about 50 g daily.
2. Protein-sparing effect (see below).
3. Antiketogenic and prevents breakdown of fatty acids (see below).

2.1.2   Digestion and Absorption

Digestion of starch (main dietary carbohydrate) is described below (Ganong 2003).

2.1.2.1 Mouth

Digestion of starch is initiated inside the mouth during mastication. Salivary α-amylase (ptyalin) hydrolyzes starch for a brief period into α-dextrin, maltotriose, and maltose.

2.1.2.2 Stomach

Salivary α-amylase cannot hydrolyze starch inside the stomach as it is inhibited by high acidic gastric juice (pH is about 1.0). The optimal pH of α-amylase is about 6.8.

2.1.2.3 Small Intestine

Pancreatic α-amylase digests starch further in the duodenum and jejunum (pH is about 6.5–7.0) into α-dextrins, maltotriose, and maltose. Products are further digested by the following brush border enzymes of the small intestine. α-Dextrin, maltose, and maltotriose are converted to glucose by the enzymes α-dextrinase (isomaltase) and maltase. Sucrose is converted to glucose and fructose by the enzyme sucrase. Lactose is converted into glucose and galactose by the enzyme lactase. Glucose and galactose are absorbed from the mucosal cells of the small intestine into the portal vein via active transport. Fructose is absorbed via facilitated diffusion into the portal vein.

2.1.3   Lactose Intolerance

Lactose is not digested due to the deficiency of lactase (alactasia). The low lactase level causes intolerance to milk sugar, i.e., lactose. Apart from lactose intolerance, certain individuals are not able to tolerate milk due to hypersensitivity of milk protein β-lactoglobulin. The incidence of lactose intolerance depends on the human races. It is common in Asian population compared with American and European population.

Low lactase level may be due to the following reasons:

1. Congenital or inherited deficiency of lactase. Normally lactase activity of the small intestine is high at birth and declines gradually with age. Susceptible elderly individuals may suffer from lactose intolerance due to reduced expression of the enzyme.
2. Surgical removal of part of the small intestine.
3. Mucosal cell defect, for example, in celiac disease and viral gastroenteritis.

2.1.3.1 Symptoms

Undigested lactose, being osmotically active, holds water and is acted upon by bacteria. Bacterial fermentation produces gases (CO_2, H_2, and methane), leading to flatulence, bloating, abdominal cramp, and diarrhea.

2.1.3.2 Diagnosis

Bacterial fermentation of lactose in the colon produces hydrogen, which is absorbed from the intestine and carried through the blood to the lungs. Hydrogen is exhaled out through the lungs. Measurement of hydrogen gas during expiration is a test for lactose intolerance.

2.1.3.3 Prevention

Lactose-free diet is recommended. Lactose intolerance can be prevented by the commercial lactase preparation. Low-income groups cannot afford this commercial preparation as it is expensive. Curd contains bacterial lactase and is better tolerated than milk. One can take soya milk.

2.1.4 Glycolysis, Vitamins, and Citric Acid Cycle (Table 2.6)

Glycolysis (Fig. 2.3) involves sequence of reactions, which converts glucose into pyruvate and lactate with simultaneous production of ATP. Because of glycolysis, ATP is continuously formed and consumed. ATP transfers energy for the formation of glycogen from glucose and various anabolic reactions of proteins and triglycerides. There are two types of glycolysis (Fig. 2.4). Recent evidence indicates that lactic acid, an end product of anaerobic glycolysis, plays an important role in immune modulation, wound repair, and cancer growth (Shiren et al. 2017).

Glucose, after entering the cells, is phosphorylated to glucose-6-phosphate. The reaction is catalyzed by the enzyme hexokinase or glucokinase. Glucokinase, present in the liver, is increased by carbohydrate-rich diets and by insulin and decreased in starvation and diabetes mellitus. Glucokinase prevents hyperglycemia after carbohydrate-rich meal by the formation of glucose-6-phosphate (Fig. 2.5).

2.1.4.1 Embden-Meyerhof Pathway

Glucose catabolism proceeds through fructose to trioses and ultimately to pyruvate. Pyruvate enters from the cytosol into the mitochondria. Acetyl-CoA is formed from pyruvate by the action of pyruvate dehydrogenase. The enzymes responsible for fatty acid synthesis are present outside the mitochondria. Acetyl-CoA cannot cross the mitochondrial membrane barrier. Acetyl-CoA is condensed with oxaloacetate to form citrate (citric acid cycle). Citrate is transported out of the mitochondria and is converted to acetyl-CoA for fatty acid synthesis. Glucose can be converted to fat by this pathway. Fats are not converted to glucose as the conversion of pyruvate to acetyl-CoA is irreversible. A negligible amount of fat is converted to glucose via glycerol of the fat depot.

2.1.4.2 Pentose Phosphate Pathway/Hexose Monophosphate Shunt

This pathway occurs in the cytosol and is important in RBC, adrenal cortex, and mammary glands. It starts with the conversion of glucose-6-phosphate to 6-phosphogluconate by the enzyme glucose-6-phosphate dehydrogenase (G6PD), leading to the formation of reduced nicotinamide adenine dinucleotide phosphate (NADPH). 6-Phosphogluconate is converted to ribulose-5-phosphate, which is isomerized to ribose-5-phosphate. Pentoses are converted to fructose-6-phosphate and phosphoglyceraldehyde by several complex reactions and shunted for glycolysis. Thus, pentose phosphate pathway is also called hexose monophosphate shunt.

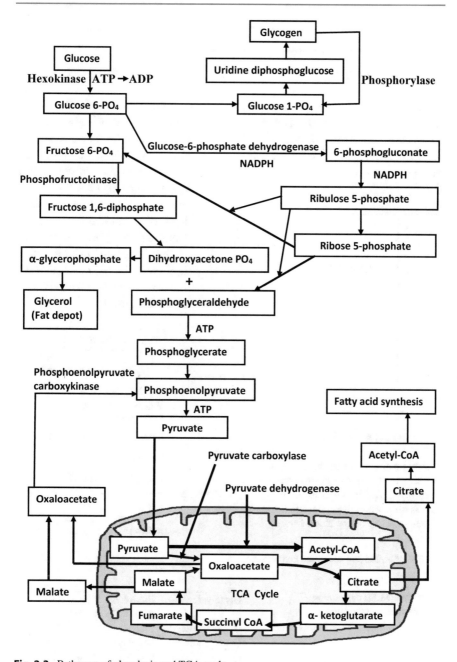

Fig. 2.3 Pathways of glycolysis and TCA cycle

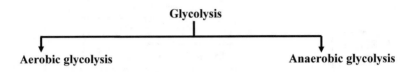

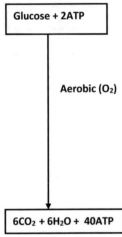

Aerobic glycolysis

Glycolysis occurs in the presence of oxygen (aerobic) and is the prelude to the citric acid cycle and electron transport chain, which together harvest energy from glucose.

Glucose + 2ATP

Aerobic (O₂)

6CO₂ + 6H₂O + 40ATP

Therefore net gain=**38ATP**

It may be noted that cardiac muscle which works under aerobiosis has poor glycolytic activity. Survival rate is poor when there is ischemia (inadequate coronary blood flow).

Anaerobic glycolysis

Glycolysis occurs in the absence of oxygen. Anaerobic glycolysis is responsible for the conversion of glucose to lactate. It allows the production of ATP in tissues that lack mitochondria (red blood cells and cornea) or in cells deprived of sufficient oxygen i.e, hypoxia (during severe exercise). Recent evidence indicates that lactic acid plays a key role in immune modulation, wound repair and growth

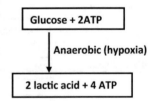

Glucose + 2ATP

Anaerobic (hypoxia)

2 lactic acid + 4 ATP

Therefore net gain=**2ATP**

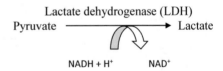

Lactate dehydrogenase (LDH)

Pyruvate ⟶ Lactate

NADH + H⁺   NAD⁺

Fig. 2.4 Aerobic and anaerobic glycolysis

Table 2.6 Terms concerning carbohydrate metabolism

Glycolysis	Breakdown of glucose to pyruvate or lactate
Glycogenesis	Process of glycogen formation
Glycogenolysis	Glycogen breakdown
Gluconeogenesis	Conversion of nonglucose molecules to glucose

Note: Insulin increases glycolysis and glycogenesis and inhibits gluconeogenesis whereas glucagon stimulates glycogenolysis and gluconeogenesis

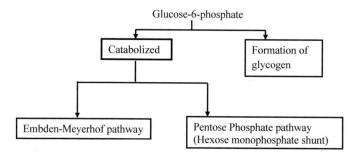

Fig. 2.5 Fate of glucose-6-phosphate

NADPH maintains reduced state of glutathione, which preserves the integrity of the membrane of RBC. RBCs have no nucleus and ribosomes. Deficiency of G6PD found in certain races causes destruction of RBCs, leading to hemolytic anemia. Furthermore, hemolysis may precipitate infection, which may generate free radicals and cause oxidative damage of RBCs. G6PD deficiency offers resistance to malarial infection. Legumes and beans are toxic to people with G6PD deficiency, and consumption causes headache, fever, and hemolysis (Colledge et al. 2010).

2.1.4.3 Role of Citric Acid Cycle [Krebs Cycle or Tricarboxylic Acid Cycle (TCA)] in Metabolism

It is called Krebs cycle as the cycle was first discovered by Krebs. It is also known as TCA as it involves acids with three carboxyl groups (–COOH). It is concerned with gluconeogenesis, transamination, deamination, and fatty acid synthesis. Citric acid cycle is a common pathway of carbohydrates, fat, and most of the amino acids for oxidation to CO_2 and H_2O. In a series of several reactions, citric acid cycle generates 12 ATP per turn of the cycle. Citric acid cycle needs oxygen and cannot operate without oxygen. Glucose catabolism, and formation of amino acids and proteins due to glycolysis is shown in Figs. 2.6 and 2.7.

2.1.4.4 Vitamins and Citric Acid Cycle

The following vitamins play a crucial role in the citric acid cycle.

1. Pantothenic acid is a part of coenzyme A, for example, succinyl-CoA of the citric acid cycle.
2. Biotin is a coenzyme in carboxylation reactions. Biotin-dependent enzymes are acetyl-CoA carboxylase, pyruvate carboxylase, and propionyl-CoA carboxylase. For example,
   (a) Propionyl CoA → Methylmalonyl-CoA, which is
            converted to succinyl CoA

      This reaction is catalyzed by propionyl-CoA carboxylase and biotin.
   (b) Pyruvate → Oxaloacetate
      This reaction is catalyzed by pyruvate carboxylase and biotin.

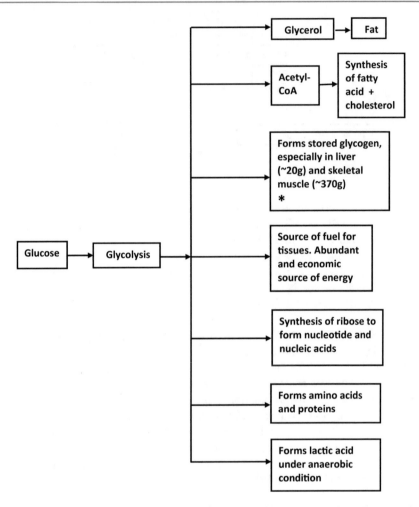

Fig. 2.6 Effects of glucose catabolism. *After about 16 h of fasting, the liver is totally depleted of glycogen, whereas skeletal muscle glycogen is depleted after exercise

3. Thiamin pyrophosphate (TPP), active form of thiamin, is a coenzyme for oxidative decarboxylation reactions. For example, pyruvate is converted to acetyl-CoA by pyruvate decarboxylase and TPP.
4. Nicotinamide adenine dinucleotide (NAD$^+$), active form of niacin, is a coenzyme for dehydrogenases in the citric acid cycle.
   (a) Pyruvic acid $\rightarrow$ Acetyl - CoA + CO$_2$
       This reaction is catalyzed by pyruvate dehydrogenase.
   (b) Malate $\rightarrow$ Oxaloacetate
       This reaction is catalyzed by malate dehydrogenase.
   (c) Isocitrate $\rightarrow$ Oxalosuccinate
       This reaction is catalyzed by isocitrate dehydrogenase.

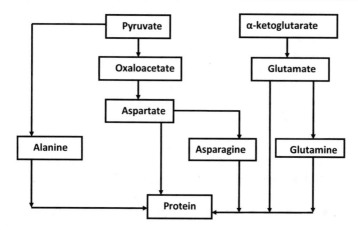

Fig. 2.7 Formation of amino acids and protein due to glycolysis

5. Flavin adenine dinucleotide (FAD), active form of riboflavin, is a cofactor of succinate dehydrogenase.

   Succinate + FAD → Fumarate + FADH$_2$.

6. Pyridoxal phosphate, active form of pyridoxine, takes part in (1) transamination reaction, for example, alanine is converted to pyruvate, and (2) decarboxylation reaction, for example, histidine is converted to histamine.
7. Vitamin B$_{12}$ converts methylmalonyl-CoA to succinyl-CoA of citric acid cycle and takes part in gluconeogenesis.

2.1.5   Dietary Sources, RDA, Deficiency, and Overconsumption

2.1.5.1 Dietary Sources (Table 2.7)
RDA for the children and adults is about 130 g/day. Pregnant women and lactating mothers require more carbohydrates. Carbohydrates are considered as protein-sparing because of intake of sufficient quantities of carbohydrate, the amino acids

Table 2.7 Digestible carbohydrate content (g/100 g)

High content >50	Sugar, jaggery, honey, rice, wheat flour, barley, pulses, peas, bajra, and jowar
Medium content <50	Banana (ripe), cashew nut, groundnut, potato, mango (ripe), grapes, apple, carrot, orange, papaya (ripe), lentils, beans, cheese, soya bean, baby corn, sweet potato, garlic, and ginger
Low content <10	Cabbage, cauliflower, cucumber, leafy vegetables, curd, mother's milk, buffalo's milk, cow's milk, lettuce, spinach, brinjal, capsicum green, ladies finger, onion, tomato ripe (hybrid), strawberry, and radish

See Chap. 7
Note: Nondigestible carbohydrate (see Chap. 5)

will not be oxidized to provide energy, rather will be utilized for repair and mainte-nance of tissue protein.

Deficiency of carbohydrate will lead to the catabolism of fat and protein for the generation of energy. If the carbohydrate intake is less than 100 g/day, increased catabolism of fat leads to the formation of ketone bodies (ketosis).

Overconsumption of carbohydrate leads to overweight and obesity as carbohy-drate is converted to fat. Excessive consumption of chewy sweets (e.g., candy) may cause dental caries.

2.2   Lipid Metabolism and Nutrition

2.2.1   Classification, Structure, Properties, and Functions

2.2.1.1 Classification
1. Simple lipids are esters of fatty acids with glycerol. Fats mainly occur as triglyc-erides (storage form of lipids in the adipose tissues). Triglycerides are esters of glycerol and three fatty acids. Fatty acids occur mainly as ester form. They also occur in the unesterified form as FFA found in plasma.
2. Complex lipids contain phosphate (e.g., phospholipids) in addition to fatty acids and alcohol. Phospholipids are esters of two fatty acids bound to two hydroxyl groups of glycerol. The third hydroxyl group of glycerol is linked to phosphate. Phospholipids are amphipathic and form the lipid bilayer of cellular membrane. The phospholipids include the following:

   > Phosphatidic acid + Choline → Phosphatidylcholine (Lecithin)
   >
   > Phosphatidic acid + Glycerol → Phosphatidylglycerol
   >
   > Phosphatidic acid + Inositol → Phosphatidylinositol

   Sphingomyelins do not have glycerol and are esters of fatty acids, phosphate, choline, and amino alcohol sphingosine. Sphingomyelins are found in brain. Phosphatidylglycerol gives rise to cardiolipin, a lipid of inner mitochondrial membranes. Lipids are complexed to other compounds, for example, lipopro-teins and glycolipids. Glycolipids (glycosphingolipids) contain fatty acid, amino alcohol, i.e., sphingosine and carbohydrate (monosaccharide or oligosaccharide) without phosphate and are present in large amounts in nerve tissues. Plasmalogens are lipids present in nervous tissue and heart. Myelin membranes contain etha-nolamine plasmalogens, and heart muscles contain choline plasmalogens (Murray et al. 1996).
3. Derived lipids are steroids (e.g., cholesterol, cortisol, sex hormones), prostaglan-dins, ketone bodies, lipid soluble vitamins, etc.
4. Neutral lipids are triglycerides, cholesterol, and cholesteryl esters (hydroxyl group of cholesterol is esterified with a long-chain fatty acid).

5. Cellular lipids consist of (1) structural lipids that are inherent part of cellular membranes and are not mobilized and (2) neutral lipids that are stored in fat depots, are mobilized during starvation, and undergo breakdown and resynthesis.
6. Miscellaneous lipids are carotenoids and squalene.

2.2.1.2 Structure

Structure of Fatty Acid and Glycerol (Fig. 2.8)
The triglycerides are made up of three fatty acids bound to glycerol. Hydroxyl groups of glycerol are bound to the carboxyl group of a fatty acid by a dehydration reaction. A part of fatty acid is nonpolar and hydrophobic or water-insoluble and the other part is polar and hydrophilic or water-soluble. Such molecules are described as amphipathic. A bilayer of amphipathic lipids forms the basic structure of biologic membranes. Naturally occurring fatty acids contain an even number of carbon atoms generally between 14 and 24. For example, number of carbon atoms in palmitic acid is 16, in linoleic acid 18, linolenic acid 18, arachidonic acid 20, etc. They may be saturated without double bonds or unsaturated, i.e., dehydrogenated with various number of double bonds. Monounsaturated fatty acids contain one double bond, whereas polyunsaturated fatty acids contain more than one double bond (Tables 2.8 and 2.9).

Structure of fatty acid		Structure of glycerol
$CH_3 (CH_2) n$ Hydrophobic hydrocarbon chain	COO^- Hydrophilic carboxyl group	CH_2OH | $CHOH$ | CH_2OH

Fig. 2.8   Structure of fatty acid and glycerol

Table 2.8   Saturated fatty acids

Common name	Number of carbon atoms	Occurrence
Acetic	2	Vinegar
Propionic	3	Product of carbohydrate fermentation in the colon of humans
Butyric	4	Butter
Caproic	6	Butter
Caprylic	8	Butter
Capric	10	Butter and coconut oil
Lauric	12	Cinnamon, coconut oil and nutmeg
Myristic	14	Butter, coconut oil and nutmeg
Palmitic	16	Adipose tissue
Stearic	18	Adipose tissue
Arachidic	20	Peanut (arachis) oil
Behenic	22	Seeds
Lignoceric	24	Peanut oil and cerebrosides

Table 2.9 Unsaturated fatty acids

Common name	Number of carbon atoms	Source
Monounsaturated		
Palmitoleic	16 (one double bond) ω 7	Adipose tissue
Oleic	18 (one double bond) ω 9	Adipose tissue
Erucic	22 (one double bond) ω 9	Mustard seed oil
Nervonic	24 (one double bond) ω 9	Cerebrosides
Polyunsaturated		
Linoleic	18 (two double bonds) ω 6	Peanut, corn, soya bean, and many vegetable oils Importance: essential fatty acid
Linolenic	18 (three double bonds) ω 3	Vegetable oils Importance: essential fatty acid
Arachidonic	20 (four double bonds) ω 6	Tropical marine fish, peanut oil Importance: precursor of prostaglandins and leukotrienes
Timnodonic	20 (five double bonds) ω 3	Cod liver oil, fatty fish oils, e.g., Salmon and Mackerel
Cervonic	22 (six double bonds) ω 3	Phospholipids of brain and fish Oils

Structure of Saturated and Unsaturated Fatty Acids (Fig. 2.9)

Saturated fatty acid

Palmitic acid contains 16 carbons atoms without double bond.

$$CH_3(CH_2)_{14} - COOH$$

Unsaturated fatty acid

Oleic acid contains 18 carbon atoms having one double bond (MUFA).

$$CH_3\text{-}(CH_2)_7\text{-} CH = CH\text{-}(CH_2)_7\text{-}COOH$$

Linoleic acid contains 18 carbon atoms having two double bonds (PUFA).

$$CH_3\text{-}(CH_2)_4\text{-} CH = CH\text{-}CH_2\text{-}CH = CH\text{-}(CH_2)_7\text{-}COOH$$

Fig. 2.9 Structure of saturated and unsaturated fatty acids

2.2.1.3 Properties

1. Sensitivity to oxidation

   Unsaturated fatty acids are oxidized in the presence of air, leading to rancidification, i.e., deterioration. Rancidification gives unpleasant smell and taste. Unsaturated fatty acids undergo auto-oxidation (peroxidation) due to the presence of double bonds. Peroxidation of lipids due to exposure to oxygen causes the formation of free radicals. Excessive and prolonged formation of free radicals may cause cancer and inflammatory diseases.

2. Saponification

Hydrolysis of triglycerides by alkali leads to the liberation of glycerol and soaps. Soaps are sodium or potassium salts of long-chain fatty acids.

3. Hydrogenation

Unsaturated fatty acids can be converted to saturated fatty acids by hydrogenation of the double bond. Trans fatty acids are formed during hydrogenation (partial) of liquid vegetable oils due to repeated cooking of vegetable oils and during the processing of fatty foods. They raise the plasma cholesterol level.

4. Melting Point

Double bond decreases melting temperature. Unsaturated fatty acids have lower melting temperature compared to saturated fatty acids. This property helps to maintain the fluid nature of membrane lipids.

5. Surfactant

Surfactant is a mixture of various lipids present in the alveoli of lungs. It reduces surface tension between air and fluid lining the alveoli and prevents the collapse of alveoli. Surfactant production begins late in gestation (after 36 weeks of gestation). Infant respiratory distress syndrome (IRDS) is a disease of preterm newborns delivered before 36 weeks of gestation. In IRDS high surface tension due to deficiency of surfactant collapses many alveoli (atelectasis). Infants die due to pulmonary edema and hypoxemia.

6. Insulation

Lipids act as an insulator. When the body is exposed to high/low environmental temperature, lipids prevent the entry of high/low temperature inside the body in order to maintain body temperature. Nerve impulse, i.e., depolarization cannot pass through myelin sheath (protein-lipid complex) of the nerve fiber as it acts as an insulator, and thereby, impulse jumps from one node of Ranvier to another, causing rapid propagation of impulse of the myelinated nerve.

7. Fatty acids

Fatty acids are weak acids. They form salts with alkali and esters with alcohols.

Fatty acid + glycerol → Monoglyceride

Monoglyceride + fatty acid → Diglyceride

Diglyceride + fatty acid → Triglyceride

8. Emulsions

These are formed by nonpolar (hydrophobic) lipids in an aqueous solution and stabilized by emulsifying agents, for example, lecithin (amphipathic lipid). Amphipathic lipids form a layer, which separates the bulk of nonpolar material from the aqueous solution.

2.2.1.4 Functions (Table 2.10) and Lipogenesis vs Lipolysis (Table 2.11)

Table 2.10 Functional role of lipids

Lipids	1. Major source of energy fuel. About 80% of the energy fuel of the body is derived from lipids. The amount of ATP formed per mole of oxidized free fatty acid is large, but varies with the size of the free fatty acid, for example, complete oxidation of one molecule of palmitic acid generates 129 ATP. The brain and RBC cannot use plasma free fatty acids for fuel
	2. Provide the hydrophobic barrier (because of insolubility in water) between the membrane associated lipids and aqueous content of cells
	3. Transported by plasma in association with protein as lipoproteins. Lipoproteins are important constituents of plasma membrane as well as mitochondrial membrane
	4. Essential fatty acids are essential for fluidity of the membrane

Table 2.11 Factors affecting lipogenesis and lipolysis

Factor	
Insulin	Inhibits lipolysis and stimulates lipogenesis
Epinephrine, norepinephrine, glucagon, growth hormone, and glucocorticoids	Stimulates lipolysis (see hormone-sensitive lipase)
Thyroid hormones	Stimulates lipolysis
Adiponectin	Inhibits lipogenesis and stimulates oxidation of fatty acids
Exercise	Enhances lipolysis through the liberation of epinephrine, norepinephrine, and adiponectin

Note: The synthesis of lipids from amino acids or glucose is called lipogenesis. The breakdown of triglycerides into glycerol and fatty acids is called lipolysis. Lipolysis mobilizes fatty acids in the circulation for oxidation

2.2.2 Digestion and Absorption

2.2.2.1 Stomach
Lingual lipase secreted from the dorsal surface of the tongue is activated by the acidic gastric juice and hydrolyzes dietary triglycerides into fatty acids and 2-monoglycerides. The rate of breakdown is slow as the lipids are not emulsified. Gastric lipase is of no physiological significance as it acts only at neutral pH. Gastric lipase can hydrolyze milk lipids in infants as pH of the infant stomach is near neutral.

2.2.2.2 Small Intestine
Lipids are digested mainly by pancreatic lipase as lipids are emulsified into small lipid droplets by the detergent action of bile salts, lecithin, and phospholipids as well as by mixing due to peristalsis. Moreover, emulsification increases the surface area of the lipid globules to facilitate the action of pancreatic lipase. Pancreatic lipase hydrolyzes triglycerides into fatty acids and 2-monoglycerides. Colipase, a protein, is secreted by the pancreatic juice and is activated by trypsin. It facilitates

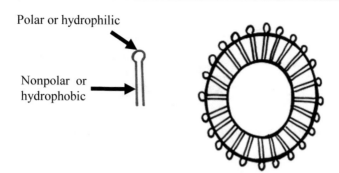

Fig. 2.10 Mixed micelle showing hydrophilic and hydrophobic groups. Bile salts, phospholipids, fatty acids, and monoglycerides are gathered with polar ends directed towards the surface and nonpolar ends directed towards the core. Cholesterol and fat-soluble vitamins are present in the core of the micelle

the action of lipase. It is amphipathic and carries the lipase on the lipid droplets. Cholesteryl esters are degraded by pancreatic cholesteryl ester hydrolase into cholesterol and fatty acids. Phospholipids are degraded by pancreatic phospholipase A_2 (activated by trypsin) into fatty acids and lysophospholipids.

Glycerol and fatty acids having chain length less than 10–12 carbon atoms pass from the mucosal cells of the intestine directly into the portal blood. All other products of digestion (2-monoglycerides, fatty acids having 10–12 carbon atoms reesterified to triglycerides, esterified cholesterol, and phospholipids) are absorbed with the help of bile salts by forming micelles. The bile salts (sodium and potassium salts of bile acids) are amphipathic having both hydrophilic and hydrophobic domains. The bile salts tend to form spherical disks called micelles with hydrophilic exterior and hydrophobic interior core. Concentration of bile salts required for the formation of micelles is known as critical micellar concentration. Above this critical concentration, all bile salts form micelles. Fatty acids, monoglycerides, and cholesterol collect in hydrophobic center, and phospholipids are lined up with their hydrophilic heads on the outside and their hydrophobic tails in the center. These molecular aggregates are called mixed micelles (Fig. 2.10). The micelle core also contains lipid-soluble vitamins. Mixed micelles having lipids are transported to the mucosal surface for absorption. Lipid molecules are released from the micelles and diffuse into the mucosal cells. Inside the mucosal cell, the fatty acids and monoglycerides are reesterified to form triglycerides. The enzymes for triglyceride synthesis are present in the smooth endoplasmic reticulum, resulting in the synthesis of triglycerides. Lipoproteins consist of a core of nonpolar hydrophobic lipids surrounded by polar hydrophilic lipids and a shell of proteins called apolipoproteins (APO) (Fig. 2.11). These lipoprotein complexes are called chylomicrons and solubilize highly hydrophobic lipids. Chylomicrons are absorbed by exocytosis from the intestinal mucosa into the blood via lymphatic vessels (lacteals) of the villi of the small intestine. Triglyceride catabolism is shown in Fig. 2.12.

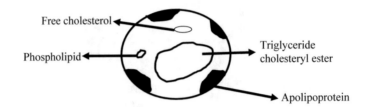

Fig. 2.11 Structure of lipoprotein. Lipoprotein is a stable structure as it is surrounded by apolipoprotein

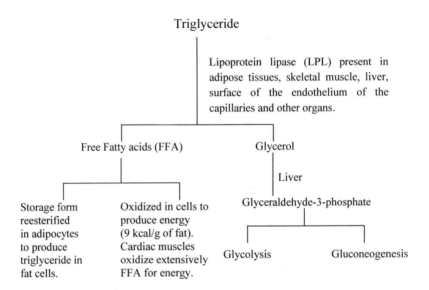

Fig. 2.12 Triglyceride catabolism

2.2.3 Steatorrhea

Steatorrhea is characterized by abnormal altered stools (stools are pale, bulky, greasy, and foul smelling). The amount of fat and protein in the stool is very high. Steatorrhea is due to (1) loss of pancreatic lipase and proteolytic enzyme after the surgical removal of pancreas or due to pancreatitis, which destroys the exocrine pancreas and thus prevents the digestion of fat and protein. (2) About 90% of bile salts are absorbed from the terminal ileum and enter the liver via portal vein (entero-hepatic circulation of bile salts). Due to the malabsorption syndrome, bile salts are not reabsorbed and excreted through the feces. Bile salts are essential for the digestion of fat. Fat digestion is impaired if bile salts are excreted through the feces. The abnormal altered stools are mainly due to defective digestion of fat (either due to the absence of pancreatic lipase or due to the absence of bile salts).

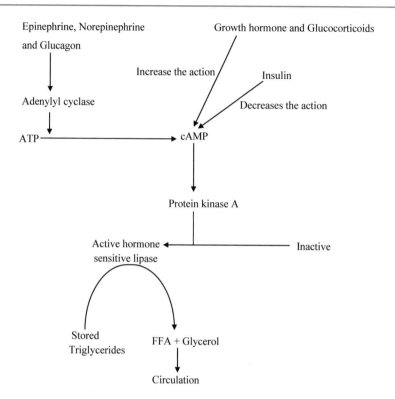

Fig. 2.13 Hormone-sensitive lipase

2.2.4   Hormone-Sensitive Lipase and Eicosanoids

2.2.4.1 Hormone-Sensitive Lipase (Ganong 2003)

Adenylyl cyclase converts ATP to cAMP. cAMP activates protein kinase A, which activates hormone-sensitive lipase. Hormone-sensitive lipase breaks down stored triglycerides into FFA and glycerol. Various hormones modify the activity of hormone-sensitive lipase (Fig. 2.13). Catabolism of fat depot is increased by stress (due to secretion of epinephrine, norepinephrine, and glucocorticoids) and diabetes mellitus due to the activation of hormone-sensitive lipase.

2.2.4.2 Eicosanoids

Prostanoids (prostaglandins, prostacyclin, and thromboxanes) and leukotrienes are called eicosanoids as they originate from 20 carbon (eicosa-) arachidonic acid. Membrane phospholipids are converted to arachidonic acid, which gives rise to arachidonic acid metabolites. Actions of arachidonic acid metabolites are mentioned in Fig. 2.14. Leukotrienes are mediators of allergic response. Thromboxane A_2 produced by platelets promotes aggregation of platelets and vasoconstriction. Platelet aggregation and vasoconstriction promote clot formation and thrombosis, which will occlude the lumen of small arteries and lead to ischemia and infarction of vital

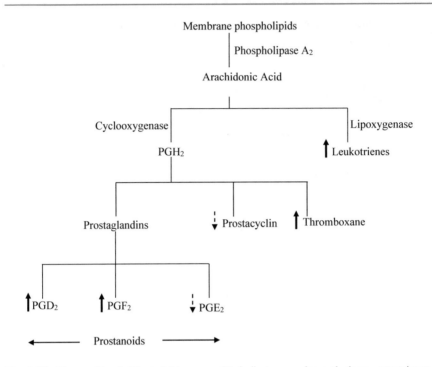

Fig. 2.14 Eicosanoids. Solid straight arrows (↑) indicate vascular and airway constrictors. Inverted dashed arrows (↓) indicate vascular and airway relaxants. PG prostaglandin

organs. Omega-3 fatty acids found in fish oils prevent the conversion of arachidonic acid to thromboxane A_2 and prevent thrombosis. Furthermore, omega-3 fatty acids reduce the development of atherosclerosis as they increase peripheral utilization of glucose by increasing insulin sensitivity and thus decrease adiposity.

2.2.5   Fatty Acid Synthesis and Oxidation (Murray et al. 1996)

2.2.5.1 Synthesis (Fig. 2.15)
The first step for fatty acid synthesis is due to carboxylation of acetyl-CoA to malonyl-CoA catalyzed by acetyl-CoA carboxylase. Insulin increases fatty acid synthesis by activating acetyl-CoA carboxylase, whereas glucagon and epinephrine decrease synthesis by inactivating the same. Acetyl-CoA cannot cross the mitochondrial membrane and is condensed with oxaloacetate to form citrate. As the enzymes responsible for fatty acid synthesis are present outside the mitochondria, citrate from citric acid cycle is transported out of the mitochondria to form acetyl-CoA. Palmitate is formed when fatty acid reaches a length of 16 carbons. Triglycerides are synthesized by inserting fatty acids to glycerol phosphate. Fatty acid synthesis is inhibited by starvation and high dietary fat, whereas it is stimulated

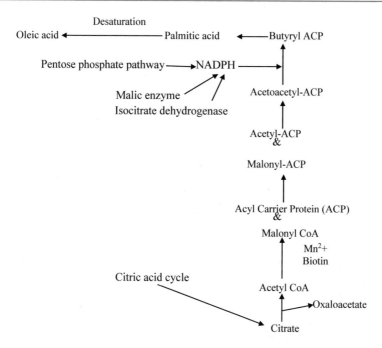

Fig. 2.15 Fatty acid synthesis

by high dietary carbohydrate and low fatty acid levels. Acyl carrier protein (ACP) contains pantothenic acid. Deficiency of biotin and pantothenic acid interferes in fatty acid synthesis.

2.2.5.2 Oxidation (Fig. 2.16)

A sequential removal of two carbon atoms at a time occurs due to oxidation of the hydrocarbon chain. β-Oxidation is termed so because fatty acids are degraded by oxidation at β-carbon. Fatty acids are activated before breakdown. Fatty acid + ATP + CoA → Acyl-CoA + PPi + AMP. This reaction is catalyzed by acyl-CoA synthetase.

Catabolism of fatty acid occurs inside the mitochondria. Long-chain fatty acids bound to carnitine (synthesized from lysine and methionine) enter inside the mitochondria. Medium- and short-chain fatty acids can enter inside the mitochondria without the help of carnitine. Complete catabolism of palmitate generates 129 ATP. Deficiencies of lysine and methionine lead to deficiency of carnitine, which results in deficient oxidation of fatty acids. As less energy is derived due to deficient oxidation of fatty acids, glucose stores are used up to provide energy, causing hypoglycemia. Hypoglycemia triggered by fasting may be fatal and may lead to coma. Hypoketonemia occurs as ketone bodies are not formed due to lack of acetyl-CoA. Due to defective synthesis of carnitine, skeletal muscle is not able to use fatty acid as fuel, resulting in weakness following muscular exercise.

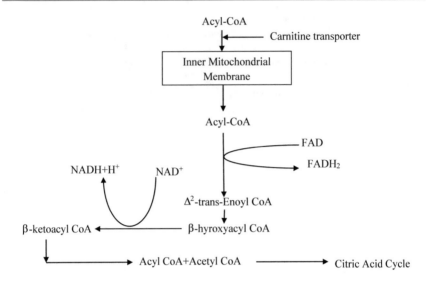

Fig. 2.16   Steps in β-oxidation of fatty acids

2.2.6   Ketone Bodies (Fig. 2.17)

Acetoacetate, β-hydroxybutyrate, and acetone are called ketone bodies. Two acetyl-CoA condense to form acetoacetyl-CoA, which further combines with acetyl-CoA to form 3-hydroxy-3-methylglutaryl-CoA (HMG-CoA). Acetyl-CoA is split off from HMG-CoA forming free acetoacetate. Acetoacetate is the primary ketone body and gives rise to β-hydroxybutyrate and acetone. Ketone bodies enter into the circulation (ketosis) as they are metabolized slowly in the liver. Acetoacetate and β-hydroxybutyrate (anions of acetoacetic and β-hydroxybutyric acids) cause metabolic acidosis, for example, diabetes mellitus and starvation (Fig. 2.18).

2.2.7   Dietary Sources, RDA, Deficiency, and Overconsumption

2.2.7.1 Dietary Sources (Fig. 2.19)

2.2.7.2 RDA (Fig. 2.20)

2.2.7.3 Deficiency

1. Essential fatty acids cannot be synthesized by human beings and are essential for fluidity of the membrane. They must be consumed in the diet as their deficiencies will lead to loss of hair, dermatitis, poor wound healing, growth retardation, failure to reproduce, and mental illness.
2. Lack of fat in the subcutaneous tissues will lead to rise or fall of normal body temperature in relation to high or low environmental temperature.

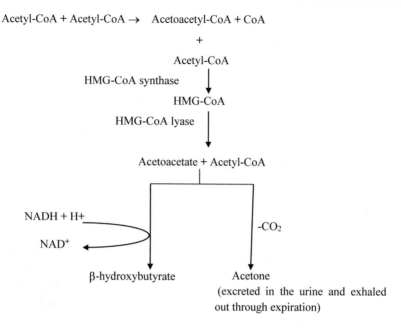

$$\text{Acetyl-CoA} + \text{Acetyl-CoA} \rightarrow \quad \text{Acetoacetyl-CoA} + \text{CoA}$$

$$+$$

$$\text{Acetyl-CoA}$$

HMG-CoA synthase

HMG-CoA

HMG-CoA lyase

Acetoacetate + Acetyl-CoA

NADH + H+

NAD$^+$

-CO$_2$

β-hydroxybutyrate

Acetone
(excreted in the urine and exhaled
out through expiration)

Fig. 2.17 Synthesis of ketone bodies in the liver

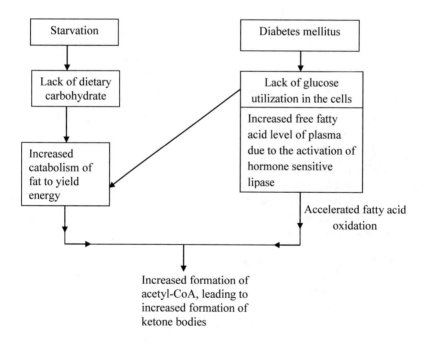

Starvation

Diabetes mellitus

Lack of dietary
carbohydrate

Lack of glucose
utilization in the cells

Increased free fatty
acid level of plasma
due to the activation of
hormone sensitive
lipase

Increased
catabolism of
fat to yield
energy

Accelerated fatty acid
oxidation

Increased formation of
acetyl-CoA, leading to
increased formation of
ketone bodies

Fig. 2.18 Formation of ketone bodies due to starvation and diabetes mellitus. Starvation due to gastroenteritis (vomiting and diarrhea) causes ketosis, resulting in acetone odor of expiration. Ingestion of carbohydrate prevents ketosis. Carbohydrate acts as an antiketogenic

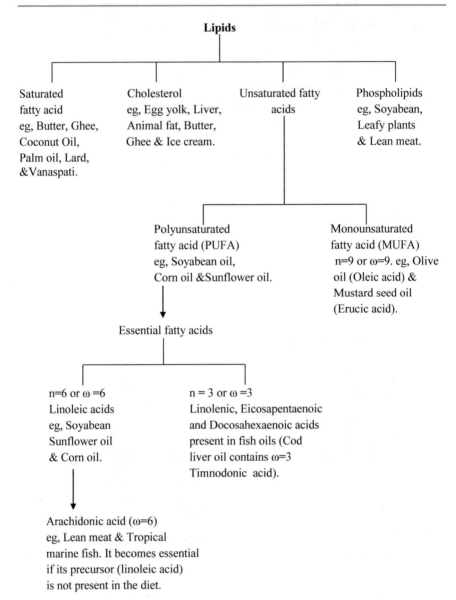

Fig. 2.19 Dietary sources of lipids. Organic eggs having omega-3 fatty acid are produced by adding algae in the chicken feed. Fatty fish such as salmon, tuna, mackerel, sardine, and herring are rich in omega-3 fatty acids

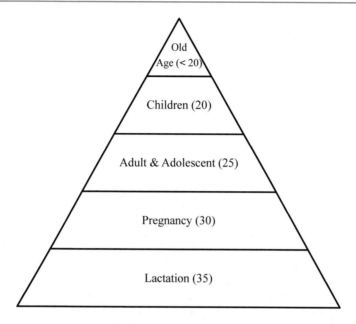

Fig. 2.20 Pyramid levels show RDA for lipids at different age groups, and during pregnancy and lactation. Figures in parenthesis indicate the amount of lipids (g/day)

3. Lowering of BMR as the lipids are the main source of energy fuel.
4. Minimizes or remediates the cognitive dysfunction.

2.2.7.4 Overconsumption
High dietary fat causes overweight and obesity despite the fact that fatty acid synthesis is inhibited by high dietary fat.

2.2.8 Saponification, Iodine, and Acid Values (Hasan et al. 2018)

Saponification value or number is the number of milligrams of potassium hydroxide required to saponify 1 g of fat. It depends on the kind of fatty acid present in the diet. It is an indicator of the molecular weight (or chain length) of fatty acids. Higher saponification indicates low molecular weight and vice versa. The long-chain fatty acids will have low saponification value. Low saponification value of the long-chain fatty acids is due to lesser number of carboxylic groups compared with short-chain fatty acids. The saponification value is determined by taking 1 g oil sample and 25 mL 0.5 N alcoholic KOH in a flask, which is heated for 1 h, and in the hot solution, phenolphthalein indicator is added. KOH is titrated with 0.25 M HCl until pink endpoint is reached. Blank experiment without oil is conducted under similar conditions.

$$\text{Saponification value} = \frac{(A-B)\times S \times 56.1}{W}$$

S = molarity of HCl used
A = volume of HCl in mL for blank
B = volume of HCl in mL for sample
W = weight of oil used in g

Saponification values (mg KOH/g) of coconut oil, butter, and human fat are about 260, 220, and 196, respectively.

Iodine value or number of a fat indicates the number of grams of iodine used by 100 g of fat. Yellow or brown color of iodine will disappear after reacting with mainly unsaturated fatty acids at a particular concentration. It is directly proportional to the content of unsaturated fatty acids. It determines the amount of unsaturation in fatty acids. Iodine monobromide converts potassium iodide to iodine. Iodine concentration is determined by titration with sodium thiosulfate.

Iodine value (g/100 g) of sunflower oil (polyunsaturated) is 120–130, groundnut oil (more monounsaturated) 85–100, butter (saturated) 25–30, and coconut oil (saturated) 5–10.

Acid value or number is the mass of potassium hydroxide in milligrams to neutralize 1 gram of oil or fat. The acid value is a measure of the carboxylic acid group in chemical compounds such as fatty acid and is used to quantify the amount of acid present. Acid value is determined by taking 25 mL of absolute ethanol and two drops of phenolphthalein indicator in a dry conical flask, containing 1 g of oil sample. The mixture is shaken in water bath at about 60 °C for 15 min. After cooling, the mixture will be titrated against 0.1 M KOH solution.

$$\text{Acid value} = \frac{V \times M \times 56}{W}$$

V = volume of KOH solution in mL
S = volume of HCl in mL for blank
W = weight of oil used in g

The maximum level of acid value for refined oil is 0.6 mg KOH/1 g oil and for natural fat is 4.0 mg KOH/1 g fat. Due to rancidity, triglycerides are converted into fatty acids and glycerol, causing an increase in acid value. Cooking oil is deteriorated by heat, light, and bacteria, causing unpleasant odor. Deteriorated cooking oil causes destruction of essential fatty acids and vitamins A, D, and E. It increases acid value due to the generation of fatty acids.

Note: Ester value is milligrams of KOH required to saponify the esters contained in 1 g of oil. Ester value = saponification value − acid value.

2.3   Protein Metabolism and Nutrition

2.3.1   Classification, Structure, Properties, and Functions

2.3.1.1 Classification
1. Based on physiological functions
   (a) Defensive proteins, for example, immunoglobulins, cytokines (hormone-like molecules that regulate immune functions) and chemokines (protective substances that attract neutrophils to areas of inflammation).
   (b) Contractile proteins, for example, myosin and actin.
   (c) Storage proteins, for example, ferritin and hemosiderin.
   (d) Transport proteins, for example, hemoglobin, B_{12}-binding protein, ceruloplasmin, and transferrin.
   (e) Regulatory proteins, for example, growth factors and DNA-binding proteins.
   (f) Structural proteins, for example, collagen, elastin, and keratin.
2. Based on shape
   (a) Globular proteins are folded and coiled due to complex interaction of secondary, tertiary, and sometime quaternary structures, for example, hemoglobin and myoglobin.
   (b) Fibrous proteins are not coiled and folded, having high tensile strength. These are formed due to the combination of specific amino acids with secondary structure proteins, for example, collagen and elastin of the skin and keratin of the skin and hair.
3. Conjugated proteins

   Based on the combination with carbohydrate (glycoproteins) or with lipids (lipoproteins) or with a specific metal, for example, Zn^{2+}, Fe^{2+}, and Cu^{2+} (metalloproteins).

2.3.1.2 Structure

Primary Structure (Fig. 2.21)
Primary structure indicates the order in which amino acids are joined together. It indicates the sequence of amino acids in a protein. Amino acids are joined by covalent bonds of the peptide. Peptide is a molecule consisting of amino acids linked by

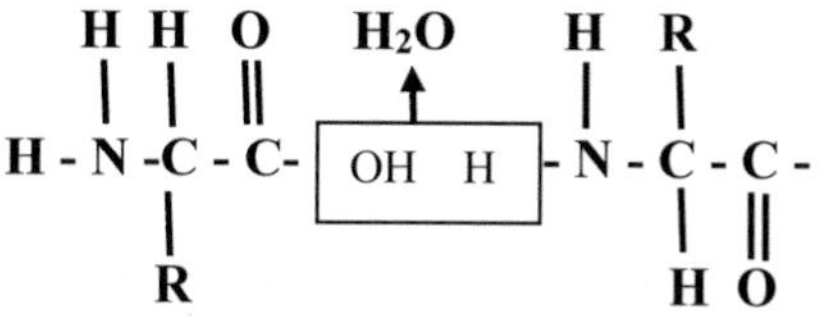

Fig. 2.21 Peptide bonds are formed with the production of H_2O. The peptide bonds of a polypeptide chain are –CO–NH– (amide) linkages. Amino acids have the general formula –RCH(NH$_2$)COOH. R indicates amino acid side chain

Fig. 2.22 Tertiary
structure. Twisted chain of
entire polypeptide (α-chain
of hemoglobin)

Heme

bonds between amino group (–NH$_2$) and carboxyl group (–COOH). There are two
terminals. At one end there is N-terminal, and at the other end there is C-terminal.
Many amino acids, usually more than 100, are joined by peptide bonds, forming an
unbranched polypeptide chain.

Secondary Structure (Voet et al. 2016)

Secondary structure consists of α-helix and β-sheet. α-Helix is a rod-like structure
and consists of a tightly packed polypeptide backbone. The side chains of amino
acids extend outward. The structure is stabilized by hydrogen bonds between NH
and CO groups in the same polypeptide chain. α-Helix is abundant in hemoglobin.
Fibrous protein keratin, found in the skin and hair, consists of entirely α-helix.
β-Sheet is another type of secondary structure. It is a sheet-like structure, not rod-
like structure. The surfaces of sheets are pleated, and that is why, it is often termed
β-pleated sheet. The polypeptide chain in the β-pleated sheet is almost fully extended
and not like tightly coiled. β-Pleated sheet is stabilized by hydrogen bonds between
NH and CO groups in neighboring polypeptide strands. β-Sheets are found in
fibrous and globular proteins. Sheets are of two varieties: (1) antiparallel β-sheets
that are neighboring hydrogen bonded polypeptide chains, which run in opposite
direction and (2) parallel β-sheets that are hydrogen-bonded polypeptide chains,
which extend in the same direction.

Tertiary Structure (Fig. 2.22)

Tertiary structure indicates the three-dimensional structure of entire polypeptide. It
is the arrangement of twisted chains into layers. Tertiary structure is composed of
α-helices (hemoglobin subunit) or mostly β-strands (immunoglobulin).

Quaternary Structure (Fig. 2.23)

Many protein molecules are made of subunits (hemoglobin consists of 2α and 2β
chains; each polypeptide chain is termed a subunit). Quaternary structure refers to
the spatial arrangement of subunits.

2.3.1.3 Properties

1. Osmotic Property

    The plasma proteins, mainly albumin, exert osmotic pressure of 25 mmHg.
    This osmotic pressure holds water inside the capillaries and thereby prevents
    water from going out of the capillaries. Water will go out of the capillaries and
    accumulate outside the vascular system (edema) due to less plasma proteins.

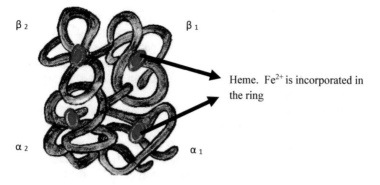

β 2   β 1

Heme. Fe^{2+} is incorporated in
the ring

α 2   α 1

Fig. 2.23 Quaternary structure. 2α and 2β chains of hemoglobin show arrangement of subunits

2. Isoelectric Point

Proteins can ionize either as acids or bases due to the fact that the side chains of amino acids contain amino groups ($-NH_2$) and carboxyl groups ($-COOH$). In alkaline solutions, the proteins ionize as acids, and free protein anions are formed. In acidic solutions, the protein amino groups act as bases and take up H^+ while the ionization of the carboxyl group is suppressed. Thus, the proteins carry a net positive charge. The method of electrophoresis depends on this principle. At an intermediate pH, the protein molecules carry equal number of positive and negative charges with net charge zero. This electrical neutrality is known as isoelectric point.

3. Buffer action

Plasma proteins act as buffers as they can act both as an acid and as a base (amphoteric property). Proteins can either donate or accept a proton (H^+).

4. Precipitation by salts

Different proteins are precipitated from solution by the addition of different concentration of salt, for example, albumin is precipitated by saturation with $(NH_4)_2SO_4$.

5. Denaturation of proteins

Protein can be denatured by heat, strong acids, or bases and by organic solvents. Protein denaturation is due to unfolding and disorganization of the structure of protein.

6. Hydrophobic and hydrophilic amino acids

Some of the amino acids (alanine, leucine, phenylalanine, tyrosine, isoleucine, methionine, tryptophan, valine, and proline) are hydrophobic. The rest are hydrophilic (serine, threonine, cysteine, asparagine, glutamine, lysine, arginine, and histidine).

2.3.1.4 Functions

Proteins play a key role in almost all functions of the human body.

1. All the enzymes are proteins. Nearly all chemical reactions are catalyzed by the enzymes. Proteins determine ultimately the pattern of various chemical transformations in the cell.

2. Amino acids are the basic structural units of proteins. All proteins are made up of amino acids. Amino acids take part in the formation of neurotransmitters, hormones, and digestive enzymes and take part in various physiological functions. Amino acids are required for the formation of new tissues during growth and pregnancy. Amino acids maintain the structure of every tissue and maintain the content of the enzyme systems of every cell.
3. Hemoglobin transports oxygen to the cells and carbon dioxide to the lungs. Each iron atom of four heme groups of hemoglobin combines with oxygen molecules and transports oxygen to the cells. Carbon dioxide combines with amino group of the proteins, mainly with hemoglobin and slightly with plasma protein and forms carbamino compounds. Carbamino compounds are transported to the lungs for the exhalation of carbon dioxide. Myoglobin stores oxygen in the muscles, and this stored oxygen will be utilized when there is an increased oxygen demand during severe exercise (Chakrabarty and Chakrabarty 2006).
4. Immunoglobulins, cytokines, and chemokines are specific proteins that take part in various immune functions.
5. Various peptide growth factors take part in the control of growth and differentiation of cells.
6. Muscular contraction is brought about by sliding of protein actin, thin filament, over the protein myosin, thick filament. The troponin-tropomyosin constitutes a relaxin protein that causes termination of interaction between actin and myosin. Myosin contains amino acid 6-N-methyllysine.
7. Fibrous protein is responsible for the tensile strength of skin. Fibrous protein, for example, collagen contains amino acids (4-hydroxyproline derived from proline and 5-hydroxylysine derived from lysine).
8. Receptors, specific proteins, play crucial roles in the action of various hormones, neurotransmitters, and other various ligands. Receptor potentials are generator potentials for the generation of action potentials of various cells. Propagated action potentials take part in the genesis of various functions, for example, vision, hearing, smell, taste, and synaptic transmission.

2.3.2   Digestion and Absorption

Pepsinogens (inactive proenzymes) secreted by peptic cells of the stomach are converted to active pepsin (endopeptidase) by HCl secreted by parietal cells of the stomach. Pepsin hydrolyzes polypeptides into smaller peptides. As the optimal pH of pepsin is about 2–3, pepsin cannot act further when the gastric contents are mixed with alkaline pancreatic secretions in the duodenum and jejunum. Proteolytic enzymes secreted by exocrine pancreas as inactive proenzymes are subsequently converted to the active form. Endopeptidases are active trypsin, chymotrypsin, and elastase. They act on the interior peptide bond.

Exopeptidases are active carboxypeptidase A and B. They cleave terminal amino acids at the carboxyl and amino end of a peptide. The final digestion of protein by brush border enzymes of the small intestine is shown in Fig. 2.24.

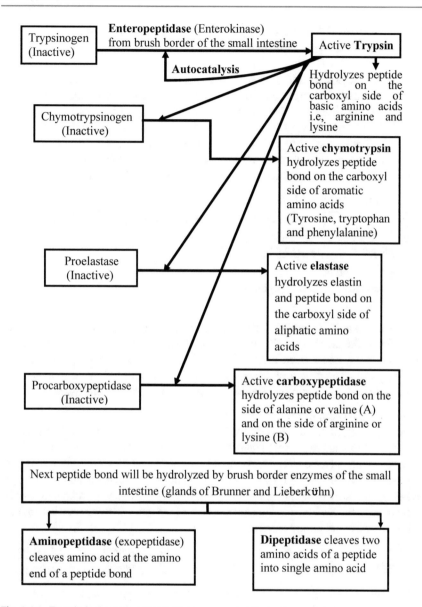

Fig. 2.24 Trypsin is the main proteolytic enzyme and activates proenzymes

The final digestion product of proteins are amino acids, which are actively transported by different transporters depending on the types of amino acids in the intestinal cells. Some of dipeptides that are absorbed by intestinal cells are hydrolyzed to amino acids in the cytosol. Amino acids enter inside the capillaries by diffusion and are finally transported to the liver by the hepatic portal system. Amino acids are metabolized by the liver. Some amino acids enter into the general circulation.

2.3.3 Essential Amino Acids, Nonessential Amino Acids, and Nitrogen Balance

Essential or indispensable amino acids cannot be synthesized by humans. Except arginine and histidine, eight amino acids must be supplied in the diet at all times. Arginine and histidine (semi-essential) are required only during tissue growth, during childhood and recovery from illness. Sufficient arginine is synthesized by the urea cycle. Lack of any one indispensable amino acids results in negative nitrogen balance characterized by anorexia, loss of body weight, fatigue, and irritability. These are essential for normal growth and development of cells at all times.

Ten essential amino acids are

1. Valine[a]	2. Isoleucine[b]	3. Phenylalanine[b]	4. Histidine[a]
5. Arginine[a]	6. Leucine[c]	7. Lysine[c]	8. Methionine[a]
9. Tryptophan[b]	10. Threonine[a]		

[a]Only gluconeogenic
[b]Both gluconeogenic and ketogenic
[c]Only ketogenic

Five amino acids (cysteine, tyrosine, arginine, proline, and glycine) are grouped under "conditionally essential" amino acids as they can be synthesized from other amino acids (cysteine from methionine and serine; tyrosine from phenylalanine; arginine from glutamine and aspartic acid; proline from glutamate; glycine from serine and choline). Common cereals such as wheat and maize are lysine-deficient.

Nonessential Amino Acids
Ten amino acids are termed nonessential as they are synthesized in vivo by metabolic pathways in amounts sufficient to meet metabolic needs, and their absence in food will not produce any adverse effect. One should not forget that maximum growth is only possible when all amino acids (both essential and nonessential) are available to the tissues simultaneously in sufficient amounts. The following ten nonessential amino acids are mentioned below. Some are gluconeogenic or both gluconeogenic and ketogenic.

The nonessential amino acids are

1. Alanine	2. Asparagine	3. Aspartic acid	4. Glutamate	5. Glutamine
6. Glycine	7. Cysteine	8. Proline	9. Serine	10. Tyrosine

Tyrosine is the only amino acid that is both gluconeogenic and ketogenic. The remaining nine are only gluconeogenic.

Nitrogen balance occurs when the nitrogen content of dietary protein is equal to loss of nitrogen in the urine, feces, and sweat. Positive nitrogen balance occurs when nitrogen intake of dietary protein is more than the loss of nitrogen. It occurs during growth or recovery from severe illness. Negative nitrogen balance occurs when nitrogen loss exceeds nitrogen intake of dietary protein. It occurs during starvation and forced immobilization due to burns, injury, surgery, etc. Burns and injury cause increased protein catabolism, resulting in nitrogen loss. Nitrogen balance becomes negative as protein is not synthesized due to lack of even a single amino acid in the diet.

2.3.4   Water-Soluble Vitamins in Gluconeogenesis and Ketogenesis (Fig. 2.25)

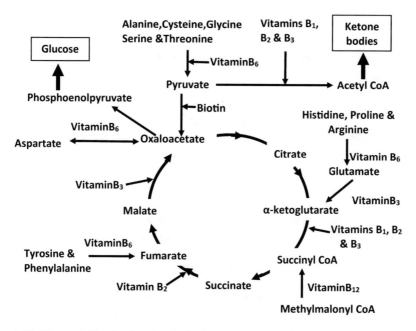

Fig. 2.25   Water-soluble vitamins take part in gluconeogenesis and ketogenesis

2.3.5   Ketogenic Amino Acids (Fig. 2.26)

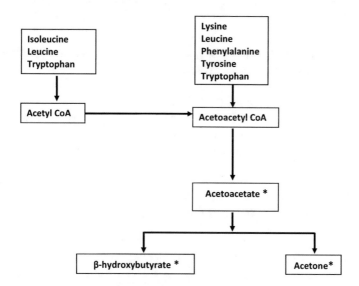

Fig. 2.26   Ketogenic amino acids. ∗ indicates ketone bodies

2.3.6   Role of Vitamins in Oxidative Deamination and Transamination (Fig. 2.27)

Oxidative Deamination

Deamination involves removal of amino group ($-NH_2$) to form a ketoacid and NH_3. For example,

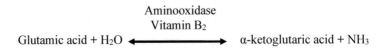

$$\text{Glutamic acid} + H_2O \underset{\text{Vitamin } B_2}{\overset{\text{Aminooxidase}}{\longleftrightarrow}} \alpha\text{-ketoglutaric acid} + NH_3$$

Transamination

Transamination involves conversion of one amino acid to a keto acid with simultaneous conversion of another keto acid to an amino acid

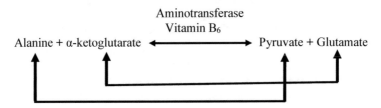

$$\text{Alanine} + \alpha\text{-ketoglutarate} \underset{\text{Vitamin } B_6}{\overset{\text{Aminotransferase}}{\longleftrightarrow}} \text{Pyruvate} + \text{Glutamate}$$

Fig. 2.27 Deamination and transamination of amino acids. Role of vitamins. α-Ketoglutarate of the Krebs cycle or pyruvate of the glycolytic pathway can be metabolized for the formation of ATP and gluconeogenesis as well as for the formation of fatty acids via acetyl coenzyme A

2.3.7   Amino Acid Metabolism, Synthesis of Nonessential Amino Acids and Protein Synthesis

2.3.7.1 Amino Acid Metabolism (Fig. 2.28)

Amino acids are derived from digestion of proteins, are synthesized in the body, and are obtained from dietary amino acids. Amino acids form various neurotransmitters, hormones, fatty acids, urea, purine, pyrimidine, creatine, nitric oxide (NO), and vitamins.

2.3.7.2 Synthesis of Nonessential Amino Acids

1. NH_4^+ enters into amino acids and forms glutamate and glutamine.

$$NH_4^+ + \alpha - \text{ketoglutarate} + NADPH \rightarrow \text{Glutamate} + NADP^+ + H_2O$$

$$\text{Glutamate} + NH_4^+ + ATP \rightarrow \text{Glutamine} + ADP + P_i + H^+$$

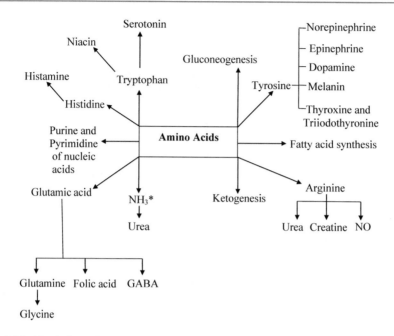

Fig. 2.28 Metabolism of amino acids. *It may be noted that NH_3 is removed by the liver under normal conditions. The concentration of NH_3 may rise to toxic levels in liver diseases and may contribute to hepatic coma. High-protein diet must be avoided in patients suffering from liver diseases

2. The carbon skeletons of amino acids are derived from glycolytic pathway and citric acid cycle.
   (a) α-Ketoglutarate is converted to glutamate by glutamate dehydrogenase. Glutamate is ultimately converted to glutamine by glutamine synthetase.
   (b) Oxaloacetate is converted to aspartate by aminotransferase. Aspartate is converted to asparagines by asparagines synthetase.
   (c) Pyruvate is converted to alanine by aminotransferase.
   (d) 3-Phosphoglycerate is converted to serine via phosphohydroxypyruvate and phosphoserine. Serine is converted to glycine catalyzed by serine hydroxymethyl transferase. Serine in combination with homocysteine gives rise to cystathionine, which gives rise to cysteine and homoserine
   Serine + Homocysteine → Cystathionine → Cysteine + Homoserine.

2.3.7.3 Protein Synthesis (Fig. 2.29)
Protein synthesis involves transcription, translation and post-translational modification.

2.3.7.4 Transcription [(DNA Is Converted to Ribonucleic Acid (RNA)]
DNA gene in the nucleus consists of coding regions (exons), which are interspersed with noncoding regions (introns). Transcription starts at the 5′ end of DNA by addition of 7-methylguanosine triphosphate, which is essential for binding to the ribosome.

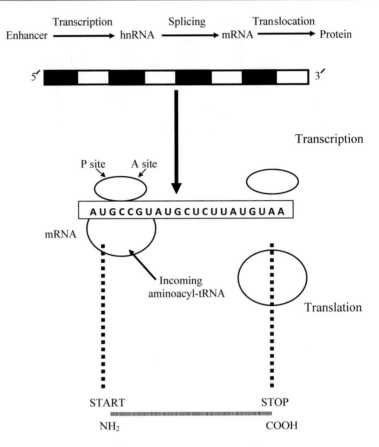

Fig. 2.29 Transcription and translation. Enhancer is a binding segment of DNA. Heterogeneous nuclear DNA (hnRNA) is a precursor of mRNA. Amino acid + ATP + tRNA ↔ aminoacyl-tRNA + AMP + pp$_i$ GTP is hydrolyzed for the binding of aminoacyl-tRNA to the ribosome

Poly (A) tail consisting of about 100 bases is added at 3′ end. The intron sequences are cleaved, and the exons are joined by interweaving strand to form the mature mRNA. Large portions of mRNA are cleaved off. mRNA from the nucleus moves to the cytoplasm. The mRNA is a transcript of a small segment of the DNA gene. The codons adenine (A), guanine (G), cytosine (C), and uracil (U) are present in the mRNA and are written from 5′ end to 3′ end. The mRNA ultimately dictates the synthesis of protein (translation, i.e., mRNA is translated into polypeptide chains).

Translation involves initiation, elongation, termination, and post-translation modification.

2.3.7.5 Initiation
In the cytoplasm, AUG codon on the mRNA initiates protein synthesis and is responsible for the interaction of the ribosome with mRNA as well as with the

tRNA. Ribosomes (protein-synthesizing machine) are large complexes of protein and rRNA. The prokaryotic 50S and 30S ribosomal subunits form 70S ribosome (S = sedimentation coefficient). The first step of translation is the binding of the small ribosomal subunit to the mRNA, which contains an AUG codon. The small ribosome has two binding sites for tRNA molecules, A (aminoacyl) and P (peptidyl) sites. During translation, A site binds an incoming aminoacyl-tRNA, which is N-formyl-methionine. P site is occupied by peptidyl-tRNA. The tRNA initiates protein synthesis and is known as adaptor molecule. The mRNA is translated from its 5′ end to 3′ end synthesizing a protein from its amino terminal to its carboxy-terminal end. Initiation factors (IF-1, IF-2, and IF-3) are protein factors associated with 30S subunits, which enhances protein synthesis. Binding of the large ribosomal subunit to the small subunit releases the initiation factors that can be reutilized.

2.3.7.6 Elongation
Peptide chain is elongated by the addition of amino acids to the carboxyl end of the growing peptide chain. During elongation the ribosomes moves from the 5′ end to the 3′ end of the mRNA.

2.3.7.7 Termination
It occurs when one of the three terminating codons (UAG, UGA, and UAA) reaches the A site. After termination, ribosome is immediately dissociated from the mRNA. The ribosomal subunits, mRNA, tRNA, and protein factors are used to synthesize another polypeptide. The ribosome repeat the synthesis several times.

2.3.7.8 Post-translation Modification
Part of the translated sequence may be removed or there may be covalent modification (e.g., phosphorylation and hydroxylation). This modification is required for the activity of protein.

2.3.7.9 Polysome
During translation, each mRNA molecule is translated by many ribosomes simultaneously forming a structure called a polysome. The purpose is to increase the speed for the synthesis of a long polypeptide chain.

2.3.8 Protein Degradation

Nonfunctional and damaged proteins (which are toxic) are degraded by three organelles: (1) lysosomes (acidic pH and contain proteolytic enzymes), (2) peroxisomes (contain oxidative enzymes), and (3) ubiquitin, a small heat-stable protein, attaches to protein and directs the protein to proteosome for degradation into smaller peptides. Ubiquitination is a common posttranslational mechanism for the degradation of zinc, iron, manganese, copper, and other trace minerals (Hennigar and McClung 2016). Proteins are constantly synthesized and degraded. The rate of protein

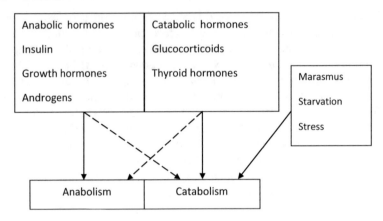

Fig. 2.30 Hormones affecting protein metabolism. Solid lines indicate stimulation. Dashed lines indicate inhibition. Anabolic hormones stimulate protein anabolism with the simultaneous inhibition of protein catabolism. Catabolic hormones will have the opposite effects. Marasmus, starvation, and stress increase the protein catabolism

synthesis is more or less equal to protein degradation so that the total amount of protein in the body remains constant. About 350 g protein per day undergoes degradation and resynthesis. Protein catabolism is stimulated by glucocorticoids and thyroid hormones, whereas protein synthesis is stimulated by insulin, testosterone, and growth hormone (Fig. 2.30).

2.3.9 Dietary Sources, Biologic Values, RDA, Deficiency, and Overconsumption

2.3.9.1 Dietary Sources (Table 2.12)

Table 2.12 Dietary sources of proteins (g/100 g)

1. High content >15	Egg, chicken, red meat, pork meat, fish (salmon, pomfret, hilsa, rohu), lobster, bengal gram dal, black gram dal, lentil dal, peas dry, soya bean white/brown, almonds, groundnut, cashew nut, Indian cottage cheese
2. Medium content between 10 and 15	Bajra, barley, jowar, whole wheat, walnut
3. Low content <10	Rice parboiled, rice flakes, rice puffed
4. Lowest content <5	Milk (buffalo, cow, and human), yoghurt, green leafy vegetables, brinjal, broad beans, green capsicum, cauliflower, green cucumber, French beans, ladies finger, onion, ripe tomato, fruits, roots and tuber, condiments

2.3.9.2 Biologic Values
1. Animal proteins with high biologic value are egg, fish, poultry, meat, and milk. Biologic value of a protein is determined by the presence of essential amino acids (Fig. 2.31).

2. Plant proteins with lower biologic value are cereals, legumes, and pulses. Proteins of plant sources may be combined in order to reach higher biologic values (that will be more or less equivalent to animal proteins). For example, wheat (lysine deficient but rich in methionine) + kidney beans (poor in methionine but rich in lysine) = animal protein. Similarly two or more pulses can be combined to reach higher biologic value.

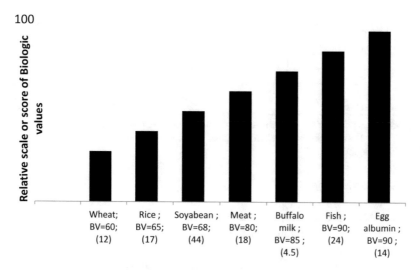

Fig. 2.31 Biologic values (BV) of animal and plant proteins. Numbers in parenthesis indicate the amount of protein in g/100 g. BV of egg albumin is highest followed by fish. Although the amount of protein of egg albumin is lesser compared with soya bean, the biologic value of egg albumin is much higher than that of soya bean

2.3.9.3 RDA (Fig. 2.32) (Table 2.13)

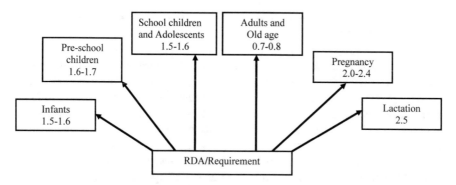

Fig. 2.32 Approximate RDA/Requirement for proteins g/kg/day at different age groups, during pregnancy and lactation. Elderly person's protein intake should not be less than 0.8 g/kg/day as they suffer from depletion of protein stores

Table 2.13 Estimates of amino acid requirements

Requirements, mg/kg per day, by age group

Amino acid	Infants, age 3–4 months	Children, age ~2 years	Children, age 10–12 years	Adults
Histidine	28	?	?	8–12
Isoleucine	70	31	28	10
Leucine	161	73	42	14
Lysine	103	64	44	12
Methionine plus cystine	58	27	22	13
Phenylalanine plus tyrosine	125	69	22	14
Threonine	87	37	28	7
Tryptophan	17	12.5	3.3	3.5
Valine	93	38	25	10
Total without histidine	714	352	214	84

Reproduced, with permission, from Recommended Dietary Allowances, Table 6-1, 10th edition, copyright © 1989 by the National Academy of Sciences, courtesy of the National Academies Press, Washington, D.C.

2.3.9.4 Deficiency

1. Hypoalbuminemia causes edema (kwashiorkor).
2. Impairment of the immune system.
3. Presence of growth retardation.
4. Presence of anemia due to low hemoglobin and transferrin.

2.3.9.5 Overconsumption
1. Protein in excess of the body's requirement will be deaminated to provide energy or will be converted to acetyl-CoA for fatty acid synthesis.
2. Urine is concentrated due to excess protein intake. Concentrated urine may result in the formation of kidney stones.

2.4   Summary

Citric acid cycle is concerned with gluconeogenesis, transamination, deamination, and fatty acid synthesis. Water-soluble vitamins play a crucial role in the citric acid cycle (carboxylation and dehydrogenation). Carbohydrates are considered as protein-sparing, because when there is intake of sufficient quantities of carbohydrates, the amino acids will be utilized for the maintenance of tissue protein. Carbohydrates are antiketogenic and prevent breakdown of fatty acids. Cardiac muscle has poor glyco-lytic activity, and survival rate is poor when there is inadequate coronary blood flow. Neutral lipids (e.g., triglycerides) from fat depots undergo breakdown and resynthe-sis. Catabolism of fat is increased by stress and diabetes mellitus. Cardiac muscles oxidize extensively FFA for energy. Lipids are the main source of energy fuel. Exercise enhances lipolysis. Lipolysis mobilizes fatty acids in the circulation for oxidation. Amino acids are the basic structural units of proteins. Essential amino acids must be supplied in the diet, whereas nonessential amino acids are synthesized in vivo. Maximum growth is possible when all amino acids (both essential and nonessential) are available to the tissues. Proteins are constantly synthesized and degraded. Protein synthesis is stimulated by insulin, androgen, and growth hormone. Protein catabolism is enhanced by marasmus, starvation, and stress.

References

Chakrabarty AS, Chakrabarty K (2006) Fundamentals of respiratory physiology, 1st edn. I.K. International Publishing House Pvt Ltd, New Delhi

Champe PC, Harvey RA (1994) Lippincott's illustrated reviews: biochemistry, 2nd edn. Lippincott Williams and Wilkins, Philadelphia

Colledge NR et al (eds) (2010) Davidson's principles and practice of medicine, 21st edn. Churchill Livingstone, London

Ganong WF (2003) Review of medical physiology, 21st edn. McGraw-Hill, New York

Hasan MK et al (2018) Comparison of some physiochemical parameters between marketed and virgin coconut oils available in Bangladesh. Int J Recent Sci Res 9(7):27979–27982

Hennigar SR, McClung JP (2016) Homeostatic regulation of trace mineral transport by ubiquitination of membrane transporters. Nutr Rev 74(1):59–67

Murray RK et al (1996) Harper's illustrated biochemistry, 24th edn. McGraw Hill, New York

Shiren S et al (2017) Lactic acid: no longer an inert and end-product of glycolysis. Physiology 32:453–463

Vasudevan DM et al (2011) Textbook of biochemistry for medical students, 6th edn. Jaypee Brothers Medical Publishers (P) Ltd, Chennai

Voet D et al (2016) Fundamentals of biochemistry: life at the molecular level, 5th edn. Wiley, Hoboken

Further Reading

Gibney MJ et al (eds) (2009) Introduction to human nutrition, 2nd edn. Wiley-Blackwell, Hoboken

Mann J, Truswell S (eds) (2017) Essentials of human nutrition, 5th edn. Oxford University Press, Oxford

Petersen MC et al (2017) Regulation of hepatic glucose metabolism in health and disease. Nat Rev Endocrinol 13:572–587

Ponziani FR et al (2015) Physiology and pathophysiology of liver lipid metabolism. Expert Rev Gastroenterol Hepatol 9:1055–1067

Reynolds A et al (2019) Carbohydrate quality and human health: a series of systematic review and meta-analyses. Lancet 393(10170):434–445

Enzymes

3

Abstract

Enzymes are specific protein catalysts in biological systems, which regulate the rate of biochemical and physiological processes. A catalyst is a substance that increases the rate of a chemical reaction and remains unchanged after the reaction. Coenzymes are essential for the activity of enzymes and are mainly derivatives of vitamins and AMP. Cofactor is a nonprotein metal ion associated with enzyme that is also essential for the activity of the enzyme. Zn^{2+} is associated with carbonic anhydrase and alcohol dehydrogenase, Cu^{2+} is associated with cytochrome oxidase, Mg^{2+} is associated with hexokinase and pyruvate kinase, etc. Carbonic anhydrase is one of the fastest enzymes. Apart from catalytic power, carbonic anhydrase takes part in the regulation of pH, carboxylation reaction, and CO_2 transport. Enzyme molecules contain active sites, are highly specific, and catalyze only one type of chemical reaction. Enzymes can be activated or inhibited. Allosteric enzymes can be inhibited by an inhibitor, which is not an analogue (e.g., cholesterol inhibits HMG-CoA reductase). Trypsin inhibitors are present in many vegetables, are denatured and inactivated by cooking. Lead inhibits ferrochelatase by noncompetitive inhibition and prevents the incorporation of ferrous ion into protoporphyrin for forming heme. Similarly, lead inhibits δ-aminolevulinate dehydrogenase and prevents the conversion of ALA into porphobilinogen. Velocity or rate of enzyme reaction depends on optimum temperature and pH. Michaelis–Menten equation explains the kinetic properties of the enzymes.

Keywords

Catalyst · Coenzyme · Cofactor · Metal ions · Carbonic anhydrase · Allosteric enzyme · Michaelis–Menten equation

© Springer Nature Singapore Pte Ltd. 2019
K. Chakrabarty, A. S. Chakrabarty, *Textbook of Nutrition in Health and Disease*,
https://doi.org/10.1007/978-981-15-0962-9_3

3.1   Classification

Six major classes of enzymes are shown below (Vasudevan et al. 2011).

1. **Oxidoreductases** catalyze oxidation–reduction reactions and will catalyze oxidation of one substrate with simultaneous reduction of another substrate or coenzyme. Oxidation is the combination of a substance with O_2 or loss of hydrogen or loss of electron. Conversion of lactic acid into pyruvic acid is an example of an oxidation reaction, which is catalyzed by lactate dehydrogenase (LDH) with loss of 2H. Reduction is the reverse of oxidation. Conversion of pyruvic acid into lactic acid is an example of a reduction reaction, which is catalyzed by LDH with the gain of 2H. Oxidation and reduction reactions are always coupled. Such paired reactions are called redox reactions, for example, the interconversion of pyruvate and nicotinamide adenine dinucleotide (NADH) reduced to lactate and nicotinamide adenine dinucleotide (NAD), which is catalyzed by LDH.
2. **Transferases** transfer one group from one substrate to another substrate. For example, CH_2 of serine is transferred to tetrahydrofolate and is converted to glycine. The reaction is catalyzed by serine hydroxymethyl transferase.
3. **Hydrolases** catalyze cleavage of bonds by the addition of water. For example, acetylcholine is converted to choline and acetate by the enzyme acetylcholine esterase.
4. **Lyases** catalyze cleavage of bonds. For example, malic acid is converted to fumaric acid by the enzyme fumarase.
5. **Isomerases** catalyze the formation of a substrate's isomer. For example, glyceraldehyde-3-phosphate is converted to dihydroxyacetone phosphate by the enzyme triose phosphate isomerase.
6. **Ligases** link two substrates with the hydrolysis of ATP. For example, acetyl CoA is converted to malonyl CoA. This reaction is catalyzed by acetyl CoA carboxylase and biotin.

3.2   Coenzymes and Cofactors

Many enzymes require a coenzyme. Coenzymes are heat stable low molecular weight organic molecules that are essential for the activity of enzymes. Whereas cofactor is a nonprotein metal ion associated with an enzyme that is also essential for the activity of the enzyme. For example, Zn^{2+} is associated with carbonic anhydrase and alcohol dehydrogenase. Cu^{2+} is associated with cytochrome oxidase. Mg^{2+} is associated with hexokinase, glucose-6-phosphate, and pyruvate kinase. K^+ is associated with pyruvate kinase, etc. Holoenzyme refers to the combination of apoenzyme and coenzyme. Apoenzyme refers to the protein portion of the holoenzyme. Most coenzymes are linked to enzymes by noncovalent bonds. A prosthetic group is a coenzyme tightly bound by covalent bond, which will not dissociate from the enzyme. Coenzymes are mainly derivatives of vitamins (e.g., pyridoxal phosphate, thiamin pyrophosphate, biotin, nicotinamide, folic acid, and vitamin B_{12}). Many coenzymes are derivatives of AMP.

Table 3.1 Coenzymes (mostly derivatives of water-soluble B-complex vitamins and AMP)

	Vitamins	Coenzymes	Mechanism of action
1.	Riboflavin	FMN, FAD	Based on transfer of hydrogen
2.	Niacin	NAD, NADP$^+$	Based on transfer of hydrogen
3.	Thiamin	Thiamine Pyrophosphate	Based on decarboxylation
4.	Pyridoxine	Pyridoxal phosphate	Based on transfer of amino group
5.	Biotin	Carboxylase	Based on carboxylation
6.	Folic acid	Tetrahydrofolic acid (active form of folic acid)	Based on a carrier of activated one-carbon unit
7.	Pantothenic acid	Coenzyme A	Role in fatty acid synthesis and oxidation. ACP contains pantothenic acid
8.	Cyanocobalamin	Methylmalonyl-CoA mutase	Converts methylmalonyl CoA to succinyl CoA, which then enters the citric acid cycle for gluconeogenesis

3.2.1 Classification of Coenzymes (Table 3.1)

1. Based on the transfer of hydrogen, i.e., by donating or accepting hydrogen, for example, NAD$^+$, NADP$^+$, flavin mononucleotide (FMN), and flavin adenine dinucleotide (FAD). For example, reactions catalyzed by LDH.
2. Based on the transfer of groups except hydrogen, for example, thiamin pyrophosphate, biotin, pyridoxal phosphate, folate coenzymes, and ATP. Amino group is transferred from pyridoxal phosphate, CO_2 is transferred from biotin, phosphate is transferred from ATP, etc. For example, glucose is converted to glucose-6-phosphate by hexokinase. ATP transfers a phosphate group to glucose to form glucose-6-phosphate (Murray et al. 1996).

3.3 Properties

1. **Catalytic power** of enzymes enhances the rate of biochemical reactions. Carbonic anhydrase is one of the fastest enzymes and catalyzes the following reaction: $CO_2 + H_2O \rightarrow H_2CO_3$. Each molecule can hydrate 10^5 molecules of CO_2 in 1 s. Most enzymes are 10^3–10^8 times faster than uncatalyzed reactions. Apart from catalytic power, carbonic anhydrase takes part in various physiological events such as pH regulation, carboxylation reactions, and CO_2 transport.
2. **Active sites** are present on enzyme molecules, which are clefts or crevices. The substrate binds to the active site and forms an enzyme-substrate (ES) complex. ES is converted to enzyme-product (EP). Finally, EP is dissociated to enzyme and product.

   Lock and Key model (Fig. 3.1) indicates the interaction of substrates and enzymes. The active site of the enzyme is complementary in shape to that of the substrate.

   Induced-fit model (Fig. 3.2) indicates a different type of interaction between substrate and enzyme. The active site will change shape complimentary to that of substrate only after the substrate is bound (Berg et al. 2015).

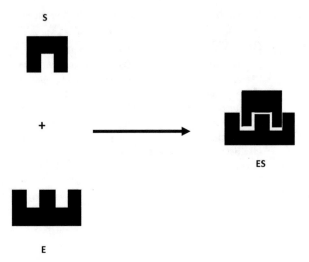

Fig. 3.1 Lock and key model

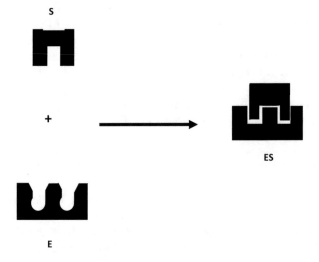

Fig. 3.2 Induced-fit model

3. **Specificity** of enzymes implies that the enzymes are highly specific and catalyze only one type of chemical reaction. Glucose oxidase will oxidize only
glucose and no other simple sugar. Trypsin hydrolyzes peptide bond on the
carboxyl side of lysine and arginine residues, whereas chymotrypsin cleaves
peptide bond on the carboxyl side of tyrosine, tryptophan, and phenylalanine.
4. **Activation or inhibition** of enzymes occurs according to the needs of the cell.
Trypsin inhibitors are widespread in vegetables and legumes, for example, oats,
maize, onion, and beetroot. Fortunately, the protease inhibitors are denatured
and inactivated by cooking.

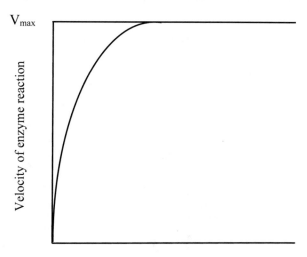

Fig. 3.3 Velocity of enzyme reaction against substrate concentration (hyperbolic shape)

5. **Velocity** or rate of enzyme reaction increases in proportion with enzyme concentration and with the availability of substrate concentration. At higher concentration of substrate, all enzyme molecules will be utilized or saturated, and afterwards, further increase of substrate will not increase the rate of enzyme reaction, and the maximum velocity will be achieved (V_{max}) (Fig. 3.3). This saturation effect will not be observed in uncatalyzed reactions.

6. **Optimum temperature** is the temperature at which velocity of reaction is maximum (maximum amount of substrate converted to the product). Increased temperature after optimum temperature will decrease reaction velocity due to denaturation of enzyme protein (Fig. 3.4).

7. **Optimum pH** depends on a particular enzyme. Each enzyme acts maximum at an optimum pH. Enzyme action will be reduced below or above optimum pH. For example, optimum pH of pepsin is between 1.0 and 2.0, whereas optimum pH of trypsin or chymotrypsin is between 6.0 and 7.0.

8. **Isoenzymes (isozymes)** are different molecular forms of the same enzyme. LDH has five isoenzymes. Isoenzymes are different forms having the same catalytic activity. They are products of closely related genes.

9. **Activation of enzyme** occurs by conversion of inactive proenzyme to the active form. For example, active trypsin is formed from inactive trypsinogen by cleaving a peptide bond and removal of peptide from trypsinogen.

10. **Inhibition** (Berg et al. 2015).

    (a) **Irreversible inhibition** occurs when the inhibitor binds to the enzyme tightly by covalent bond, preventing dissociation. For example, cholinesterase (which breaks down acetylcholine into acetate and choline) is inhibited by diisopropylphosphofluoridate (DIPF), so that acetylcholine is able to transmit impulse through the neuromuscular junction for a longer duration. Tacrine, galantamine, and donepezil are used clinically as cholinesterase inhibitors (Yamali et al. 2018).

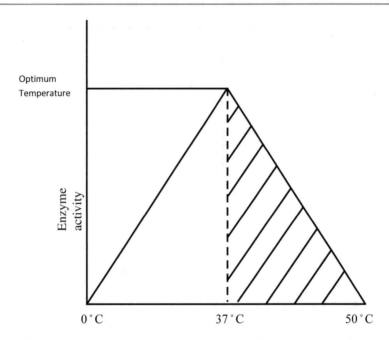

Fig. 3.4 Enzyme activity against temperature. Optimum temperature of enzymes is about 37 °C. Shaded areas indicate denaturation of enzyme

(b) **Reversible inhibition**

   Competitive inhibition occurs when an inhibitor resembles the substrate and binds to the active site of the enzyme, preventing the substrate from binding. However, excess substrate can abolish the inhibition. Malonate, because of structural resemblance, binds to the active site of succinate dehydrogenase and prevents succinate to bind. Dicoumarol, because of structural resemblance, will occupy the active site of vitamin K in the liver and will act as an anticoagulant (Fig. 3.5).

   Noncompetitive inhibition occurs when an inhibitor does not prevent the substrate from binding. The inhibitor binds simultaneously at a different site of the same enzyme. No competition occurs between substrate and inhibitor. There is no structural resemblance between substrate and inhibitor. For example, lead inhibits ferrochelatase by noncompetitive inhibition, preventing the incorporation of ferrous ion into protoporphyrin for ultimately forming heme (Fig. 3.6). Similarly, lead inhibits δ-aminolevulinate dehydratase, preventing the conversion of δ-aminolevulinic acid into porphobilinogen.

11. **Allosteric enzymes** are those in which one active site in the enzyme can influence another active site on the same enzyme. Allosteric enzymes can be inhibited by the inhibitor, which is not an analogue. For example, cholesterol inhibits HMG-CoA reductase. Allosteric enzymes do not exhibit Michaelis–Menten equation.

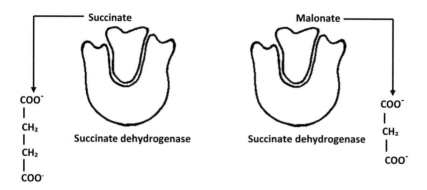

Fig. 3.5 Competitive inhibition

Fig. 3.6 Noncompetitive inhibition

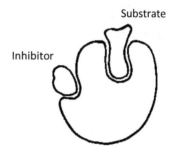

3.4 Michaelis–Menten Equation or Model

Michaelis–Menten equation or model explains the kinetic properties of most enzymes. The equation relates to initial velocity of a simple enzyme-catalyzed reaction to initial substrate concentration. These are measured under steady-state condition.

$$E + S \underset{K_1}{\overset{K_2}{\longleftrightarrow}} ES \overset{K_3}{\longrightarrow} E + P$$

where
E = enzyme
S = substrate
ES = enzyme-substrate complex
P = product
K_1, K_2, and K_3 = rate constants

ES can dissociate to E and S or can proceed further to form E and P. Once E and P are formed, the reaction is irreversible.

$$K_m \left(\text{Michaelis constant} \right) = \frac{K_2 + K_3}{K_1}$$

It was further proved that

$$V = \frac{V_{max}[S]}{K_m + [S]} \qquad (3.1)$$

V = reaction velocity
V_{max} = maximal velocity

When the amount of S is equal to K_m, i.e., $[S] = K_m$, Eq. (3.1) becomes

$$V = \frac{V_{max}[S]}{[S]+[S]} = \frac{V_{max}[S]}{2\,[S]} = \frac{V_{max}}{2}$$

$$\therefore V = 1/2V_{max}$$

When K_m is equal to [S], V, i.e., reaction velocity, is half of maximal velocity (Fig. 3.7).

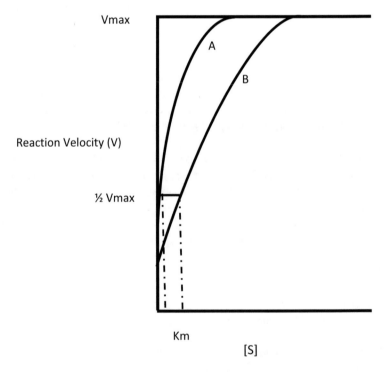

Fig. 3.7 Reaction velocity is plotted against substrate concentration. Reaction velocity of enzyme A rises immediate and rapidly reaches maximum, giving hyperbolic shape of the plot. K_m indicates the affinity of enzyme for the substrate. Low K_m of enzyme A indicates a high affinity of the enzyme for the substrate. Low concentration of substrate is required to half-saturate the enzyme A. High K_m of enzyme B indicates a low affinity of the enzyme for the substrate, and thereby, higher concentration of substrate is required to half-saturate the enzyme B

3.5   Clinically Important Enzymes (Table 3.2)

Table 3.2   Clinically important enzymes

	Enzymes	Diagnostic use
1.	Aspartate amino transferase (AST) or serum glutamate oxaloacetate transaminase (SGOT)	Myocardial infarction Liver disease (hepatitis)
2.	Alanine amino transferase (ALT) or serum glutamate pyruvate transaminase (SGPT)	Hepatitis
3.	Lactate dehydrogenase (LDH)	Myocardial infarction Hepatocellular damage Hemolytic anemia
4.	Creatine phosphokinase (CPK)	Myocardial infarction Muscular dystrophies
5.	Alkaline phosphatase	Cholestasis
6.	Gamma-glutamyl transpeptidase (GGT)	Alcoholism
7.	Acid phosphatase	Prostate cancer
8.	Lipase	Pancreatitis
9.	Amylase	Pancreatitis

Note: It has been suggested that modification of gut bacterial flora using probiotics and synbiotics may improve serum concentration of liver enzymes and thus may improve liver function (Khalesi et al. 2018)

3.6   Summary

Enzymes, protein catalysts, regulate the physio-biochemical functions. Almost all the chapters of our book involve knowledge of enzymes to explain the biochemical and physiological processes. The role of various major classes of enzymes is discussed. Coenzymes and cofactors are essential for the activity of enzymes. Coenzymes are mostly derivatives of water-soluble vitamins and AMP. Cofactors are needed for the activity of various enzymes such as carbonic anhydrase, alcohol dehydrogenase, cytochrome oxidase, hexokinase, and pyruvate kinase. Various models are discussed to explain the interaction of substrates and enzymes. Most enzymes exhibit Michaelis–Menten equation, which explain the kinetic properties of enzymes. Clinically important enzymes are described.

References

Berg JM et al (2015) Biochemistry, 8th edn. W.H. Freeman and Company, New York
Khalesi S et al (2018) Effect of probiotics and synbiotics consumption on serum concentrations of liver function test enzymes: a systematic review and meta-analysis. Eur J Nutr 57:2037–2053
Murray RK et al (1996) Harper's illustrated biochemistry, 24th edn. McGraw Hill, New York
Vasudevan DM et al (2011) Textbook of biochemistry for medical students, 6th edn. Jaypee Brothers Medical Publishers (P) Ltd, Chennai
Yamali C et al (2018) Synthesis, molecular modeling and biological evaluation of 4-[5-aryl-3(thiophen-2-yl)-4,5-dihydro-1H-pyrazol-1-yl] benzenesulfonamides toward acetylcholinesterase, carbonic anhydrase I and II enzymes. Chem Biol Drug Des 91:854–866

Further Reading

Campbell E et al (2016) The role of protein dynamics in the evolution of new enzyme functions. Nat Chem Biol 12:944–950

Cuesta SM et al (2015) The classification and evolution of enzyme function. Biophys J 109(6):1082–1086

Ianiro G et al (2016) Digestive enzyme supplementation in gastrointestinal diseases. Curr Drug Metab 17(2):187–193

Micronutrients

4

Abstract

Vitamins are organic compounds that are required by the body to perform various cellular functions. These are essential for the maintenance of normal health, growth, and perform a variety of physio-biochemical functions. These cannot be synthesized by the body and must be supplied by the diet in order to prevent deficiencies of vitamins. The precursors of vitamins in the diet are called provitamins. For example, provitamin beta-carotene present in carrot and spinach produces vitamin A in the body. Vitamins do not provide energy. However, some vitamins can provide energy indirectly, since they act as coenzymes for carbohydrate, lipid, and amino acid metabolism. Vitamin D can be synthesized by the skin in the presence of sunlight, and niacin can be synthesized in the body from tryptophan. Vitamin D and niacin may not be strictly vitamins. Rather vitamin D may be termed prohormone, since it gives rise to hormone calcitriol. Minerals must also be supplied in the diet and consist of macrominerals and microminerals. Deficiencies and toxicities of minerals are highlighted in the respective sections. Genetic diseases due to mutation of genes will cause deficiencies or toxicities of micronutrients.

Keywords

Micronutrients · Vitamins · Provitamins · Macrominerals · Microminerals · Genetic diseases

© Springer Nature Singapore Pte Ltd. 2019
K. Chakrabarty, A. S. Chakrabarty, *Textbook of Nutrition in Health and Disease*,
https://doi.org/10.1007/978-981-15-0962-9_4

4.1   Fat-Soluble Vitamins (Longo et al. 2011)

4.1.1   Vitamin A

4.1.1.1 Structure (Fig. 4.1)

Vitamin A is a polyisoprenoid compound having a cyclohexenyl ring. Retinoids (compounds with vitamin A activities) are retinol (vitamin A alcohol), retinal (vitamin A aldehyde) and retinoic acid (vitamin A acid).

4.1.1.2 Physio-biochemical Role

1. Rhodopsin (visual purple) is the photosensitive pigment of the rods and is responsible for night-light or dim light (scotopic) vision. Vitamin A is necessary for the synthesis of rhodopsin (Fig. 4.2). Whereas cones are responsible for photopic vision (daylight vision), visual acuity, and color vision.

Fig. 4.1 Structure of vitamin A

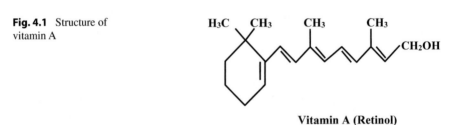

Vitamin A (Retinol)

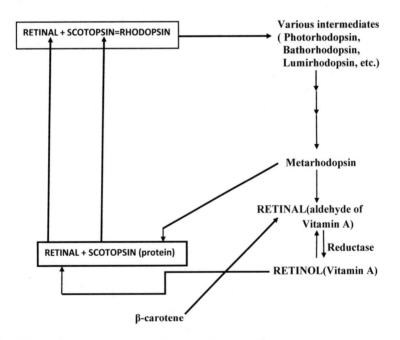

Fig. 4.2 Visual cycle, showing role of vitamin A in the synthesis of rhodopsin

2. Vitamin A is required for the normal differentiation of epithelial tissues to ultimately produce mature epithelium.
3. Retinoic acid derived from oxidation of retinal takes part in the synthesis of glycoproteins and GAGs. This may explain, in part, the action of retinoic acid in promoting growth and differentiation of tissues. Retinoic acid also takes part in lubrication between the joints.
4. Carotenoids (a variety of carotenes) are cleaved by oxidation to give rise to retinaldehyde and thus retinol. Carotenoids are antioxidants, protect the macula cells from oxidative damage, and have anticancer property. Lutein and xeaxanthine are carotenoids and form the yellow macular pigment that minimize chromatic aberration on visual acuity. They may play a preventive role in fundus diseases (Zhao et al. 2019).
5. Vitamin A prevents drying of the skin and eyes.
6. Vitamin A and beta-carotene are needed for reproduction.

4.1.1.3 Dietary Sources
Yellow and green vegetables, green leafy vegetables (e.g., spinach), carrot, fruits (papaya, mango, pumpkin, etc.), fish, liver, egg yolk, cream, and cheese are the major sources. Liver is the richest source. Yellow vegetables, carrots, and fruits contain the yellow pigment beta-carotene, which serves as a precursor of vitamin A. Colostrum (yellowish fluid secreted from the mammary gland for about 2–3 days after the delivery) contains high amount of vitamin A.
Note: Thyroid hormones convert carotene to vitamin A in the liver.

4.1.1.4 RDA (µg/day) at Different Age Groups and During Pregnancy and Lactation
For infants, it is about 350–400. Whereas it is about 600 from the schoolchildren to elderly person. It is about 800 for pregnant women. Excessive intake must be avoided during pregnancy as it may cause developmental abnormalities of the fetus. The lactating mothers need more amount (not less than 1000).

4.1.1.5 Deficiencies
1. From the visual cycle, it is evident that an individual due to vitamin A deficiency will suffer from night blindness known as nyctalopia. A person suffering from nyctalopia will not be able to drive any vehicle at night and will not be able to read in dim light.
2. If a person spends a considerable length of time in bright light and then moves to dim light (e.g., entering a cinema hall), he/she will not be able to see anything immediately as bright light causes depletion of stores of rhodopsin in the rods. However, after a few minutes, rhodopsin is re-synthesized and vision is improved. This period is called dark adaptation time. Vitamin A-deficient persons will have an increased dark adaptation time (Ganong 2003).
3. Pathological changes of the eye (Fig. 4.3).
   (a) Xerophthalmia is a condition in which the conjunctiva (mucus membrane that lines the front of the eye and lines the inside of the eyelids) becomes

Fig. 4.3 Illustration of pathological changes of the eye due to vitamin A deficiency (described in detail in the text)

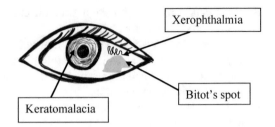

　　　dry, thickened, and wrinkled. The conjunctiva loses its normal transparency, since it gets keratinized (keratin belongs to a family of proteins, which is fibrous in nature).

(b) Bitot's spots are grayish-white triangular spots on the surface of conjunctiva. They consist of fragments of keratinized epithelium.

(c) Keratomalacia is a condition in which the corneal epithelium degenerates and will be vascularized. As a result, corneal opacities will develop. Corneal opacities, necrosis, ulceration, and perforation of the cornea can ultimately lead to blindness. The children of age group up to 5 years are commonly affected.

4. Metaplasia (abnormal cell growth) is caused due to vitamin A deficiency, leading to keratinized stratified squamous epithelium in the respiratory passages and kidneys. It is suggested that vitamin A controls gene expression and thus cellular growth and differentiation. Loss of mucociliary epithelium of the airways causes infection of the lungs. Desquamation of keratin and debris predisposes to the renal and urinary bladder stones.

5. Rough skin and papules termed "goose bump rash" (raised spots) of the skin are due to hyperkeratinization of the epithelium lining the follicles.

6. Structure degeneration of the retina may occur.

7. Appetite may be reduced, probably due to keratinization of taste buds.

8. Bone growth and maturation is slow, probably due to defective synthesis of chondroitin sulfate (present in ground substance of the bone and cartilage) with degeneration of cells in the epiphyseal cartilage. Retinoic acid inhibits osteoblast activity, accelerates osteoclast formation, and stimulates bone resorption (Hamishehkar et al. 2016).

9. Anemia may occur, but the cause is not understood.

10. Vitamin A plays a role in the differentiation of immune system cells (humoral immunity and T-cell-mediated immunity). It takes part in phagocytosis. Vitamin A deficiency leads to increased susceptibility to infections.

11. Retinitis pigmentosa may be aggravated. It is a rare hereditary retinal disease due to degeneration of rods and cones. It is characterized by nyctalopia, diminished vision, visual field defects, pallor of the optic disc, and intrarenal pigment around mid-periphery (Zhao et al. 2019).

　　　Note: Low-fat diet leads to vitamin A deficiency, since retinol is absorbed from the small intestine dissolved in lipids. People suffering from malabsorption of fat are susceptible to developing vitamin A deficiency. Prolonged use of barbiturates may lead to vitamin A deficiency due to the catabolism of retinol. Zinc deficiency augments vitamin A deficiency by interfering with the mobilization of vitamin A from the liver.

4.1.1.6 Prevention

An individual with any sign of vitamin A deficiency, especially appearance of xerophthalmia must be advised to take diet rich in vitamin A. The pregnant women should take more vitamin A to build up stores of retinol in the fetal liver. Excessive intake of vitamin A may cause a serious problem in the pregnant woman as retinol may be teratogenic (developmental abnormalities of the fetus). The pregnant woman should not consume more than 3000 µg/day (recommended by the American Pediatric Association). When keratomalacia is present, oral dose of retinol must be given to prevent blindness. According to National Blindness Control Programme, oral dose of vitamin A should be given to the children below the age of 5 years, irrespective of vitamin A deficiency. WHO report indicates that about 250 million children below the age of 5 suffer from subclinical deficiency. Prevention of vitamin A deficiency is one of the main priorities (other than iron and iodine) of WHO.

4.1.1.7 Hypervitaminosis

Hypervitaminosis may occur due to excessive intake of vitamin A. It was first recognized among the arctic explorers, because of eating polar bear liver, containing high amount of vitamin A. Toxicity is characterized by scaly dermatitis, anorexia, headache, aching joints and muscles, alopecia, dizziness, hepatomegaly, and even cirrhosis of the liver. Hypervitaminosis may lead to bone resorption, bone fragility, and spontaneous fracture. Other features are thickness of long bones, hypercalcemia, and calcification of soft tissues.

4.1.2 Vitamin D

4.1.2.1 Structure

Vitamin D refers to a group of closely related steroid prohormone (various intermediates of cholesterol synthesis). Structure of calcitriol (active form) is shown in Fig. 4.4.

4.1.2.2 Physio-biochemical Role

Vitamin D-binding proteins are synthesized in liver cells and transport active form to various target organs. Synthesis and action of active form is shown in Fig. 4.5.

In addition to the main action of calcitriol on increased plasma levels of calcium and phosphate, calcitriol may have a role for the synthesis and secretion of insulin and parathyroid hormone. Vitamin D may activate tryptophan hydroxylase-2, which catalyzes the synthesis of serotonin and thus may minimize the adverse consequences of mental disorders such as attention deficit hyperactive disorder (ADHD), bipolar disorder, schizophrenia, major depression, and autism (Blumberg et al. 2018). An association between vitamin D and allergic diseases has been implicated (Searing and

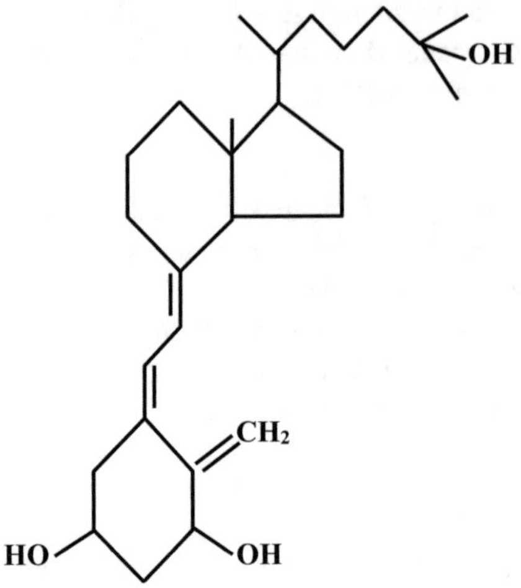

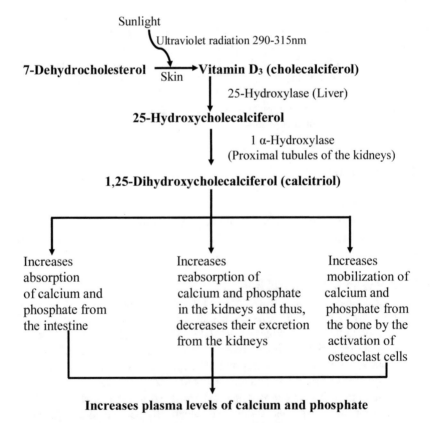

Fig. 4.4 Structure of 1,25-dihydroxycholecalciferol (calcitriol)

Sunlight

Ultraviolet radiation 290-315nm

7-Dehydrocholesterol $\xrightarrow[\text{Skin}]{}$ **Vitamin D₃ (cholecalciferol)**

25-Hydroxylase (Liver)

25-Hydroxycholecalciferol

1 α-Hydroxylase
(Proximal tubules of the kidneys)

1,25-Dihydroxycholecalciferol (calcitriol)

| Increases absorption of calcium and phosphate from the intestine | Increases reabsorption of calcium and phosphate in the kidneys and thus, decreases their excretion from the kidneys | Increases mobilization of calcium and phosphate from the bone by the activation of osteoclast cells |

Increases plasma levels of calcium and phosphate

Fig. 4.5 Synthesis and action of calcitriol

Leung 2010; Abuzeid et al. 2012; Osborne et al. 2012). It has been reported that patients suffering from chronic urticaria had significantly lower serum vitamin D levels (Abdel-Rehim et al. 2014; Tuchinda et al. 2018; Nasir-Kalmarzi et al. 2018). Indeed, vitamin D could reduce IgE production by inhibiting B-cell function. Various allergic diseases such as atopic dermatitis (Heine et al. 2013), psoriasis (Bergler-Czop and Brzezinska-Wcislo 2016), and acne (Lim et al. 2016) had low serum 25-hydroxyvitamin D levels. It may be noted that population of the USA and Australia have an increased risk of allergic diseases as they are far away from the equator (Poole et al. 2018). Significant association of food allergy and ultraviolet radiation intensity due to summer or due to winter season was observed (Mullins et al. 2011).

4.1.2.3 Dietary Sources
Dietary sources are fish cod liver oil, liver, egg yolk, and milk. Dietary sources are required in individuals with limited exposure to sunlight.

4.1.2.4 RDA (µg/day) at Different Age Groups and During Pregnancy and Lactation
The requirement of different age groups is about 10. Pregnant women, lactating mothers, and elderly persons should have an additional requirement. After several pregnancies, women are susceptible to suffer from osteomalacia due to inadequate exposure to sunlight and low reserve of calcium. Higher incidence of hypocalcemia and defective dental elements have been reported in the infants of pregnant mothers. Elderly people suffer from osteomalacia due to inadequate exposure to sunlight and due to less formation of 7-dehydrocholesterol in the skin.

4.1.2.5 Deficiencies
Deficiency of vitamin D causes defective calcification of the bone matrix due to failure in delivering adequate amounts of calcium and phosphate at the site of mineralization. The defective calcification leads to the following:

1. Rickets in children mostly occurs between the age of 1 and 3 years (Fig. 4.6). Incomplete calcification of bone results in soft and pliable bones. Weight-bearing bones are bent, for example, bowlegs or knock-knees. Knock-knee or genu valgum is an abnormal curving of the legs when the knees are in contact. This is due to a gap between the ankles (Fig. 4.7).

   Prominence of the sternum produces pigeon-like chest. Swelling of the ribs at the costochondral junction is prominent. The process of ossification at the epiphyses (end of long bones) takes place in an abnormal manner, which can be felt as a marked projection on the surface. Vitamin D-resistant rickets is caused by mutations of the enzyme 1-α hydroxylase. Mutation in PHEX gene leads to lowering of plasma calcium and phosphate, resulting in rickets.
2. Osteomalacia in adults is a form of adult rickets. Many bones become soft and deformed. Due to progressive loss of bone matrix (osteoporosis), the incidence of fractures is increased. Chronic renal failure causes rickets and osteomalacia due to the loss of 1α-hydroxylases from proximal tubules of the kidneys.

Fig. 4.6 Rickets. The
bowing of legs in a toddler
due to the formation of
poorly mineralized bone is
evident. (Reproduced, with
permission, from Elsevier,
Book Title: Robbins and
Cotran Pathologic basis of
Disease, 2005, Author:
Vinay Kumar, Abul K
Abbas, Nelson Fausto
(Chapter 9, Title:
Environmental and
Nutritional Pathology,
Fig. 9–26, Pages 505 to
505) Copyright © Elsevier)

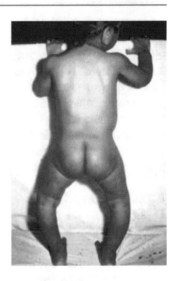

Fig. 4.7 Illustration of
genu valgum (knock-knee).
Note abnormal curving of
the legs and a gap between
the ankles when the knees
are in contact

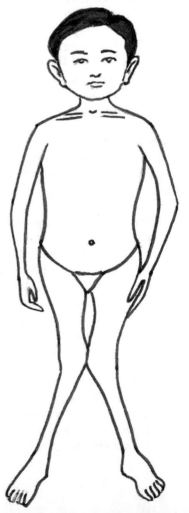

3. Vitamin D level is correlated with high HDL cholesterol level. Vitamin D supplementation may cause a decrease in blood cholesterol level. Vitamin D deficiency may result in low HDL cholesterol.
4. Vitamin D deficiency may lead to hypertension, which may be due to activation of RAAS (Garach et al. 2019).

Note: Vitamin D is absorbed in lipid micelles. People on a low fat diet may have vitamin D deficiency. It has been reported that low vitamin D leads to insulin resistance with impaired glucose tolerance and thus predisposes type 2 diabetes mellitus. Use of barbiturates and isoniazid over a long period may lead to the development of osteomalacia, since barbiturates causes catabolism of calcitriol and isoniazid inhibits 25-hydroxylase in the liver. Paget's disease is characterized by bowing of the femur and tibia due to increased osteoclastic activity. Injection of calcitonin will be useful in treatment of Paget's disease.

4.1.2.6 Hypervitaminosis
High level of vitamin D is toxic as increased mobilization of calcium from the bone causes hypercalcemia, which causes calcification of the tissues (e.g., the kidneys and arteries). Calcification of the arteries can lead to hypertension.

4.1.3 Vitamin E (Tocopherols)

4.1.3.1 Structure
Tocopherols consist of 6-hydroxy chroman ring with an isoprenoid side chain (Fig. 4.8).

4.1.3.2 Physio-biochemical Role
Vitamin E consists of several tocopherols of which α-tocopherol is the most active. Vitamin E acts as an antioxidant to prevent oxidation of polyunsaturated fatty acids (PUFA) in the cell membrane by free radicals. Free radicals may lead to the development of cancer and may increase the incidence of coronary ischemic heart disease (CIHD) by lipid peroxidation and DNA damage. Vitamin E accumulates in the mitochondria and plasma membrane to prevent oxidation by free radicals. Vitamin E can inhibit HMG-CoA reductase and thus may prevent the incidence of atherosclerosis.

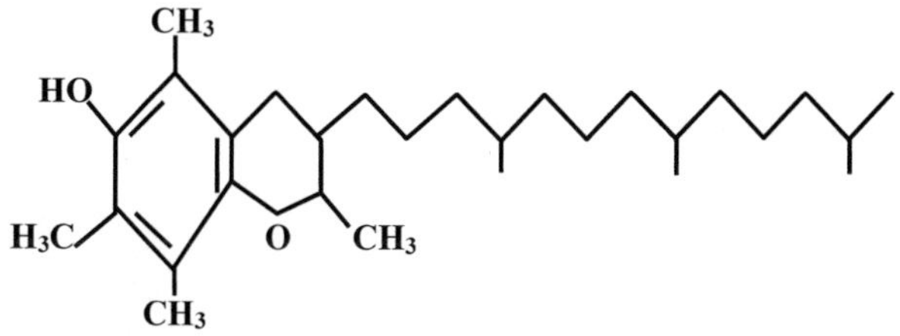

Fig. 4.8 Structure of α-tocopherol

Furthermore, vitamin E prevents the formation of atherosclerotic plaque by inhibiting oxidation of low-density lipoprotein (LDL). It may protect the structure of rods and cones from oxidative damage.

4.1.3.3 Synergistic Action of Vitamin E and Selenium
Selenium present in glutathione peroxidase destroys free radicals and removes the cause of lipid peroxidation. Vitamin E prevents lipid peroxidation. Thus, vitamin E and selenium act synergistically to prevent lipid peroxidation.

4.1.3.4 Dietary Sources
Vegetable oils (e.g., sunflower oil, soya bean oil, etc.) are rich sources of vitamin E. Wheat germ, tomatoes, carrots, leafy vegetables, nuts, seeds, fish, liver, and eggs contain significant amount. Vitamin E is destroyed by cooking and by food processing.

4.1.3.5 RDA (mg/day) at Different Age Groups and During Pregnancy and Lactation
In infants, preschool children, and schoolchildren, it is about 4–6. Whereas adults, elderly persons, and pregnant women require about 10. The lactating mothers require slightly more, i.e., about 12. Requirement depends on the intake of PUFA.

4.1.3.6 Deficiencies
Defective fat absorption leads to vitamin E deficiency as vitamin E is dissolved in dietary fat and incorporated into chylomicrons and absorbed during fat digestion. Deficiency of vitamin E may occur due to insufficient dietary intake. Deficiency may cause increased fragility of RBCs, resulting in hemolytic anemia. This may be due to abnormal cellular membrane caused by oxidative radicals. Vitamin E deficiency causes degeneration of the axons of spinal cord and cerebellar dysfunction (e.g., ataxia).

4.1.4 Vitamin K

4.1.4.1 Structure
Vitamin K is a derivative of naphthoquinone, having an isoprenoid side chain (Fig. 4.9). There are several forms of vitamin K, for example, phylloquinone (vitamin K_1), menaquinone (vitamin K_2), menadione (vitamin K_3). Phylloquinone is

Fig. 4.9 Structure of vitamin K

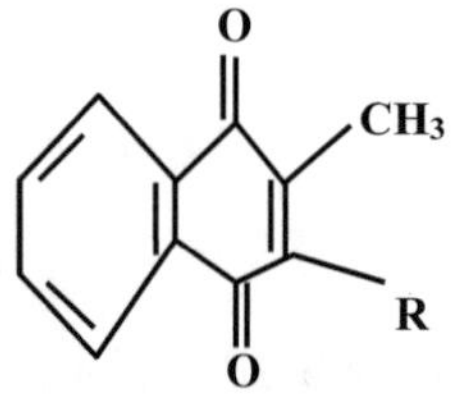

present in green leafy vegetables. Menaquinone is synthesized by the intestinal bacteria. Menadione, a synthetic compound, is metabolized to phylloquinone.

4.1.4.2 Physio-biochemical Role

1. Vitamin K is essential for the synthesis of blood clotting factors. It catalyzes γ-carboxylation of glutamic acid residues on various clotting factors. The liver synthesizes inactive precursors of clotting factor II (prothrombin), factor VII (proconvertin or stable factor), factor IX (Christmas factor), and factor X (Stuart-Prower factor), containing glutamic acid residues. Glutamic acid residues are converted to γ-carboxyglutamate (Gla) of the clotting factors (active form) by the enzyme vitamin K-dependent carboxylase with the incorporation of CO_2. Active Gla is released into the circulation and takes part in the cascade of clotting reactions. In brief, prothrombin binds with calcium and platelet phospholipid and is converted to thrombin that converts soluble fibrinogen into insoluble fibrin, resulting in clot formation.
2. Osteocalcin of the bone matrix is essential for the bone mineralization and contains Gla. Newborn children of pregnant women treated with warfarin may suffer from bone abnormalities ("fetal warfarin syndrome").

4.1.4.3 Vitamin K Antagonists

Coumarin derivatives dicumarol and warfarin inhibit Gla and thus act as anticoagulants. Furthermore, dicumarol is having chemical resemblance of vitamin K and prevents the action of vitamin K by occupying the site of action of vitamin K in the liver (substrate competition).

4.1.4.4 Dietary Sources

Spinach, broccoli, peas, green beans, cabbage, cauliflower, egg yolk, liver, soya bean, and olive oil are rich in vitamin K. Mother's milk contains low amount of vitamin K and is less than cow's milk. Vitamin K is synthesized by the intestinal bacteria. This synthesis is sufficient for the body's requirement, provided the intestinal absorption is normal.

4.1.4.5 RDA (μg/day) at Different Age Groups and During Pregnancy and Lactation

In infants, preschool children, and schoolchildren, it is about 15–20. Adults and elderly persons require about 60. Pregnant women and lactating mothers require more, i.e., about 70–80.

4.1.4.6 Deficiencies

Deficiency may be due to (1) extensive use of antibiotic and (2) fat malabsorption. Vitamin K is absorbed into the lymphatics via chylomicrons (therapeutic use of menadione, being water soluble, is absorbed directly into the hepatic portal vein). Vitamin K deficiency will lead to hypoprothrombinemia and will cause hemorrhage due to the defect of clotting mechanism. The newborn infants are very much vulnerable to vitamin K deficiency and may suffer from hemorrhagic disease due to the following reasons: (1) Vitamin K cannot cross the placental barriers efficiently to reach the fetus. Reserve of vitamin K of the liver is low in the fetus. (2) The intestines are sterile after birth. (3) Human milk is a poor source of vitamin K. In view of the above reasons, all newborn infants should be given a prophylactic dose of vitamin K in order to prevent hemorrhagic disease.

4.2   Water-Soluble Vitamins (Gibney et al. 2009)

4.2.1   Vitamin B₁ (Thaimin)

4.2.1.1 Structure
Contains a substituted pyrimidine ring joined to a substituted thiazole ring by a methylene bridge (Fig. 4.10).

4.2.1.2 Physio-biochemical Role
Thiamin pyrophosphate is the active form of vitamin B_1 formed by the transfer of a pyrophosphate group from ATP to thiamin. Thiamin is converted to thiamin pyrophosphate (TPP) by the enzyme thiamin diphosphotransferase with the conversion of ATP to AMP. TPP is the coenzyme for oxidative decarboxylation reactions. For example, (1) pyruvate is converted to acetyl-CoA by pyruvate dehydrogenase and TPP with the liberation of CO_2. (2) α-Ketoglutarate is converted to succinyl-CoA by α-ketoglutarate dehydrogenase and TPP with the liberation of CO_2. TPP takes part in carbohydrate, lipid, and amino acid metabolism. It takes part in gluconeogenesis. TPP is coenzyme of transketolase in the hexose monophosphate shunt pathway of glycolysis and provides NADPH for fatty acid synthesis. It releases energy from carbohydrate. Ketogenic amino acids (leucine, isoleucine, etc.) are metabolized by α-ketoglutarate dehydrogenase (TPP is the coenzyme) with the formation of NADH (see Chap. 2). TPP takes part in phosphorylation of sodium transporter of a nerve membrane and thus takes part in nerve conduction.

4.2.1.3 Dietary Sources
The best sources are yeast, pulses, unrefined cereals, nuts, meat, fish, egg, and liver. Thiamin is almost absent in polished rice. Aleurone layer of food grains contains thiamin. When the grain is boiled along with husk, aleurone layer will be fixed with the grain. By parboiling process, thiamin is fixed in the grain. Whole-wheat bread is

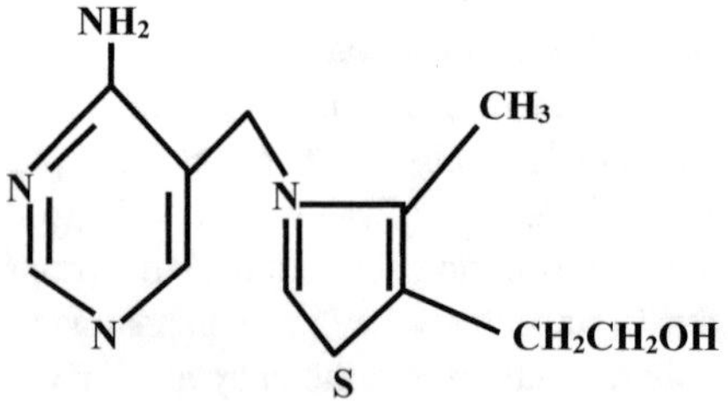

Fig. 4.10 Structure of thiamin

a good source compared with white bread. Thiamin is destroyed by heat and sunlight. Thiaminase present in certain fishes, red cabbage, tea, and coffee destroys thiamin. Apart from consumption of polished rice, continuous chewing of betel nuts can precipitate thiamin deficiency. Polyphenols and thiaminase present in betel nuts destroy thiamin.

4.2.1.4 RDA (mg/day) at Different Age Groups and During Pregnancy and Lactation

Infants need about 0.3. In adults, elderly persons, pregnant and lactating mothers, the requirement is about 1.5. The RDA at different age groups and during pregnancy/lactation may vary as it depends on the pattern of dietary intake.

4.2.1.5 Deficiencies

Thiamin is synthesized by the intestinal bacteria. Thiamin deficiency can be precipitated by the oral administration of antibiotics, which kill intestinal bacteria. One should take B-complex tablet, containing thiamin along with antibiotic. Thiamin deficiency results in a decreased production of ATP. Thus, cellular functions are impaired. Carbohydrate utilization is impaired with the accumulation of pyruvic and lactic acid. It is well established that all nervous tissues, including brain, use blood glucose as their primary source of energy. Thiamin deficiency leads to beriberi. It occurs mainly in areas where polished rice and white flour are the major components of the diet.

Adult beriberi is of three types

1. Neuritic (dry form) is characterized by the inflammation of the spinal cord and peripheral nerves. Neuritis is present. Muscles are weak and patient is unable to walk.
2. Cardiac (wet form) is characterized by palpitation, dyspnea, and enlarged heart (Fig. 4.11). Accumulation of lactic and pyruvic acid causes dilatation of the arterioles, which causes edema of the legs and face.
3. Cerebral form is characterized by loss of appetite, nausea, vomiting, insomnia, depression, recent memory loss, nystagmus, and ophthalmoplegia.

Infantile Beriberi

It occurs in the nursing infants (usually 2–6 months of age) if the mothers are suffering from thiamin deficiency. The disease is found in underdeveloped countries among the malnourished infants. It is characterized by anorexia, weight loss, tachycardia, vomiting, diarrhea, peripheral edema (lower legs), and convulsion. In the acute form, the baby suffers from dyspnea on exertion and the baby may die of heart failure.

4.2.1.6 Alcoholism and Thaimin Deficiency

Thiamin is absorbed in the duodenum and jejunum and then is absorbed directly into the portal circulation by an active transport that is inhibited by alcohol. Chronic alcoholics generally suffer from thiamin deficiency either due to dietary insufficiency or due to impaired absorption of thiamin. Alcohol interferes not only with the intestinal

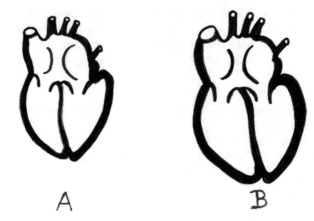

Fig. 4.11 Illustration of normal heart (**A**) and enlarged heart (**B**) in adult beriberi (cardiac form)

absorption of thiamin but also with the synthesis of TPP. Chronic alcoholics suffer from Wernicke's encephalopathy due to thiamin deficiency, which is characterized by loss of recent memory (past memory unimpaired), nystagmus, ataxia, neuritis and ophthalmoplegia. The syndrome is known as Wernicke-Korsakoff syndrome if there is an associated loss of recent memory.

4.2.2   Vitamin B$_2$ (Riboflavin)

4.2.2.1 Structure
Riboflavin consists of three-ring cyclic structure, isoalloxazine attached to the ribitol, which is alcohol of ribose sugar (Fig. 4.12).

4.2.2.2 Properties
It is soluble in water and heat stable. It is a fluorescent pigment and decomposes in the presence of light. Due to its yellow color, it is used as a food color. Riboflavin is converted to an inactive lumiflavin (in alkaline solution) and lumichrome (in acidic or neutral solution) due to photolysis. These inactive compounds (due to oxidation of lipids and methionine) produce unpleasant smell when milk and fish are exposed to sunlight ("Sunlight flavor").

4.2.2.3 Physio-biochemical Role
Active forms of riboflavin are FMN and FAD. Enzymes containing riboflavin are called flavoproteins. Flavoproteins act as electron carriers in the mitochondrial electron transport chain. FAD and FMN are the coenzymes, which are tightly bound and essential for the activity of various enzymes. FMN-dependent enzymes, for example, amino acid oxidase takes part in oxidative deamination.

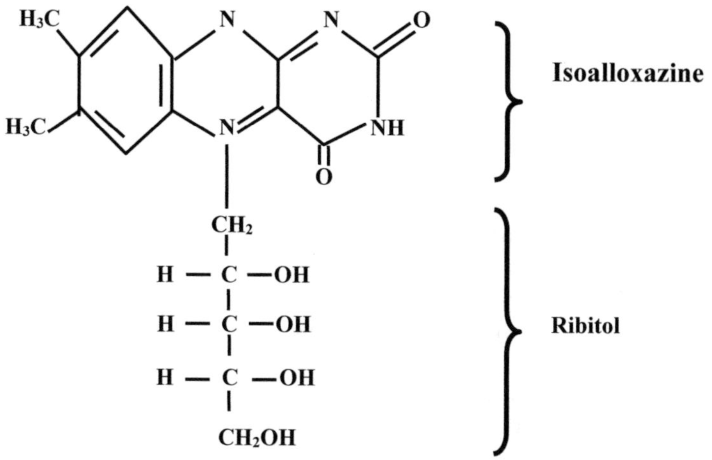

Fig. 4.12   Structure of riboflavin

$$\text{Amino acid} + \text{FMN} \rightarrow \text{Imino acid} + \text{FMNH}_2$$

FAD-dependent enzymes, for example, succinate dehydrogenase catalyzes the following reaction.

$$\text{Succinate} + \text{FAD} \rightarrow \text{Fumarate} + \text{FADH}_2$$

FAD takes part in fatty acid synthesis and gluconeogenesis and is one of the key coezymes of the TCA cycle.

4.2.2.4 Dietary Sources
Dietary sources are milk, meat, cheese, yeast, liver, fish, eggs, whole cereals, green leafy vegetables, and broccoli. Germination of cereals increases the content of riboflavin.

4.2.2.5 RDA (mg/day) at Different Age Groups and During Pregnancy and Lactation
For infants, it is about 0.3. For other age groups and during pregnancy/lactation, it varies from 1.4 to 1.6.

4.2.2.6 Deficiencies
Changes observed clinically are dermatitis, cheilosis (fissures at the corners of mouth) and glossitis (tongue appears purplish known as "magenta tongue"), angular stomatitis (inflammation at the corners of mouth), photophobia (discomfort of the eye due to exposure of light), and seborrhea (excessive secretion by the sebaceous glands, especially of the nose). Other features are conjunctivitis with vascularization of the cornea and opacity of the lens. Glutathione maintains the normal clarity of lens, since glutathione reductase is a flavoprotein. Riboflavin-deficient subjects are resistant to malaria, the cause of which is not known.

Fig. 4.13 Structure of
nicotinic acid

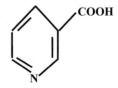

4.2.3   Vitamin B$_3$ (Niacin)

4.2.3.1 Structure (Fig. 4.13)
Niacin is a derivative of pyridine having carboxylic acid.

4.2.3.2 Physio-biochemical Role
Niacin and nicotinamide are the precursors of coenzymes NAD$^+$ and NADP$^+$, which take part in oxidation and reduction reactions. NAD$^+$ and NADP$^+$ play an important role as coenzymes to many dehydrogenase enzymes, for example, lactate dehydrogenase, pyruvate dehydrogenase, malate dehydrogenase, etc. and take part in carbohydrate, amino acid, and lipid metabolism. Niacin takes part in gluconeogenesis. Tryptophan in lactalbumin can be converted readily to niacin derivatives.

4.2.3.3 Dietary Sources
Dietary sources are legumes, peanuts, liver, milk, eggs, fish, and lean meat. Niacin, to a certain extent, can be obtained from the metabolism of tryptophan (pyridoxal phosphate is involved in the synthesis of tryptophan). Diet low in tryptophan and pyridoxine (vitamin B$_6$) will aggravate deficiency of niacin. Niacin in cereals (especially bran) cannot be utilized as niacin is bound as niacytin. Deficiency of niacin may occur in India among people whose main diet cereal is jowar (*Sorghum vulgare*) as leucine present in jowar can inhibit the synthesis of NAD$^+$ from tryptophan. Isoniazid prevents the conversion of tryptophan to NAD$^+$ by inhibiting pyridoxal phosphate.

4.2.3.4 RDA (mg/day) at Different Age Groups and During Pregnancy and Lactation
For infants, it is about 0.5, and for other groups, it is about 15. Pregnant women and lactating mothers need more than 15.

4.2.3.5 Deficiencies
Deficiency of niacin causes pellagra, which is characterized by the following.

1. Dermatitis characterized by skin lesions with exudation. The skin exposed to sunlight will look like severe sunburn. Butterfly like pattern of dermatitis distribution occurs over the face. Rash occurs around the neck in the form of a ring ("Casal's necklace") (Fig. 4.14).
2. Dementia occurs leading to deterioration of higher intellectual functions, poor memory, and deterioration of personal care.

Fig. 4.14 Pellagra in a girl of 5 years, showing skin lesions on the neck (Casal's collar) is pathognomonic. (Reproduced, with permission, from Elsevier, Book Title: Davidson's Principles and Practice of Medicine A textbook for Students and Doctors, 1991, Edited by Christopher R.W. Edwards and Ian A.D. Boucher (Chapter number 3, Title: Nutritional factors in disease, Fig 3.8, Pages 63 to 63) Copyright © Elsevier)

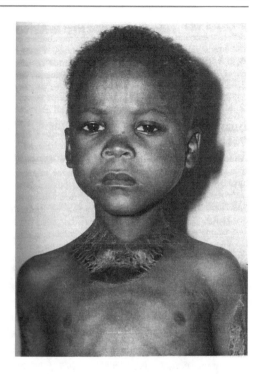

3. Delirium results in mental disturbances, disorientation, and hallucination. Extreme excitement can occur.
4. Depressive psychosis may be due to reduced synthesis of serotonin from tryptophan.
5. Diarrhea may be accompanied by nausea and anorexia.
6. Dysphagia.

Students can remember six D's due to niacin deficiency (dermatitis, dementia, delirium, depressive psychosis, diarrhea, and dysphagia).

Note: Hartnup disease causes defective intestinal absorption and impaired reabsorption of tryptophan from the renal tubules, leading to the deficiency of niacin. Tryptophan deficiency may also occur in carcinoid tumor (tumor of argentaffin cells in the glands of the intestinal tract).

4.2.3.6 Hypervitaminosis
Excessive intake of niacin (more than 500 mg/day) may cause liver damage. Prolonged use can result in liver failure.

4.2.3.7 Clinical Importance
Due to inhibition of lipolysis in adipose tissue, it causes hypocholesterolemia and hypolipidemia.

4.2.4 Vitamin B₅ (Pantothenic Acid)

4.2.4.1 Structure
Pantothenic acid is formed by combination of pantoic acid and β-alanine.

4.2.4.2 Physio-biochemical Role
Pantothenic acid is a part of coenzyme A. In brief, CoA takes part in fatty acid synthesis, fatty acid oxidation, cholesterol synthesis, and synthesis of steroid hormones. ACP takes part in fatty acid synthesis and contains pantothenic acid.

4.2.4.3 Dietary Sources
Dietary sources are yeast, liver, eggs, whole grain cereals, meat, etc.
    Daily dietary intake for children and adults is about 4–6 mg.

4.2.4.4 Deficiencies
Deficiency of pantothenic acid is rare as it is widely distributed in foods and synthesized by the intestinal bacteria. The name pantothenic acid is derived from the Greek meaning "from everywhere." "Burning foot syndrome" (burning and lightning pain of the lower extremities) has been ascribed to the deficiency of pantothenic acid. Deficiency may cause insomnia, muscle spasm, paresthesia of the hand and feet, and hyperactive tendon reflexes, for example, knee reflex. Impaired steroidogenesis decreases secretion of the adrenal cortex hormones and serum cholesterol level. Deficiency can interfere with fatty acid synthesis. The individuals may suffer from mental depression. Deficiency impairs acetylcholine synthesis from acetyl-CoA and impairs threonine acyl esters formation in myelin. Neuromuscular disorders of "burning foot syndrome" may be due to impaired synthesis of acetylcholine and myelin.

4.2.5 Vitamin B₆ (Pyridoxine)

4.2.5.1 Structure (Fig. 4.15)
Vitamin B₆ consists of three related pyridine derivatives (pyridoxine, pyridoxal, and pyridoxamine). Pyridoxine is the primary alcohol, pyridoxal is an aldehyde, and pyridoxamine contains an amino methyl group. Pyridoxal phosphate, coenzyme, is the active form and is synthesized by pyridoxal kinase and ATP.

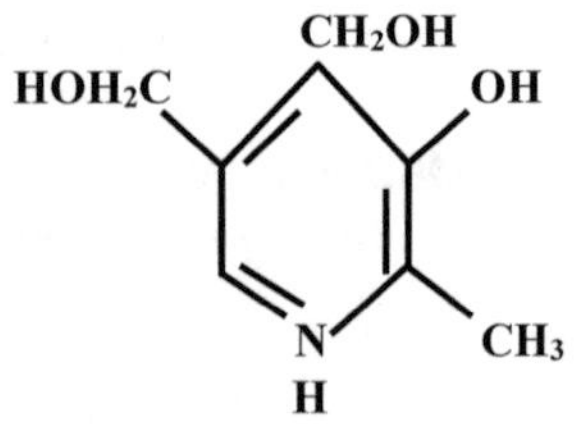

Fig. 4.15 Structure of pyridoxine

4.2.5.2 Physio-biochemical Role

Pyridoxal phosphate takes part in the following biochemical reactions:

1. Transamination

   Oxaloacetate + Glutamate $\leftrightarrow$ Aspartate + α-Ketoglutarate
2. Decarboxylation

   Histidine $\rightarrow$ Histamine + CO_2

   Glutamate $\rightarrow$ Gamma amino butyric acid (GABA) + CO_2
3. Glycogenolysis

   Phosphorylase requires pyridoxal phosphate and breaks down glycogen to glucose-1-phosphate.
4. Condensation

   Pyridoxal phosphate acts as a coenzyme for the condensation of glycine and succinyl-CoA to form δ-aminolevulinic acid (ALA), which forms heme via porphobilinogen and protoporphyrin IX (Champe and Harvey 1994).

   Glycine + Succinyl-CoA $\rightarrow$ δ-aminolevulinic acid

4.2.5.3 Dietary Sources

Dietary sources are liver, banana, egg yolk, wheat, corn, yeast, and green leafy vegetables.

4.2.5.4 RDA (mg/day) at Different Age Groups and During Pregnancy and Lactation

For infants, it is about 0.3–0.4, for adults, elderly persons, pregnant women, and lactating mothers, it varies from 2.0 to 2.5.

4.2.5.5 Deficiencies

Deficiency may occur in alcoholics due to the hydrolysis of phosphate. Isoniazid, a drug used for the treatment of tuberculosis, can cause B_6 deficiency by converting pyridoxal phosphate into an inactive derivative. Deficiency, especially in children, may lead to neurological symptoms like hyperirritability and convulsion due to decreased formation of GABA. Deficiency causes microcytic hypochromic anemia due to diminished synthesis of heme as pyridoxal phosphate is essential for the synthesis of heme via ALA. Deficiency can cause hypersensitivity of steroid hormones as pyridoxal phosphate can remove the hormone-receptor complex from DNA binding to terminate the action of steroid hormones like cortisol, androgens, estrogens, etc. It has been reported that prolonged use of oral contraceptives may cause B_6 deficiency. It has also been reported that deficiency of vitamin B_6 may lead to the development of cancer of the breast, uterus, and prostate.

4.2.5.6 Toxicity

Prolonged use of high dose of vitamin B_6 may cause peripheral neuritis.

4.2.6   Biotin

4.2.6.1 Structure (Fig. 4.16)
An imidazole ring combined with a thiophene ring.

4.2.6.2 Physio-biochemical Role
Biotin is a coenzyme in carboxylation reactions (biotin-dependent enzymes are acetyl-CoA carboxylase, pyruvate carboxylase, and propionyl-CoA carboxylase). Biotin takes part in gluconeogenesis and fatty acid synthesis. It induces the synthesis of glucokinase and pyruvate kinase (enzymes of glycolysis) as well as phosphoenolpyruvate carboxykinase (key enzyme of gluconeogenesis). Because of increased synthesis of glycolytic enzymes, biotin administration may improve glucose tolerance in diabetes mellitus.

4.2.6.3 Dietary Sources
Dietary sources are liver, soya bean, yeast, egg yolk, tomato, milk, and meat. **Dietary intake** should be between 100 and 200 µg/day.

4.2.6.4 Deficiencies
Biotin deficiency is rare as it is present in almost all foods. Moreover, a large amount of biotin can be synthesized by the intestinal bacteria. Prolonged consumption of raw eggs can cause biotin deficiency. Egg white contains heat-labile protein avidin, which tightly combines with biotin and thus, prevents absorption of biotin from the intestine. Avidin is denatured by boiling and cannot bind biotin. Prolonged use of oral antibiotics and prolonged consumption of raw eggs can cause biotin deficiency characterized by depression, dermatitis, glossitis, loss of appetite, muscle pain, nausea, loss of hair (alopecia), and rash around the eyes and mouth. Hyperglycemia may occur due to reduced synthesis of glucokinase. As biotin is a coenzyme necessary for the conversion of acetyl-CoA to malonyl-CoA, biotin deficiency interferes with fatty acid synthesis.

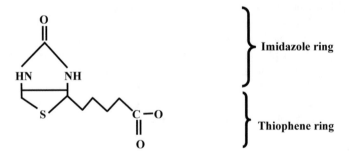

Fig. 4.16 Structure of biotin

4.2.7   Vitamin B$_{12}$

4.2.7.1 Structure

Vitamin B$_{12}$ (cobalamin) has a complex corrin structure. It has the most complex structure of any vitamin known. Four pyrrole (Fig. 4.17) rings with a cobalt ion at the center is called a corrin ring. Cobalt is held in the center of the corrin ring by four bonds from the nitrogen of the pyrrole groups. The commercial preparation for oral use is cyanocobalamin (cyanide is added to cobalamin ring to get stable crystals). Vitamin B$_{12}$ is a red crystalline substance, which is water soluble and heat stable.

4.2.7.2 Physio-biochemical Role

1. Methyl group bound to cobalamin (methylcobalamin) is transferred to homocysteine to form methionine. Cobalamin removes methyl group from methyltetrahydrofolate to form tetrahydrofolate (active folate). Tetrahydrofolate takes part in the synthesis of purine, pyrimidine, and nucleic acid (Fig. 4.18).
2. Fatty acids are converted to methylmalonyl-CoA. Vitamin B$_{12}$ converts methylmalonyl-CoA to succinyl-CoA of citric acid cycle and thus takes part in gluconeogenesis (Fig. 4.19).
3. Methylmalonyl-CoA accumulates due to deficiency of vitamin B$_{12}$ as it cannot be converted to succinyl-CoA. It is hydrolyzed to methylmalonic acid, which is excreted in the urine (methylmalonic aciduria). Homocystinuria occurs due to

Fig. 4.17 Structure of a pyrrole ring

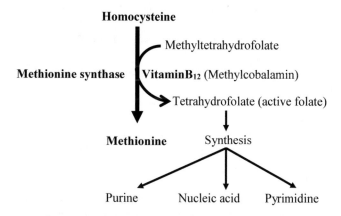

Fig. 4.18 Formation of methionine and active folate is catalyzed by vitamin B$_{12}$-dependent enzyme methionine synthase

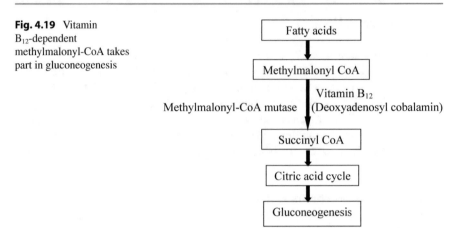

Fig. 4.19 Vitamin B$_{12}$-dependent methylmalonyl-CoA takes part in gluconeogenesis

lack of cystathionine beta synthase (vitamin B$_6$ dependent) enzyme. Homocysteine cannot be converted to cystathionine. As a result, homocysteine accumulates in the blood, and homocysteine is excreted in the urine. Thus, homocystinuria is characterized by high plasma level of homocysteine and high urinary excretion of homocysteine. The individuals (mostly infants and young adults) are mentally retarded, very tall with long fingers (because of overgrowth of bones). Dislocation of lens occurs, leading to myopia. It may cause thromboembolism (Edelstein and Sharlin 2009), causing early death. Dietary protein restriction with vitamin B$_6$, vitamin B$_{12}$, and folate supplementation is essential for treatment with an early diagnosis (Fig. 4.20).

4.2.7.3 Dietary Sources
Dietary sources are found in food from animals and absent in vegetables. Richest sources are found in the liver, milk, and curd (curd contains more vitamin B$_{12}$ than milk as lactobacillus can synthesize vitamin B$_{12}$). Egg, fish, and meat also contain good amount. Vitamin B$_{12}$ is synthesized by the intestinal bacteria.

4.2.7.4 RDA (µg/day) at Different Age Groups and During Pregnancy and Lactation
Infants need about 0.2. Other age groups, pregnant women, and lactating mothers need about 1.5.

4.2.7.5 Absorption
Intrinsic factor of the stomach is essential for the absorption of vitamin B$_{12}$, which is transported to the tissues as well as is stored in the liver (Fig. 4.21).

4.2.7.6 Deficiencies
Lack of intrinsic factor (due to autoimmune destruction of gastric parietal cells or after total gastrectomy) causes deficiency of vitamin B$_{12}$ (Fig. 4.22).

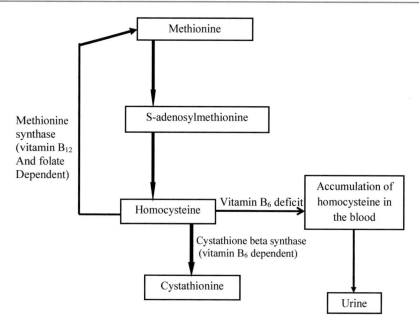

Fig. 4.20 Methionine is converted to *S*-adenosylmethionine, which is subsequently hydrolyzed to homocysteine. Homocysteine is metabolized to cystathione. Homocysteine can be converted back to methionine by methionine synthase, which is a vitamin B_{12}- and folate-dependent enzyme

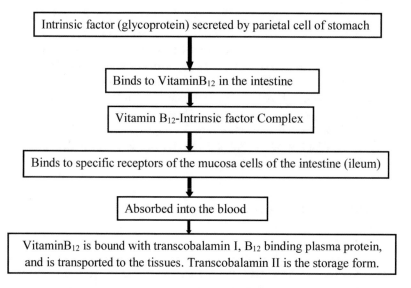

Fig. 4.21 Role of intrinsic factor in the absorption of vitamin B_{12}

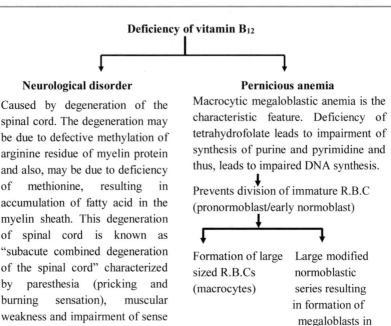

Fig. 4.22 Vitamin B_{12} deficiency causes neurological disorder and pernicious anemia

Elderly people may have atrophic gastritis and may suffer from B_{12} deficiency due to defective absorption. Strict vegetarians of low economic group who cannot afford milk or curd are likely to suffer from vitamin B_{12} deficiency. In the stomach, vitamin B_{12} binds to a protein cobalophilin. In the duodenum, cobalophilin is hydrolyzed and releases vitamin B_{12} to bind to intrinsic factor. Thus, pancreatic disease can cause B_{12} deficiency as cobalophilin cannot be hydrolyzed. Vitamin B_{12} is absorbed from the terminal ileum. Individuals suffering from Crohn's disease affecting the ileum are at higher risk of developing B_{12} deficiency. Surgical removal of the stomach, which produces the intrinsic factor essential for B_{12} absorption is likely to precipitate B_{12} deficiency.

4.2.8 Folic Acid/Folate (Pteroylglutamic Acid)

4.2.8.1 Structure
Folic acid consists of the base pteridine, which is attached to para aminobenzoic acid (PABA) to form pteroic acid. Pteroic acid is attached to glutamic acid to form pteroylglutamic acid, i.e., folic acid.

Pteridine + PABA = Pteroic acid + glutamic acid = Folic acid

It is insoluble in water and is destroyed when exposed to light.

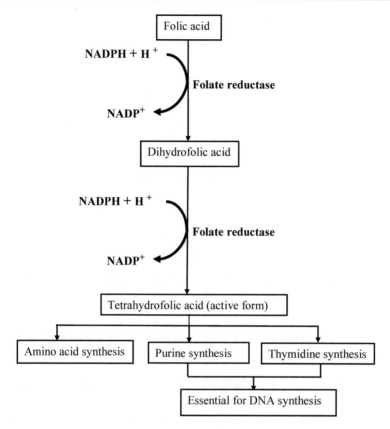

Fig. 4.23 Formation of active folate is catalyzed by folate reductase, which reduces folic acid to dihydrofolic acid and ultimately to tetrahydrofolic acid

4.2.8.2 Physio-biochemical Role

1. Tetrahydrolate is the active form of folic acid/folate and is a carrier of activated one-carbon units. It takes part in amino acid and nucleotide metabolism (Fig. 4.23).
2. Tetrahydrofolate takes part in conversion of one amino acid to another. Histidine → Formiminoglutamic acid (Figlu) → Glutamic acid Catabolism of serine and glycine forms methyltetrahydrofolate via methylene tetrahydrofolate, which in turn forms methionine (Fig. 4.24).
3. The beneficial effect of folate intake on the polymorphism of methylene tetrahydrofolate reductase gene (MTHFR) in overweight women has been reported. The beneficial effect of dietary folate intervention on oxidative stress in terms of total antioxidant capacity has also been observed (Ribeiro et al. 2018).

4.2.8.3 Dietary Sources

Mammals are unable to synthesize a pteridine ring. They obtain folic acid from the diet. Folic acid can be synthesized by the intestinal bacteria. Folic acid is found in

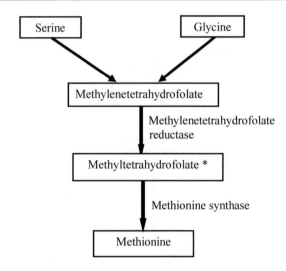

Fig. 4.24 Methylenetetrahydrofolate reductase and methionine synthase catalyzes serine and glycine to methionine. ∗Methyltetrahydrofolate is the main circulating form and is taken up by the tissues

green leafy vegetables, whole grain cereals, liver, egg, and yeast. Liver contains large amount of folic acid.

4.2.8.4 RDA (µg/day) at Different Age Groups and During Pregnancy and Lactation

The infants need about 25–30, and other age groups need about 100. Pregnant women and lactating mothers need about 300–400.

4.2.8.5 Interaction Between Folic Acid and Vitamin B$_{12}$

Methionine synthase is the key enzyme for the functions of folate and B$_{12}$. In B$_{12}$ deficiency when methionine synthase activity is impaired, uptake of folate into the tissues is impaired from the circulating methyltetrahydrofolate (folate is trapped as methyltetrahydrofolate). B$_{12}$ deficiency, therefore, causes simultaneous folate deficiency. Both tetrahydrofolate and vitamin B$_{12}$ are concerned with the conversion of homocysteine to methionine. Both promote DNA synthesis. Like vitamin B$_{12}$, folic acid deficiency causes macrocytic megaloblastic anemia. However, folic acid deficiency will not cause any neurological disorder. Thus, administration of folic acid in patients suffering from vitamin B$_{12}$ deficiency will correct anemia, but not neurologic disorder (for unknown reason it may aggravate neurological disorder) (Table 4.1).

4.2.8.6 Clinical Implication

1. Pregnant women should take adequate amount of folate (not less than 400 µg/day) in order to prevent abnormal fetal development, for example, spina bifida (the baby is born with the spinal cord and its covering exposed through a

Table 4.1 Comparison between folic acid and vitamin B_{12} deficiencies

Folic acid deficiency	Vitamin B_{12} deficiency
1. Presence of macrocytic megaloblastic anemia	1. Presence of macrocytic megaloblastic anemia
2. Absence of neurological disorder	2. Presence of neurological disorder
3. Absence of methylmalonic aciduria	3. Presence of methylmalonic aciduria
4. Presence of Figlu in the urine after administration of histidine (Figlu test)	4. Figlu test may be positive in some cases

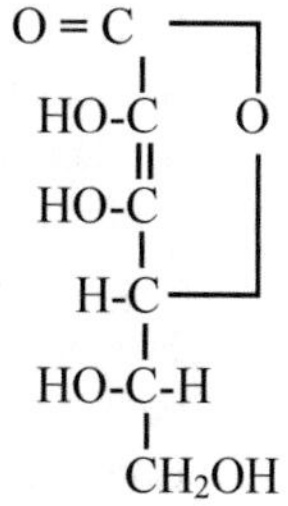

Fig. 4.25 Structure of vitamin C

$$
\begin{array}{l}
O=C \\
HO-C \quad O \\
HO-C \\
H-C \\
HO-C-H \\
CH_2OH
\end{array}
$$

gap in the backbone). Folate supplementation during pregnancy prevents neural tube defects (Blumberg et al. 2018).

2. Hyperlipidemia, hypercholesterolemia and higher incidence of cardiovascular diseases such as hypertension and CIHD are linked to folate deficiency.

3. Folate toxicity occurs due to consumption of excessive amount of folic acid, which may lead to an increase in epileptic attacks (folic acid antagonizes the effects of anticonvulsants). Higher dose for many years may precipitate cancerous growth. It has been reported that increased incidence of prostate cancer may occur in men supplemented with high dose for many years. Excessive consumption over a long period may have adverse effect on cognitive functioning.

4.2.9   Vitamin C (Ascorbic Acid)

4.2.9.1 Structure (Fig. 4.25)
The structure resembles simple sugar ($C_6H_8O_6$) (Fig. 4.25).

4.2.9.2 Properties
Vitamin C is a white substance, which is soluble in water and is easily destroyed by heat. It is a moderately strong acid. It is stable in the crystalline state, but in solution is readily oxidized by atmospheric oxygen. The oxidized form is dehydroascorbic acid, which is readily reduced to reform ascorbic acid. This property of easily reversible oxidation and reduction occurs in the tissues. Vitamin C is a strong reducing agent. The reducing property depends on the double-bonded carbons.

4.2.9.3 Physio-biochemical Role

Vitamin C can be synthesized from glucose by all the species except primates, including humans, monkeys, and guinea pigs. Primates cannot synthesize vitamin C due to the lack of gulonolactone oxidase (Fig. 4.26) (Murray et al. 1996). To prevent its deficiency human beings must consume vitamin C-containing diets.

Vitamin C takes part in the following biochemical and physiological functions:

1. It is required for the synthesis of intercellular cement substance of capillaries.
2. It is required for hydroxylation of proline, collagen synthesis, and connective tissue formation. Proline and lysine hydroxylases are required for collagen formation. Proline hydroxylase is also required for the formation of osteocalcin.
3. It takes part in bile acid formation.
4. It takes part in the intracellular electron transfer.
5. It takes part in the synthesis of epinephrine and norepinephrine from tyrosine.
6. It increases the absorption of iron from the intestine as it is a reducing agent and converts ferric form to ferrous form, which can be absorbed from the intestine.
7. Like vitamin E and β-carotene, it acts as an antioxidant. It is able to inactivate toxic oxygen free radicals (radical-trapping antioxidant). It may prevent the incidence of cancer.
8. Adrenal cortex contains large amount of vitamin C. ACTH administration causes depletion of vitamin C, indicating that vitamin C may have some role in steroidogenesis.

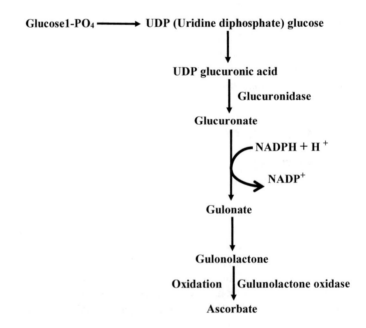

Fig. 4.26 Synthesis of ascorbic acid

9. It is suggested that high dose increases resistance to common cold and improves immune function.
10. Nitrites, used as preservatives in food, can be converted to nitrosamines. It can prevent the formation of nitrosamines, which are potent carcinogens.

4.2.9.4 Dietary Sources

Dietary sources are citrus fruits [orange, lemon, guava, amla (Indian gooseberry)], tomatoes, broccoli, and green leafy vegetables. Among all these, amla contains large amount of vitamin C. Outer covering of citrus fruits contains bioflavonoid, which is an antioxidant as well as an anti-inflammatory and enhances immunity.

4.2.9.5 RDA (mg/day) at Different Age Groups and During Pregnancy and Lactation

Infants should have 25, and other age groups including pregnant women need about 50. The lactating mothers need more than 50 as breast milk contains less amount of vitamin C.

4.2.9.6 Deficiencies

Deficiency of vitamin C causes scurvy characterized by

1. Gum bleeding from the tender swollen gums.
2. Capillaries are fragile. Tendency to bleed under minor pressure, resulting in subcutaneous hemorrhage. It reduces platelet aggregation as normal platelets contain high amount of vitamin C.
3. Delayed wound healing.
4. Painful swelling of the joints and bones due to hemorrhage.
5. Bone fracture with minimal trauma occurs since, the bones are rarified. The signs from 1 to 5 are due to the defective collagen formation and brittle intercellular cement substance (Fig. 4.27).
6. Impaired activity of lysine hydroxylase causes depletion of muscle carnitine, which may lead to fatigue of the scorbutic individuals.

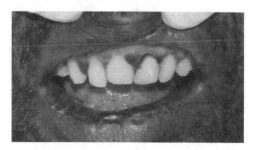

Fig. 4.27 Scurvy. A gingival swelling and bleeding. (Reproduced, with permission, from Elsevier, Book Title: Davidson's Principles and Practice of Medicine, 2010, Edited by Nicki R. Colledge, Brian R. Walker and Stuart H. Ralston (Chapter number 5, Title: Environmental and Nutritional factors in disease, Fig 5.19, Pages 128 to 128) Copyright © Elsevier)

7. Dimorphic anemia consists of (1) microcytic hypochromic anemia, which occurs due to deficient iron absorption. Hemorrhage also contributes. (2) Macrocytic anemia may occur as vitamin C can reduce folic acid to tetrahydrofolic acid.

4.3   Macrominerals

4.3.1   Calcium

4.3.1.1 Physio-biochemical Role

1. Ionic form (Ca^{2+}) takes part in blood coagulation, vasoconstriction, muscular contraction, nerve impulse transmission, hormone action, neurotransmitter action, endocytosis, exocytosis, cellular motility, and glycogen metabolism.
2. Found as calcium crystals in bone and teeth. Calcium takes part in mineralization of bone, i.e., deposition of inorganic minerals in the organic matrix of bone, leading to the formation of bone and teeth.

4.3.1.2 Regulation (Table 4.2)

It is regulated by vitamin D, parathyroid hormone (PTH), and calcitonin of the thyroid gland. 1,25-Dihydroxycholecalciferol increases absorption of dietary calcium from the intestine. PTH increases plasma calcium level. Calcitonin decreases plasma calcium by inhibiting bone resorption. Carbohydrate such as lactose augments passive diffusion across the intestine and thus facilitates calcium absorption.

   Calcium deficiency is rare due to the actions of PTH (Fig. 4.28).

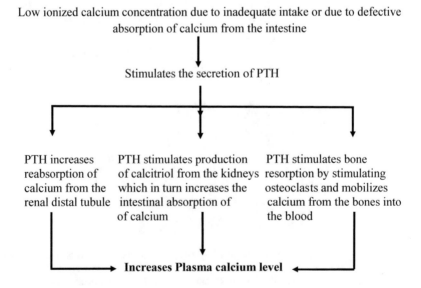

Fig. 4.28 Actions of PTH

Table 4.2 Facilitation and inhibition of calcium absorption

Calcium absorption	
Facilitation	Inhibition
1. Adequate intake of vitamin D	1. Deficiency of vitamin D
2. Deficiency of calcium	2. Deficiency of magnesium
3. Deficiency of phosphorus	3. Excess intake of calcium
4. Pregnancy	4. Menopause
5. Lactation	5. Old age
6. Hyperparathyroidism (due to increased PTH)	6. Phosphate, phytate, oxalate, and dietary fiber
7. 1,25-Dihydroxycholecalciferol (Calcitriol)	7. Hypoparathyroidism
8. Idiopathic hypercalciuria	8. Malabsorption syndrome
9. Carbohydrate such as lactose	9. Celiac disease
	10. Crohn's disease
	11. Diabetes
	12. Renal disease

4.3.1.3 Dietary Sources
Dietary sources are milk and milk products (yoghurt and cheese). Calcium content of cow's milk is more than that of mother's milk. Egg yolk, leafy green vegetables, soya beans, and fish contain good amount of calcium.

4.3.1.4 RDA (mg/day) at Different Age Groups and During Pregnancy and Lactation
Different age groups from infants to children require about 400–600. Adolescents, adults, and elderly persons require about 800–1000. Pregnant women and lactating mothers require about 1200.

4.3.1.5 Deficiencies
1. Magnesium deficiency causes calcium deficiency, since magnesium is a cofactor for PTH secretion.
2. Low-ionized calcium due to impaired secretion of PTH causes tetany, which is characterized by neuromuscular hyperexcitability, leading to spasm of skeletal muscle, especially the muscles of extremities and larynx. Spasm of larynx causes obstruction of the airway, resulting in death due to asphyxia (lack of O_2 and excess of CO_2 in the blood).
3. Rickets in the children and osteomalacia in the adults due to defective calcification of bone matrix (see vitamin D).
4. Osteoporosis in the elderly people, especially women leads to bone fracture. Osteoporosis, i.e., loss of bone mass (loss of both matrix and minerals) is due to increased osteoclastic activity. Osteoclasts cause bone resorption by eroding and absorbing formed bone (whereas osteoblasts are the cells concerned with bone formation). Loss of bone matrix due to osteoporosis causes increased incidence of fractures. Fractures are common in the distal forearm known as Colles' fracture, hip, and vertebrae. Fracture of vertebrae causes kyphosis characterized by

hunching of the back due to compression of the vertebrae and outward curvature of the spine. Osteoporosis is more common in elderly women, especially after menopause due to the deficiency of estrogen. Normally estrogen inhibits osteo-clastic function by increasing apoptosis of osteoclasts. Decreased absorption of calcium in elderly woman aggravates osteoporosis. Osteoporosis is also com-mon in Cushing's syndrome as excessive glucocorticoid secretion leads to loss of bone mass by inhibiting bone formation and stimulating bone resorption. Glucocorticoids inhibit protein synthesis of osteoblasts as well as inhibit the absorption of calcium and phosphate from the intestine. Parathyroid hormone and thyroid hormones cause osteoporosis due to increased bone resorption and decreased bone mass. Low plasma concentration of insulin, androgen, and calci-tonin causes decreased bone formation, leading to osteoporosis. Increased intake of calcium and vitamin D with moderate exercise will slow the progression of osteoporosis.

4.3.1.6 Calcium Toxicity
Excess calcium may be deposited in the kidney, resulting in stone formation (neph-rolithiasis), and may cause constipation. Hypercalcemia increases hydrogen ion secretion and causes alkalosis. Consumption of excessive amount of dietary cal-cium reduces zinc absorption and increases the zinc requirement. Consumption of large quantities of calcium carbonate in antacids for the treatment of peptic ulcer may override the renal ability to excrete excess calcium, causing toxification. This is known as milk alkali syndrome.

4.3.2 Phosphorus

4.3.2.1 Physio-biochemical Role
Phosphorus occurs as hydroxyapatite in the bone (about 80% of total body phos-phorus present in the bone) and as phospholipids. Phosphate is a constituent of ATP, creatine phosphate, cAMP, 2,3-diphosphoglycerate, and many proteins. Phosphorylation and dephosphorylation of proteins take part in various cell func-tions. Phosphate crystals along with calcium take part in the bone formation. Phosphate buffers are involved in the acid-base regulation of blood. Phosphorus is a component of DNA and RNA. In contrast to plasma calcium level, PTH decreases plasma phosphate by increasing phosphate excretion in the urine.

4.3.2.2 Dietary Sources
Dietary sources are milk, nuts, cereals, meat, poultry, and phosphate food additives.

4.3.2.3 RDA (mg/day) at Different Age Groups and During Pregnancy and Lactation
Similar to calcium (see before).

4.3.2.4 Deficiencies

Include anorexia, paresthesia, ataxia, confusion, muscle weakness, bone pain, and rickets in children or osteomalacia in adults. In both rickets and osteomalacia, there is a deficit in the formation of plate cartilage or bone matrix, which leads to impairment of the formation of chondroblasts and osteoblasts. Phosphorus deficiency occurs in patients with Fanconi's syndrome. Fanconi's syndrome (autosomal dominant disorder) is a disorder of proximal convoluted renal tubule, leading to deficient phosphate reabsorption and consequently large amount of urinary excretion of phosphate as well as large amount of amino acids and glucose.

4.3.3 Sodium, Potassium, and Chloride

4.3.3.1 Physio-biochemical Role

1. These three minerals take part in the regulation of acid-base balance, maintain the osmolarity and water balance of the body.
2. Sodium is the principal cation of extracellular fluid, whereas potassium is the principal cation of intracellular fluid. Chloride is found mainly in the extracellular fluid.
3. Na^+ influx and K^+ efflux are responsible for the generation of action potential of nerves, skeletal, and cardiac muscles. Thereby, they take part in the transmission of nerve impulse and contraction of skeletal and cardiac muscles. $Na^+K^+ATPase$ of the plasma membranes extrude three Na^+ from the cell and take two K^+ into the cell due to hydrolysis of ATP. Various hormones, for example, thyroid hormones, insulin, and aldosterone increase the activity of $Na^+K^+ATPase$.
4. Chloride forms hydrochloric acid of the stomach. Chloride is concerned with chloride shift of RBCs and takes part in the transport of carbon dioxide.

4.3.3.2 Dietary Sources

Dietary sources of potassium are banana, apple, bean, potato, nut, meat, fish, and poultry. Coconut water is a good source of potassium. An individual consumes about 8–10 g of salt daily in food and salt added at the table. Recommended intake of sodium is about 2 g/day and of potassium is about 3 g/day.

4.3.3.3 Deficiencies

Deficiencies may occur due to vomiting and diarrhea. Salt intake should be restricted to prevent hypertension. Insulin causes potassium to enter inside the cells, resulting in hypokalemia. Bartter syndrome is an inherited condition due to mutation in the NaK_2Cl cotransporter gene on chromosome 15. It is characterized by hypokalemia, alkalosis, hypercalciuria, short stature, and hypovolemia. Sodium depletion liberates renin from the kidneys, which forms angiotensin II. Angiotensin II is released in the circulation, increases blood pressure due to vasoconstriction, and stimulates aldosterone release from the adrenal cortex. ANP counteracts sodium retention by inhibiting renin and aldosterone release. It decreases blood pressure and antagonizes angiotensin II (see more details in Chaps. 10 and 16).

4.3.3.4 Toxicity

Excessive salt intake may cause ischemic coronary heart disease, stroke, and hypertension. High-sodium diet increases calcium excretion and bone resorption, suggesting role of high salt intake in the development of osteoporosis. It has been suggested that high salt intake may predispose to gastric cancer. Hyperkalemia due to metabolic acidosis, Addison's disease and chronic renal failure is detrimental on heart and may even lead to cardiac arrest.

4.3.4 Magnesium

It activates many enzymes, for example, hexokinase, fructokinase, adenylyl cyclase, etc. It takes part in glycolysis. It enhances the condensation of chromatin in the regulation of gene expression. It is a constituent of bones and teeth, and about 60% of total body magnesium is present in bones. It is essential for the secretion of PTH. Fructose may facilitate magnesium absorption.

4.3.4.1 Dietary Sources

Dietary sources are cereals, pulses, nuts, leafy vegetables, fish, and chicken.

4.3.4.2 RDA (mg/day) at Different Age Groups and During Pregnancy and Lactation

In infants, it is about 40–60. Schoolchildren need about 200. Adults, elderly people, pregnant women, and lactating mothers need about 400.

4.3.4.3 Deficiencies

Deficiency is due to malnutrition. Deficiency causes hypocalcemia and hypokalemia. Hypocalcemia occurs due to the effect of magnesium depletion on PTH. Hypokalemia occurs due to excessive potassium excretion. Magnesium depletion causes neuromuscular excitability.

4.3.4.4 Toxicity

Toxicity of excessive magnesium consumption may cause nausea, diarrhea, and pain in the abdomen.

4.4 Microminerals

4.4.1 Iron

The iron content of an adult man is about 4–5 g. Iron is present mainly in hemoglobin (70%). A certain amount is present in myoglobin, cytochromes, catalase, NADH dehydrogenase, succinic dehydrogenase, aconitase, storage forms (ferritin and hemosiderin), and transport iron, i.e., transferrin.

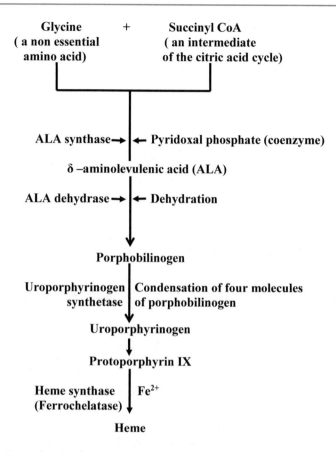

Fig. 4.29 Synthesis of Fe^{2+}-protoporphyrin complex (heme). Lead inhibits zinc-containing enzyme ALA dehydrase and also ferrochelatase (catalyzes the insertion of Fe^{2+} into protoporphyrin) and causes anemia

4.4.1.1 Physio-biochemical Role

Ferrous form is incorporated into protoporphyrin for hemoglobin synthesis (Fig. 4.29). Iron is an important constituent of the cytochromes, which are concerned with the internal respiration and energy production. It is required for intracellular electron transport. Iron generates ATP during the oxidation of substances of the macronutrients. Gastric secretion dissolves the iron, permitting vitamin C to convert ferric to ferrous form.

4.4.1.2 Dietary Sources

Dietary sources are leafy vegetables, pulses, beans, meat, and liver. Jaggery is a good source of iron, whereas milk, vegetables, and fruits are poor sources of iron. Ferrous iron is only absorbed from the intestine. Diets rich in vitamin C increase the absorption of iron by converting ferric to ferrous form, whereas phytate (present in cereals) and oxalates (in leafy vegetables) inhibit the absorption of iron by forming

Table 4.3 Facilitation and inhibition of iron absorption

Iron absorption	
Facilitation	Inhibition
1. Low iron intake	1. High iron intake
2. Vitamin C	2. Calcium
3. Chronic blood loss (hookworm infection, bleeding piles, etc.)	3. Phytate, oxalate, and phosphate
4. Repeated pregnancy	4. Phenolic compound (iron binding)
5. Dietary intake of fish, meat, and seafood	5. Achlorhydria
6. Aplastic anemia	6. Tannin of tea
7. Hemolytic anemia	7. Old age
8. Hereditary hemochromatosis	

insoluble iron salts. Polyphenols present in fruits such as berries and grapes reduce the absorption of iron. The tannin of tea reduces absorption of iron. Heme iron from meat is rapidly absorbed (Table 4.3).

4.4.1.3 RDA (mg/day) at Different Age Groups and During Pregnancy and Lactation

Infants need about 8. Adult female and elderly persons require more than adult males, i.e., about 20. Pregnant women and lactating mothers require about 30.

4.4.1.4 Iron Deficiency Anemia

It is the most common type of anemia in Asia and is characterized by microcytic hypochromic anemia. It is most common in infants, preschool children, adolescents, and pregnant women. People suffering from iron deficiency anemia are susceptible to infections due to low T-lymphocyte and B-lymphocyte counts.

Causes

1. Dietary deficiency of iron.
2. Chronic blood loss due to hookworm infection in rural areas, where sanitation is poor. Bleeding piles also cause blood loss.
3. Poor dietary habits of the elderly people.
4. Repeated pregnancy as blood loss occurs at parturition.
5. Achlorhydria.
6. Partial or complete gastrectomy.

Note: Adult female requires more iron to compensate loss of blood during menstruation. Menstruation causes average loss of 20 mg of iron per month.

Symptoms

Tiredness, fatigue, dyspnea on exertion, impaired work performance, pale conjunctiva, spooning of nails or koilonychias (Fig. 4.30), glossitis, and angular stomatitis are common features. Premature labor may occur. Iron deficiency anemia can have an effect on mental development in the children.

Fig. 4.30 Illustration of
normal nail (**A**) and
koilonychia (**B**)

Prevention

Iron deficiency anemia can be prevented by iron containing diets along with the diets containing vitamin C. Repeated pregnancy should be avoided. During pregnancy, mother's food must contain surplus quantities of iron as the fetus is dependent on the iron reserve of the mother. This demand rises as pregnancy progresses and is greatest in the second half of pregnancy. Iron deficiency anemia enhances the mortality and morbidity of the mother and infant. Sanitation should be improved in order to prevent hookworm infection.

Diagnostic Criteria of Iron Deficiency Anemia

Anemia occurs when there is a reduction in the number of circulating RBCs or a decrease in the content of hemoglobin or both (normal average RBC count is 5.5 million/cumm in adult males and 4.8 million/cumm in adult females. Normal average hemoglobin is 15.5 g% in adult males and 14 g% in adult females). Size of RBC is less than normal [average diameter of normal RBC is 7.3 μ and MCV (mean corpuscular volume), i.e., the volume of individual RBC is less than normal in microcytic anemia (normal MCV is about 85 μm^3)]. Normal color of RBC is pink or orange color due to the presence of hemoglobin. In hypochromic anemia pink color is very less due to low content of hemoglobin. [MCH (mean corpuscular hemoglobin), i.e., amount of hemoglobin per RBC is less than normal in hypochromic anemia (normal MCH is about 30 pg) and MCHC (mean corpuscular hemoglobin concentration), i.e., amount of hemoglobin per unit volume of RBC is less than normal (normal MCHC is about 35%)]. Low serum iron is below 50 μg/100 mL. Low serum iron-binding capacity is below 250 μg/100 mL (normal serum iron varies from 60 to 140 μg/100 mL and normal serum iron-binding capacity varies from 250 to 350 μg/100 mL). Thus, iron deficiency causes microcytic hypochromic anemia.

4.4.1.5 Toxicity Due to Excess Iron Intake

Iron is absorbed according to the body's demand (mucosal block theory). In contrast to the mucosal block theory, excess iron intake is stored in the liver, spleen, and bone marrow in the form of hemosiderin. In the plasma, Fe^{2+} is converted to Fe^{3+} that is incorporated by transferrin (TF). TF transports iron to the bone marrow, liver, and spleen (Fig. 4.31). Hemosiderin is an aggregation of ferritin (iron-binding protein). Hemosiderosis (excessive deposition of hemosiderin) due to repeated blood transfusion or excessive iron intake damages tissues, causing hemochromatosis. Hemochromatosis is characterized by pigmentation of skin (bronze color), bronze

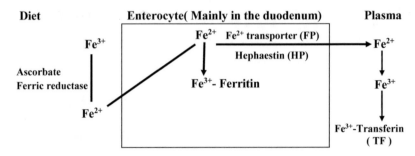

Fig. 4.31 Fe^{3+} present in the diet is converted to Fe^{2+} by ascorbate and ferric reductase. Fe^{2+} is transported by iron transporter into the enterocyte. Fe^{2+} is converted to Fe^{3+}, which is incorporated into ferritin. Fe^{2+} is transported to the blood by FP. This transport is facilitated by HP. In the plasma Fe^{2+} is converted to Fe^{3+} that is incorporated into TF. TF transports iron to the bone marrow, liver, etc. for heme biosynthesis

diabetes (due to the deposition of hemosiderin in the pancreas), and cirrhosis of the liver. South African Bantu people prepare food in iron utensils and may suffer from hemochromatosis. Mutations in the HFE gene located on the short arm of chromosome 6 (inherited as an autosomal trait) cause hereditary hemochromatosis due to iron overload.

4.4.2   Copper

4.4.2.1 Physio-biochemical Role

Copper is a component of coenzymes in the electron transport chain and is essential for various biochemical reactions. The following important enzymes contain copper:

1. Cytochrome oxidase → Essential for internal respiration. Involved in the electron transport chain and formation of ATP.
2. Tyrosinase → Essential for the formation of melanin.
3. Lysyl oxidase → Essential for the maintenance of elasticity of elastin.
4. ALA synthase → Essential for heme synthesis.
5. Monoamine oxidase → Inactivates catecholamines.
6. Ferroxidase → Catalyzes oxidation of iron.
7. Diamine oxidase → Inactivates histamine.
8. Dopamine β-hydroxylase → Converts dopamine to norepinephrine.
9. Zinc/Copper superoxide dismutase (Zn/Cu SOD) → Prevents oxidative damage by converting superoxide ion to hydrogen peroxide.

Ceruloplasmin is a globulin of plasma, which binds six atoms of copper and oxidizes ferrous form to ferric form, which is incorporated into transferrin. Transferrin transports iron to the bone marrow and liver for the synthesis of heme.

4.4.2.2 Dietary Sources

Dietary sources are green leafy vegetables (e.g., spinach), cereals, nuts, bran, liver, meat, egg, and fish. Copper content is less in milk.

4.4.2.3 RDA (μg/day) at Different Age Groups and During Pregnancy and Lactation

In adults, elderly people, pregnant women, and lactating mothers, it is about 1000. Infants and schoolchildren require less 200–400 to prevent copper toxicity.

4.4.2.4 Deficiencies

1. Copper is essential for heme synthesis. Copper deficiency will result in microcytic hypochromic anemia like iron deficiency anemia. Deficiency is common in infants getting only milk.
2. Copper deficiency causes weakening of the walls of blood vessels as copper is essential for the maintenance of elasticity of blood vessels.
3. Copper deficiency causes hypopigmentation of the skin and hair due to lack of formation of melanin.
4. Prolonged use of antacids can cause copper deficiency.

4.4.2.5 Copper Toxicity

1. Wilson's disease is a genetic disease due to mutation in the gene of copper transporting ATPase (autosomal recessive disorder), which causes low plasma ceruloplasmin. As a result, more copper is delivered and deposited in various organs, for example, the liver and brain. Deposition of copper will cause degeneration of the liver and lenticular nucleus of basal ganglia of the brain (hepatolenticular degeneration) and thus will lead to liver cirrhosis and serious neurological symptoms like defective movement, slurred speech, and muscle spasm.
2. Menkes syndrome is also a genetic disease (X-linked copper deficiency), which is fatal in the infants. Ceruloplasmin is low. Signs and symptoms are mental retardation due to defective myelination, fragile abnormal (kinky) hair, convulsion due to neurodegeneration of the brain, twisted arteries, and deformities in the skull and ribs. Death usually occurs before 3 years of age.
3. Excess copper enhances oxidation of proteins and lipids and enhances the production of free radicals. Therefore, it is necessary to have normal amount of copper in the diet.
4. Storage and boiling of food and water in brass vessels cause accumulation of copper in the liver. The infants usually suffer from copper toxicity characterized by abdominal distension, anorexia, hepatitis, jaundice, and even cirrhosis of the liver.
5. Aceruloplasminemia due to mutation in the ceruloplasmin gene causes accumulation of ferrous iron in the brain. Symptoms are dementia, poor muscle tone, and retinal degeneration.

4.4.3   Iodine

Total body contains about 25 mg of iodine. It is stored mainly in the thyroid gland (about 80%). Rest is stored in the salivary glands, gastric mucosa, choroid plexus of the eyes, and lactating mammary glands. Ingested iodine is converted to iodide and is absorbed from the intestine. The thyroid gland daily secretes 80 µg of iodine (as part of the thyroid hormones). Iodide takes part in the synthesis of thyroid hormones. This trace element is a scarce component of the soil, and therefore, there is little in the food. It is present in the seafood (especially, shellfish and sea fish) and the seaweeds, eggs, and meat. Iodine is lacking in the mountain areas due to constant erosion of soil. Such areas are called goiterous belts like Himalayan region and Alps. It is low in the soil of frequently flooded areas.

4.4.3.1   RDA (µg/day) at Different Age Groups and During Pregnancy and Lactation

In infants, it is about 40, and in adults, it is about 150. Pregnant women and lactating mothers require more than the adults to prevent congenital hypothyroidism and cretinism.

4.4.3.2   Goiter

Enlargement of thyroid gland is called goiter (Fig. 4.32). It is of the following types:

1. Iodine deficiency goiter occurs when dietary iodine intake falls below 50 µg/day, which hampers the synthesis of triiodothyronine (T_3) and tetraiodothyronine (T_4)/thyroxine. Normally secretion of thyroid hormones is controlled by thyroid-stimulating hormone (TSH)/thyrotropin of the anterior pituitary. TSH stimulates the secretion of thyroid hormones (by increasing various steps of synthesis) and is subject to negative feedback control by high circulating level of thyroid hormones, acting on the anterior pituitary. High level of thyroid hormone inhibits the secretion of TSH from the anterior pituitary (negative feedback). Thus, TSH stimulates the secretion of thyroid hormones, which inhibits the secretion of TSH (Fig. 4.33). A balance of this feedback is maintained by appropriate adjustment in order to have normal secretion of thyroid hormones. Low thyroid hormones due to iodine deficiency will lead to unchecked excessive secretion of

Fig. 4.32 Illustration of goiter

TSH. Thyroid hypertrophies (increase in size of thyroid cells) due to excessive secretion of TSH leads to an enlargement of thyroid gland.

2. Cabbage goiter occurs due to excessive consumption of vegetables of the Brassicaceae family. Cabbage and turnip contain progoitrin, which is converted to antithyroid agent, i.e., goitrin. Goitrin inhibits the synthesis of thyroid hormones and causes enlargement of the thyroid gland.

3. Pendred syndrome is an inherited disorder caused by mutation of pendrin gene present on chromosome 7. It causes deficiency of peroxidase and thus inhibits the secretion of thyroid hormones from the thyroid gland, resulting in the development of goiter. It is associated with congenital deafness.

4. Congenital hypothyroidism may be due to maternal iodine deficiency. The secretion of thyroid hormones is inhibited and causes the development of goiter in the children. Furthermore, maternal antithyroid antibodies, if present, can cross the placenta and inhibit fetal thyroid function. Placental transfers of maternal antibodies may cause hypothyroidism in infants and children. During pregnancy, iodine deficiency induces cerebral dysfunction in the children, leading to cretinism, and an increased risk of spontaneous abortion.

Mechanism of formation of goiter under two, three, and four are same as iodine deficiency goiter. Apart from goiter, iodine deficiency may cause increased stillbirth and infant mortality.

4.4.3.3 Iodine Toxicity

Iodine toxicity occurs due to excessive intake of iodized salt or seafoods and seaweeds. Near the sea areas, cow or buffalo milk may contain high amount of iodine due to iodine-enriched cattle feed. Excessive intake of iodine causes iodine-induced hyperthyroidism due to excessive secretion of the thyroid hormones.

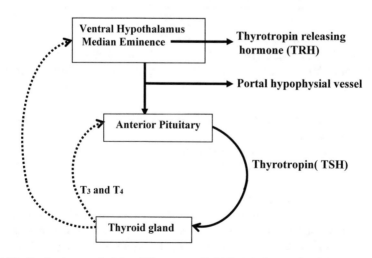

Fig. 4.33 Feedback control of thyroid hormones. Solid lines indicates stimulation. Dashed lines indicate inhibition

4.4.3.4 Prevention
About one billion people throughout the world used to suffer from iodine deficiency disorder. Government programs of addition of iodine in common salt has prevented the incidence of goiter. Goiter has almost disappeared due to use of iodized salt.

4.4.4   Zinc

Zinc is a component of metalloenzyme Zn/CuSOD, carbonic anhydrase, alcohol dehydrogenase, RNA nucleotide polymerase, alkaline phosphatase, and ALA dehydrase.

4.4.4.1 Physio-biochemical Role
1. Required by protein kinases, which regulate gene expression.
2. Essential for spermatogenesis.
3. Being component of carbonic anhydrase, takes part in the transport of carbon dioxide.
4. Plays a role in immune function.
5. Essential for taste sensation, growth, and wound healing.
6. Takes part in protein digestion as it is a component of peptidases.
7. Essential for the activities of the enzyme lactate dehydrogenase as it contains zinc.
8. Is a coenzyme for the conversion of retinol to retinaldehyde by retinol dehydrogenase.
9. ALA dehydrase converts ALA to porphobilinogen, which is converted by series of reactions to heme.
10. Zinc finger motif is formed by binding of Zn^{2+} between cysteine and histidine. Transcription factors, containing zinc finger, mediate the binding to DNA.

4.4.4.2 Dietary Sources
Dietary sources are meat, liver, cheese, nuts, beans, wheat bran, and oatmeal. Oysters contain high amount of zinc. Colostrum, compared with milk, contains high amount of zinc.

4.4.4.3 RDA (mg/day) at Different Age Groups and During Pregnancy and Lactation
At different age groups (except infants and preschool children) and during pregnancy and lactation, it is about 15. Infants and preschool children require about 4.

4.4.4.4 Deficiencies
Mutations in the gene for the intestinal zinc importer ZIP_4 may lead to zinc deficiency. It causes poor wound healing, diarrhea, impaired taste sensation (hypogeusia), alopecia, "flaky paint" rash of lower extremities, hemorrhagic dermatitis (Fig. 4.34), growth retardation, hypogonadism, sexual immaturity, impaired immune function, and increased susceptibility to infections. Low serum level during pregnancy causes

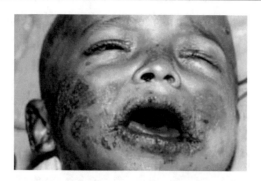

Fig. 4.34 Zinc deficiency with hemorrhagic dermatitis around the mouth and eyes. (Reproduced, with permission, from Elsevier, Book Title: Robbins and Cotran Pathologic basis of Disease, 2005, Author: Vinay Kumar, Abul K Abbas, Nelson Fausto (Chapter 9, Title: Environmental and Nutritional Pathology, Fig. 9–31, Pages 509 to 509) Copyright © Elsevier)

preterm delivery. Poor zinc absorption due to an inherited recessive condition causes acrodermatitis enteropathica characterized by eczematous skin on the hands, feet, and other parts of the body, growth retardation, low immunity, and chronic diarrhea. Excessive intake of copper may precipitate zinc deficiency due to competition between zinc and copper for the intestinal absorption.

4.4.4.5 Toxicity
Inhalation of zinc oxide fumes by welders causes toxicity characterized by nausea, vomiting, headache, and even fibrosis of the lungs. Excessive zinc inhalation causes irritation of the gastrointestinal tract and vomiting.

4.4.5 Fluoride

Fluoride forms calcium fluorapatite in teeth and bones. It increases hardness of bones and teeth. It prevents dental caries as it increases enamel resistance to acid. High level of fluoride intake (due to water intake from deep wells) may cause discoloration and surface irregularities of teeth, anorexia, and loss of body weight. Excessive intake of fluoride may cause sclerosis of the bone, hypercalcification of the ligaments and vertebrae. It may cause genu valgum (knock-knee) characterized by curving of the legs at the knees. Fluoride deficiency increases the incidence of dental caries. Fluoridated toothpaste may prevent dental caries.

Requirement of fluoride (mg/day) for the adults is about 2–4.

4.4.6 Manganese

The total amount of manganese in adult is about 15 mg. It is present in bone, liver, and intestine. It activates many enzymes, for example, pyruvate carboxylase, phosphoenolpyruvate carboxylase, glutamine synthetase, arginase, and mitochondrial superoxide dismutase. It is essential for the synthesis of glycoproteins and chondroitin sulfate. It is essential for urea formation, fatty acid synthesis, and release of insulin.

4.4.6.1 Dietary Sources
Dietary sources are nuts, ginger, seafood, and dairy products. Tea leaves contain large amount of manganese.

4.4.6.2 RDA (mg/day) at Different Age Groups and During Pregnancy and Lactation
In infants, preschool children, and schoolchildren, it is about 0.5–1.5, and for adults, elderly person, pregnant women and lactating mothers, it is about 2.5.

4.4.6.3 Deficiency
Deficiency may lead to defective formation of the bone and cartilage due to lack of chondroitin sulfate. Deficiency may be more common in infants due to low concentration of manganese present in mother's milk.

4.4.6.4 Toxicity
Toxicity occurs due to inhalation of manganese fumes and manganese ore dust by mine workers, resulting in psychosis and features of parkinsonism (cogwheel or leadpipe rigidity, resting tremor, etc.)

4.4.7 Selenium

Selenium is an antioxidative and anti-inflammatory micromineral. It is present inside the cells as selenocysteine and selenomethionine. It is a cofactor for glutathione peroxidase, which serves to protect cell membrane. It is a synergistic antioxidant with vitamin E. Deiodinase enzyme (which causes deiodination of thyroxine, i.e., T_4 to T_3) contains selenium. Selenium is essential for the motility of sperm and immune function. It prevents chromosome breakage. Selenium supplementation lowers incidence of cancer, particularly prostate cancer. It reduces DNA damage, enhances apoptosis, and inhibits angiogenesis.

4.4.7.1 Dietary Sources
Dietary sources are cereal grains (jowar and bajra), Bengal gram, meat, nuts, shellfish, chicken, egg yolk, and garlic.

4.4.7.2 RDA (µg/day) at Different Age Groups and During Pregnancy and Lactation
In infants, preschool children, and schoolchildren, it is about 15–20. Adults, elderly people, pregnant women, and lactating mothers require about 60.

4.4.7.3 Deficiency
Keshan disease causes cardiomyopathy, resulting in enlargement of heart and heart failure. It occurs in certain areas of China due to deficiency of selenium in soil.

4.4.7.4 Toxicity (Selenosis)

High intake of selenium causes brittle hair and nails, skin infection, and garlic smell of breath due to the expiration of dimethyl selinide.

4.4.8 Molybdenum

It is a cofactor of xanthine oxidase, aldehyde oxidase, and sulfate oxidase. Xanthine oxidase takes part in conversion of purine bases to uric acid. Aldehyde oxidase detoxifies purine and pyrimidine. Sulfite oxidase converts sulfite to sulfate.

4.4.9 Cobalt

Nutritional importance of cobalt is due to its presence in vitamin B_{12}. It takes part in erythropoiesis.

4.4.10 Sulfur

Sulfur is present as chondroitin sulfates in the cartilage and bone and is present in the keratin of hair and nail. Glutathione, thiamin, and biotin contain sulfur group. Bile salts contain sulfur group. Sulfur containing amino acids (cysteine and methionine) constitute body proteins. It is a component of insulin. Dietary sources are fish, eggs, liver, poultry, cheese, and beans.

4.4.11 Chromium

The kidney, spleen, liver, and skeletal muscle contain chromium. It is present in yeast, wine, black pepper, mushrooms, and nuts. It is absorbed mainly in the jejunum. Absorption is facilitated by vitamin C. It enhances biotin synthesis by the intestinal bacteria. Chromium deficiency causes insulin resistance as chromium facilitates the binding of insulin to its receptor of peripheral cells and potentiates the actions of insulin. Chromium deficiency causes impaired glucose tolerance.

4.5 Summary

The chapter deals with structure, physio-biochemical role, dietary source, RDA, features of deficiency and toxicity for each micronutrient. The fat-soluble vitamins except vitamin D are isoprenoid compounds. Vitamin D refers to a group of closely related prohormone containing various intermediates of cholesterol synthesis. Vitamin A synthesizes rhodopsin, which is responsible for scotopic vision. Vitamin A controls gene expression and thus cellular growth and differentiation. Important characteristic features of vitamin A deficiency are nyctalopia, increased dark adaptation time, and pathological changes of the eyes. Vitamin D increases plasma levels of calcium and phosphate. Deficiency of vitamin D causes defective calcification of bone matrix, leading to rickets in children and osteomalacia in adults. Vitamin E acts as an antioxidant. Vitamin K is essential for the synthesis of blood clotting factors. Water-soluble vitamins acts as coenzymes or cofactors and take part in various chemical reactions by binding with various enzymes such as amino acid oxidase, dehydrogenase, carboxylase, etc. B_{12} and folate promote DNA synthesis. Thereby, deficiency of both the vitamins causes macrocytic megaloblastic anemia. In addition, B_{12} deficiency (not folate) causes neurological factors. B-complex vitamins except folate enter TCA cycle and take part in amino acid synthesis, fatty acid synthesis, and gluconeogenesis.

Calcium takes part in various physio-biochemical reactions such as blood coagulation, muscular contraction, nerve impulse transmission, and neurotransmitter actions. Calcium deficiency is rare due to the actions of PTH. Low ionized calcium due to impaired secretion of PTH causes tetany. Phosphorylation and dephosphorylation of proteins take part in various cell functions. Phosphate buffers are involved in the acid-base regulation of blood. Phosphorus deficiency leads to impairment of the formation chondroblasts and osteoblasts. Na^+, K^+, and Cl^- take part in the regulation of acid-base balance and maintain the osmolarity and water balance of the body, and Na^+ influx and K^+ efflux are responsible for the generation of action potential of nerves, skeletal and cardiac muscles. Mg^{2+} is a cofactor for hexokinase, pyruvate kinase, and adenyl cyclase and is essential for the secretion of PTH. Ferrous ion is incorporated into protoporphyrin for hemoglobin synthesis. Copper is a cofactor in the electron transport chain. Iodide takes part in the synthesis of T_3 and T_4. Zinc is a component of Zn/CuSOD, carbonic anhydrase, and ALA dehydrase.

Fluoride prevents dental caries. Manganese is essential for urea formation, fatty acid synthesis and release of insulin. Selenium is a synergistic antioxidant with vitamin E. Sulfur is a component of insulin. Nutritional importance of cobalt is due to its presence in vitamin B_{12}. Chromium facilitates the binding of insulin to its receptors of peripheral cells and potentiates the action of insulin. Molybdenum is a cofactor of xanthine oxidase, which takes part in the conversion of purine bases to uric acid.

Various genetic diseases due to mutation of genes are highlighted, for example, (1) mutation in PHEX gene causes rickets, (2) mutation in the HFE gene causes hereditary hemochromatosis due to iron load, (3) mutation in the gene of copper transporting ATPase causes copper toxicity, (4) mutation in the ceruloplasmin gene causes aceruloplasminemia, leading to the accumulation of ferrous ion in brain, (5) mutation of pendrin gene results in the development of goiter, and (6) mutation in the gene for the intestinal zinc importer ZIP_4 leads to zinc deficiency. Poor zinc absorption due to an inherited recessive condition causes acrodermatitis enteropathica.

Table 4.4 refers to antioxidant micronutrients. Table 4.5 reviews clinical signs due to deficiencies of micronutrients.

Table 4.4 Sources of antioxidants

Antioxidant	Sources
1. Vitamin C and E, provitamin β-carotene, selenium, and phenolic compounds	Present in whole grains, soya beans, apples, and walnuts are antioxidants as they can inactivate oxygen free radicals
2. Tomatoes	Good antioxidants as they contain vitamin C as well as carotenoid lycopene
3. Melatonin (Reiter et al. 2000)	Present in vegetables, fruits, flowers, seeds, walnuts, and a variety of herbs and is a potent free radical scavenger (hydroxyl radical, hydrogen peroxide, etc.)
4. Flavonoids (Scalbert and Zamora-Ros 2015)	Present in a variety of foods such as onions, berries, apples, tea, etc.

Table 4.5 Clinical signs due to deficiencies of micronutrients

Clinical signs	Deficiencies
1. Glossitis	Present due to deficiencies of biotin, riboflavin, and iron
2. Peripheral neuropathy	Features of the deficiencies of vitamin B_{12}, thiamin, and vitamin E
3. Edema	Present due to the deficiency of thiamin
4. Dermatitis	Due to deficiencies of riboflavin, niacin, biotin, and zinc
5. Microcytic hypochromic and normoblastic anemia	Caused by the deficiencies of iron, copper, pyridoxine, and vitamin C
6. Macrocytic megaloblastic anemia	Caused by the deficiencies of vitamin B_{12} and folic acid
7. Hemorrhagic diseases	Due to the deficiencies of vitamin K, vitamin E, and vitamin C

References

Abdel-Rehim AS et al (2014) Vitamin D level among Egyptian patients with chronic spontaneous urticaria and its relation to severity of the disease. Egypt J Immunol 21:85–90

Abuzeid WM et al (2012) Vitamin D and chronic rhinitis. Curr Opin Allergy Clin Immunol 12:13–17

Bergler-Czop B, Brzezinska-Wcislo L (2016) Serum vitamin D level-the effect on the clinical course of psoriasis. Postepy Dermatol Alergol 33:445–449

Blumberg JB et al (2018) The evolving role of multivitamin/multimineral supplement use among adults in the age of personalized use among adults in the age of personalized nutrition. Nutrients 10:248. (17 p)

Champe PC, Harvey RA (1994) Lippincott's illustrated reviews: biochemistry, 2nd edn. Lippincott Williams and Wilkins, Philadelphia

Edelstein S, Sharlin J (eds) (2009) Life cycle nutrition. Jones Bartlett Publishers LLC, Burlington

Ganong WF (2003) Review of medical physiology, 21st edn. McGraw-Hill, New York

Garach AM et al (2019) Vitamin D status, calcium intake and risk of developing type 2 diabetes: an unresolved issue. Nutrients 11:642

Gibney MJ et al (eds) (2009) Introduction to human nutrition, 2nd edn. Wiley-Blackwell, Hoboken

Hamishehkar H et al (2016) Vitamins, are they safe? Adv Pharm Bull 6(4):467–477

Heine G et al (2013) Association of vitamin D receptor gene polymorphisms with severe atopic dermatitis in adults. Br J Dermatol 168:855–858

Lim SK et al (2016) Comparison of vitamin D levels in patients with and without acne: a case-control study combined with a randomized controlled trial. PLoS One 11:0161162

Longo DL et al (eds) (2011) Harrison's principles of internal medicine, 18th edn. McGraw-Hill, New York

Mullins RJ et al (2011) Season of birth and childhood food allergy in Australia. Pediatr Allergy Immunol 22:583–589

Murray RK et al (1996) Harper's illustrated biochemistry, 24th edn. McGraw-Hill, New York

Nasir-Kalmarzi R et al (2018) Evaluation of 1,25 dihydroxyvitamin D3 pathway in patients with chronic urticaria. QJM 111(3):161–169

Osborne NJ et al (2012) Prevalence of eczema and food allergy is associated with latitude in Australia. J Allergy Clin Immunol 129:865–867

Poole A et al (2018) Cellular and molecular mechanisms of vitamin D in food allergy. J Cell Mol Med 22:3270–3277

Reiter RJ et al (2000) Actions of melatonin in the reduction of oxidative stress a review. J Biomed Sci 7(6):444–458

Ribeiro MR et al (2018) Influence of the C677T polymorphism of the MTHFR gene on oxidative stress in woman with overweight or obesity: response to a dietary folate intervention. J Am Coll Nutr 37:677–684

Scalbert A, Zamora-Ros R (2015) Bridging evidence from observational and intervention studies to identify flavonoids most protective for human health. Am J Clin Nutr 101:897–898

Searing DA, Leung DY (2010) Vitamin D in atopic dermatitis, asthma and allergic diseases. Immunol Allergy Clin N Am 30:397–409

Tuchinda P et al (2018) Relationship between vitamin D and chronic spontaneous urticaria: a systematic review. Clin Transl Allergy 8:51

Zhao Y et al (2019) Vitamins and mineral supplements for retinitis pigmentosa. J Ophthalmol 2019:8524607. 11 p

Further Reading

Bailey RL et al (2015) The epidemiology of global micronutrient deficiencies. Ann Nutr Metab 66(Suppl 2):22–33

Blumberg JB et al (2018) The evolving role of multivitamin/multimineral supplement use among adults in the age of personalized use among adults in the age of personalized nutrition. Nutrients 248(17 pages):10

Elia M et al (eds) (2013) Clinical nutrition, 2nd edn. Wiley-Blackwell, Hoboken

Gibney MJ et al (eds) (2009) Introduction to human nutrition, 2nd edn. Wiley-Blackwell, Hoboken

Hamishehkar H et al (2016) Vitamins, are they safe? Adv Pharm Bull 6(4):467–477

Hennigar SR, Mcclung JP (2016) Homeostatic regulation of trace minerals transport by ubiquitination of membrane transporters. Nutr Rev 74(1):59–67

Zhao Y et al (2019) Vitamins and mineral supplements for retinitis pigmentosa. J Ophthalmol 2019:8524607 11 p

Dietary Fibers

5

Abstract

Nondigestible dietary fibers are classified according to water solubility properties. Water-soluble fibers include pectin, gums, mucilage, β-glucans, and fructans including inulin and fructooligosaccharides. On the other hand, water-insoluble fibers include lignin, cellulose, hemicellulose, whole-grain foods, bran, seeds, and nuts. Dietary fibers are complex carbohydrates except lignin (noncarbohydrate). They are derivatives of plants. However, nondigestible chitin and chitosans are derived from the shells of shrimps and crabs. The algal polysaccharides are extracted from seaweeds or algae. Fibers are non-starch polysaccharides and resistant oligosaccharides. Resistant starch escapes digestion in the small intestine and enters the colon. Prebiotic fibers are nondigestible food ingredients and include galacto-oligosaccharides, fructose-oligosaccharides (fructosans), and lactulose. They stimulate the activity or growth of intestinal bacteria. They probably increase resistance to pathogen invasion. The beneficial effects of fibers are discussed.

Keywords

Water-soluble fibers · Water-insoluble fibers · Resistant starch and prebiotic fibers

5.1 Properties

Soluble fibers are gel-forming viscous in nature. They increase transit time, delay gastric emptying and decrease nutrient absorption. They are easily fermentable by the gut bacteria and thus have some prebiotic functions. Dietary fibers cannot be digested by humans and ingested dietary fibers that reach the large intestine are in an unchanged state. For example, resistant starch present in unripe bananas will not be digested and will pass unchanged into the colon. Colonic flora ferments

© Springer Nature Singapore Pte Ltd. 2019

K. Chakrabarty, A. S. Chakrabarty, *Textbook of Nutrition in Health and Disease*,

https://doi.org/10.1007/978-981-15-0962-9_5

carbohydrates and releases short-chain fatty acids (SCFAs). SCFAs are acetate, propionate, and butyrate. They maintain the integrity of colonocyte DNA. They are rapidly absorbed from the large intestine. SCFAs decrease the solubility of free bile acids due to acidic pH and increase the excretion of bile. It has been reported that propionic acid decreases cholesterol synthesis in the liver, leading to decreased blood cholesterol. SCFAs result from fermentation of fibers by the gut microbiota and reduced risk of type 2 diabetes mellitus, metabolic syndrome, inflammatory diseases, CIHD, and obesity. They bind with the receptor of the intestine and produce PYY and glucagon-like peptide-1 (GLP-1) hormones, which may reduce appetite and food intake. Furthermore, SCFAs may reduce gastric emptying. Insoluble fibers are nonviscous in nature. They have rapid gastric emptying, decrease the intestinal transit time, and increase the fecal bulk, which may help to relieve constipation.

5.2   Biochemical Features

Dietary fibers are non-starch polysaccharides and resistant oligosaccharides. The American Association of Cereal Chemists defines dietary fibers as "Dietary fiber is the edible parts of plants or analogous carbohydrates that are resistant to digestion and absorption in the human small intestine with complete or partial fermentation in the large intestine. Dietary fiber includes polysaccharides, oligosaccharides, lignin, and associated plant substances. Dietary fibers promote beneficial physiological effects including laxation, and/or blood cholesterol attenuation, and/or blood glucose attenuation" (AACC Report 2001). Cellulose is the main carbohydrate of plant. It is a polymer of glucose and is made of glucose units linked by β-1,4 linkage. Hemicellulose contains a mixture of hexose and pentose with the side chains of galactose, arabinose, and glucuronic acid. Hemicellulose contains uronic acid and forms salts with calcium and zinc. The hemicellulose of wheat increases the water-binding capacity of the bowel contents and thus increases the bulk of feces. Pectins are galacturonic acid polymers having a chain of pentose and hexose. The gel-forming properties of pectins are utilized to make jam. Gums secreted at the site of plant injury are exudates of plants and contain galactose backbone with the side chain of arabinose, glucuronic acid, and galactose. They are used in food industry as thickener and food additive. Mucilage found in the plant psyllium is structurally similar to gums and is viscous gel-forming water-soluble fiber. They are used as thickener for creams and soups. The algal polysaccharides are extracted from seaweeds or algae, and because of gel-forming properties, they are utilized in dairy products. β-Glucans are formed of polymers of glucose subunits. Fructans are polymers of fructose. Fructans include inulin and fructooligosaccharides. Inulin is a long-chain D-fructose unit, and due to low molecular weight, it is used for measuring glomerular filtration rate. Lignin is not a carbohydrate and is a high molecular weight polymer of phenylpropane derivatives, some of which have methoxy side chains. Lignin is found in the spaces of cell wall between cellulose and hemicelluloses and prevents absorption of water into the cell wall. The prebiotic lactulose consists of unbranched fructose and galactose.

5.3   Food Sources

Food sources are listed in Table 5.1 (Longvah et al. 2017). In brief, soluble fibers are present in some foods (bananas, berries, apples, and pears), vegetables (carrot, broccoli, onion, etc.), legumes, oats, and barley. Insoluble fibers are present in whole-grain foods, wheat, bran, nuts, and seeds. Inulin is present in oats, onion, garlic, chickory, and tubers. Bran (grain husk separated from the flour) and dry outer covering of some fruits and seeds are good sources of fibers, but they have less water.

Table 5.1 Approximate content of dietary fibers in various foods, expressed as g/100 g edible portion

		Total	Insoluble	Soluble
1.	Bajra	15.6	10.0	5.6
2.	Barley	10.2	8.5	1.7
3.	Jowar	12.2	11.3	0.9
4.	Parboiled rice	3.7	3.0	0.7
5.	Wheat flour	11.4	9.7	1.7
6.	Bengal gram	15.2	12.7	2.5
7.	Soyabean white	22.6	17.0	5.6
8.	Cabbage green	2.8	1.9	0.9
9.	Lettuce	1.8	1.3	0.5
10.	Spinach	2.4	1.5	0.9
11.	Broad beans	8.6	6.6	2.0
12.	Capsicum green	2.0	1.3	0.7
13.	Cauliflower	3.7	2.7	1.0
14.	Babycorn	6.1	4.5	1.6
15.	Cucumber green, elongate	2.1	1.6	0.5
16.	Ladies finger	3.0	1.7	1.3
17.	Peas fresh	6.3	5.0	1.3
18.	Tomato ripe hybrid	1.6	1.3	0.3
19.	Apple	2.6	1.4	1.2
20.	Apricot dried	3.3	2.7	0.6
21.	Dates dry dark brown	9.1	7.6	1.5
22.	Guava white	8.6	7.1	1.5
23.	Carrot red	4.5	3.1	1.4
24.	Beetroot	3.3	2.6	0.7
25.	Radish elongate white	2.7	2.0	0.7
26.	Coriander leaves	4.7	3.2	1.4
27.	Onion	2.4	1.9	0.5
28.	Almond	13.1	10.6	2.5
29.	Cashew nut	3.8	2.2	1.6
30.	Coconut kernel dry	15.9	14.6	1.3
31.	Walnut	5.4	4.7	0.7
32.	Mushroom oyster dry	39.1	35.6	3.5

Reproduced with permission from Longvah T, Ananthan R, Bhaskarachary K, Venkaiah K (2017)

5.4   RDA

The RDA in children aged 1–3 years is 20 g/day and aged 4–8 years is 25 g/day. Recommendation for boys aged 9–13 years is about 30 g/day and for boys aged 14–18 years is about 38 g/day. Men aged 19–50 years require 38 g/day, whereas men more than 51 years require about 30 g/day. Recommendation for girls aged 9–18 years is about 26 g/day. Women aged 19–50 years require 25 g/day, whereas women more than 51 years require about 21 g/day (Soliman 2019). Overconsumption (more than recommended RDA) may cause bloating and abdominal discomfort.

5.5   Beneficial Effects

1. Effects on blood glucose: Normally glucose enters the intestinal cell-coupled Na^+ transport with the help of the cotransporter sodium glucose-linked transporter (SGLT) 1. Na^+ is actively transported out of the cell, and glucose diffuses into the blood by facilitated diffusion via interstitium. Viscosity of non-starch polysaccharides may act as a barrier to SGLT 1 and hence preventing the entry of glucose into the blood (Qi et al. 2018). Furthermore, higher viscosity induced by fiber delay gastric emptying, which, in part, reduces blood glucose after meal (Abbasi et al. 2016). Thus, increase in fiber intake improves glycemia and insulin sensitivity in both nondiabetic and diabetic individuals.

2. The "statin" drugs, for example, lovastatin and atorvastatin inhibit HMG-CoA reductase and decrease cholesterol level. However, statin treatment is not recommended as it is costly and is associated with side effects. Gel-forming viscous soluble fiber prevents the absorption of cholesterol because of the binding property and thus prevents hypercholesterolemia and atherosclerosis. Especially, pectin reduces the absorption of bile salts and thus contributes in reducing hypercholesterolemia (increased fiber intake lowers blood pressure and serum cholesterol level). Furthermore, absorption of SCFA such as propionic acid has been reported to decrease blood cholesterol level due to decreased synthesis of cholesterol in the liver (Anderson et al. 2009).

3. Effects on satiety: Viscous fibers increase retention time in the stomach and cause distension of the stomach, which will increase satiety by stimulating satiety center of the hypothalamus (see Chap. 1). Fibers having large particles may slow gastric emptying, thus increasing satiety. Both non-starch polysaccharides and resistant oligosaccharides produce SCFAs in the gut. They bind with the receptor of the intestine and produce PYY and GLP-1 hormones that may reduce hunger and appetite (Beglinger and Degen 2006). Furthermore, PYY_{3-36} is a peptide hormone secreted by the small intestine and colon and reduces food intake by acting at the level of the hypothalamus (Nelson and Cox 2008). It has been reported that inoculation with *Bifidobacterium pseudocatenulatum* (promoter of SCFA production) could reduce weight gain and body fat (Zhao et al. 2018). Food present in the gastrointestinal tract releases

the hormone cholecystokinin which, inhibits feeding by activating satiety center of the hypothalamus (see Chap. 1). SCFAs may reduce gastric emptying and may reduce appetite (Cherbut 2003).

4. Effects on carcinoma: Dietary fibers reduce the contact of gastrointestinal epithelium with carcinogenic chemicals and prevent the absorption of carcinogens. Chronic inflammation [due to bacterial overgrowth (e.g., *Helicobacter pylori* and *Fusobacterium nucleotum*) or viral infections (e.g., human papilloma virus (Tulay and Serakinci 2016) and Epstein–Barr virus (Farrell 2019))] is one of the factors for the development of cancer. Inflammation of the stomach due to *Helicobacter pylori* causes gastric carcinoma (Paula et al. 2018), whereas inflammation of the colon and rectum by bacterial outgrowth, especially by *Fusobacterium nucleotum*, promotes colorectal carcinoma (Liu et al. 2018). Butyrate plays an anti-inflammatory role as it can inhibit proinflammatory cytokines such as interleukin-1 (IL-1) and tumor necrosis factor (TNF) (Encarnação et al. 2018) and thus may prevent the incidence of cancer. Higher fiber intake decreases the incidence of gastric cancer and pancreatic cancer (Veronese et al. 2018), colorectal cancer (McRae 2018), and ovarian cancer (Huang et al. 2018). Colorectal cancer is common in developed countries due to high intake of nonvegetarian diet, whereas the low incidence of the same is observed in vegetarian population.

5. Effects on allergy: It has been suggested that Western diets may contribute to the development of allergic diseases due to reduced intake of dietary fibers (Folkerts et al. 2018). They further observed that supplementation of fiber and its metabolites prevent the development of immune disorders through the regulation of B- and T-cell activation. Wang et al. (2017) demonstrated that sodium butyrate pretreatment decreased the percentage of degranulated mast cells as well as the content of mast cell mediators. Zhang et al. (2016) also reported that butyrate pretreatment inhibited mast cell release of TNF and IL-6.

6. Other beneficial functions:
   (a) Significant fiber is correlated with longer telomeres and lower biologic aging. Shorter telomeres have been attributed for CIHD, cancer, and infectious disease, leading to shorter lives (Tucker 2018).
   (b) Fibers prevent constipation as they provide a large surface volume of nondigestive material in the intestine and increase peristalsis, especially of the colon. Moreover, insoluble fibers have rapid gastric emptying, decrease the intestinal transit time, and increase the fecal bulk.
   (c) Review indicates that fibers inhibit macronutrient absorption and inhibit starch (main dietary carbohydrate) digestion by digestive enzymes (Qi et al. 2018). Thus, fibers decrease the energy intake through inhibiting digestion and absorption of energy providing macronutrients in the diet. As a result, fibers have "low energy density" and negative energy balance (Hervik and Svihus 2019). As discussed earlier, satiety-inducing effects further reduces energy intake. It has been demonstrated that inoculation with *Bifidobacterium pseudocatenulatum* strain C95 (promoter of SCFA production) could significantly

reduce weight gain and body fat (Zhao et al. 2018). It is clear from the literature that high fiber intake in obese individuals will enhance weight loss as they
have practically no caloric value.

(d) A mixture of different strains of lactobacilli and dietary fibers reduces the
effect of pathogenic organisms.

(e) Increased fiber intake is beneficial for gastrointestinal diseases such as gastrointestinal reflux disease, diverticulitis, and hemorrhoids.

(f) Type 2 diabetes or non-insulin-dependent diabetes mellitus (NIDDM). It is
due to insulin resistance, resulting in hyperglycemia, impaired glucose tolerance, visceral/central obesity and dyslipidemia (high level of LDL cholesterol, low level of HDL (high density lipoprotein) cholesterol,
hypertriglyceridemia). Deficiency in SCFAs production from carbohydrate
fermentation is associated with type 2 diabetes. Indeed, promoter of SCFAs
production by the gut microbiota significantly reduces hyperglycemia,
insulin resistance, and postprandial glycemic response (Zhao et al. 2018).
Metabolic syndrome is a combination of the findings as mentioned in type
2 diabetes insulin resistance, abdominal obesity, dyslipidemia, etc. that
increases the risk of atherosclerosis and NIDDM. Dietary fiber intake is
found to be inversely associated with the risk of metabolic syndrome
(Chen et al. 2018).

(g) Dyslipidemia due to high LDL causes infiltration of cholesterol into the
macrophages of connective tissue of the arterial wall. This, in turn, forms
foam cells (lipid-laden macrophages). Foam cells release growth factors and
inflammatory mediators, which stimulate the proliferation of smooth muscle
and cause narrowing of the arterial wall. Narrowing of the arterial wall with
the rupture of atherosclerotic plaque causes the formation of clot (thrombosis) in blood vessels and predisposes myocardial infarction, cerebral thrombosis, and hypertension.

5.6 Summary

Apart from well-documented beneficial effects, dietary fibers because of gel-
forming properties are used as thickener and food additives, particularly in dairy
products. It is well documented that individuals with high intake of dietary fibers are
at significantly lower risk of developing CIHD, stroke, hypertension, diabetes mellitus, obesity, and certain gastrointestinal disorders (e.g., diverticulitis and hemorrhoids). Probiotic fibers are nondigestible food ingredients and consist mainly of
inulin, fructose-oligosaccharides, and galacto-oligosaccharides. Probiotic fibers
increase resistance to pathogen invasion and enhance immune function. Lastly one
should be careful not to take excessive amount of dietary fibers (>30 g/day) as they,
because of binding properties, may prevent absorption of micronutrients.

References

AACC Report (2001) The definition of dietary fiber. Cereal Foods World 46(3):112–126

Abbasi NN et al (2016) Oat β-glucan depresses SGLT1- and GLUT2- mediated glucose transport in intestinal epithelial cells (IEC-6). Nutr Res 36(6):541–552

Anderson JW et al (2009) Health benefits of dietary fiber. Nutr Rev 67(4):188–205

Beglinger C, Degen L (2006) Gastrointestinal satiety signals in humans-physiologic roles for GLP-1 and PYY? Physiol Behav 89(4):460–464

Chen JP et al (2018) Dietary fiber and metabolic syndrome: a meta-analysis and review of related mechanisms. Nutrients 10(24):17 p

Cherbut C (2003) Motor effects of short-chain fatty acids and lactate in the gastrointestinal tract. Proc Nutr Soc 62(1):95–99

Encarnação JC et al (2018) Butyrate, a dietary fiber derivative that improves irinotecan effect in colon cancer cells. J Nutr Biochem 56:183–192

Farrell PJ (2019) Epstein-Barr virus and cancer. Annu Rev Pathol 14:29–53

Folkerts J et al (2018) Effect of dietary fiber and metabolites on mast cell activation and mast cell-associated diseases. Front Immunol 9(1067):13 p

Hervik KA, Svihus B (2019) The role of fiber in energy balance. J Nutr Metab 2019(4983657):11 p

Huang X et al (2018) Association between dietary fiber intake and risk of ovarian cancer: a meta-analysis of observational studies. J Int Med Res 46(10):3995–4005

Liu L et al (2018) Diets that promote colon inflammation associate with risk of colorectal carcinomas that contain Fusobacterium nucleatum. Clin Gastroenterol Hepatol 16:1622–1631

Longvah T, Ananthan R, Bhaskarachary K, Venkaiah K (2017) Indian food composition tables. National Institute of Nutrition (Indian Council of Medical Research), Hyderabad

McRae MP (2018) The benefits of dietary fiber intake on reducing the risk of cancer: an umbrella review of meta-analyses. J Chiropr Med 17(2):90–96

Nelson DL, Cox MM (2008) Lehninger principles of biochemistry, 5th edn. W.H. Freeman and Company, New York

Paula D et al (2018) Helicobacter pylori and gastric cancer: adaptive cellular mechanisms involved in disease progression. Front Microbiol 9:5 p

Qi X et al (2018) Dietary fiber, gastric emptying, and carbohydrate digestion: a mini review. Starch-Stärke 70(1700346):5 p

Soliman GA (2019) Dietary fiber, atherosclerosis and cardiovascular disease. Nutrients 11(1155):11 p

Tucker LA (2018) Dietary fiber and telomere length in 5674 U.S. adults: an NHANES study of biological aging. Nutrients 10(4):400

Tulay P, Serakinci N (2016) The role of human papillomaviruses in cancer progression. J Cancer Metastasis Treat 2:201–213

Veronese N et al (2018) Dietary fiber and health outcomes: an umbrella review of systematic reviews and meta-analyses. Am J Clin Nutr 107(3):436–444

Wang CC et al (2017) Sodium butyrate enhances intestinal integrity, inhibits mast cell activation, inflammatory mediator production and JNK signaling pathway in weaned pigs. Innate Immun 24:40–46

Zhang H et al (2016) Butyrate suppresses murine mast cell proliferation and cytokine production through inhibiting histone deacetylase. J Nutr Biochem 27:299–306

Zhao L et al (2018) Gut bacteria selectively promoted by dietary fibers alleviate type 2 diabetes. Science 359:1151–1156

Further Reading

Chen JP et al (2018) Dietary fiber and metabolic syndrome: a meta-analysis and review of related mechanisms. Nutrients 10(24):17 p

Drew JE et al (2018) Dietary fibers inhibit obesity in mice, but host responses in the cecum and liver appear unrelated to fiber-specific changes in cecal bacterial taxonomic composition. Sci Rep 8(15566):11 p

Hervik KA, Svihus B (2019) The role of fiber in energy balance. J Nutr Metab 2019(4983657):11 p

Shichijo S, Hirata Y (2018) Characteristics and predictors of gastric cancer after Helicobacter pylori eradication. World J Gastroenterol 24(20):2163–2172

Zhao L et al (2018) Gut bacteria selectively promoted by dietary fibers alleviate type 2 diabetes. Science 359:1151–1156

Food Hypersensitivity

6

Abstract

Food hypersensitivity may be caused either by food allergy or by food intolerance. Food allergy is an abnormal reaction mediated by antigens, i.e., allergens (glycoprotein) present in certain foods. Food allergic reaction is mediated by IgE. IgE antibodies bind to mast cells. Activation of mast cells releases mediators (histamine, leukotrienes, prostaglandins, thromboxanes, and platelet-activating factors) causing allergic reactions. Food allergy may lead to life-threatening anaphylaxis, which may lead to anaphylactic shock. Severe allergic reactions may be caused by food-borne parasites. Food intolerance caused by celiac disease is due to the activation of T cells. Celiac disease is caused by prolamin present in wheat (gliadin, main component of gluten fraction), barley (hordein), rye (secalin), and oat (avenin), resulting in severe inflammation of the intestinal mucosa. Food intolerance due to non-IgE-mediated hypersensitivity occurs in individuals suffering from lactose intolerance, galactosemia or fat (present in milk) intolerance. Various nonallergic food hypersensitivities such as oral allergy syndrome, gastrointestinal infections, and gastroesophageal reflux diseases cause adverse food reactions.

Keywords

Food allergy · Anaphylaxis · Food intolerance · IgE · T cell · Non-IgE-mediated food hypersensitivity and nonallergic food hypersensitivity

6.1 Activation of B Cells

1. Food allergy is due to hypersensitive immune reaction in response to antigen (allergen) mediated by IgE antibodies. Allergens are low molecular weight proteins or glycoproteins, which are soluble in aqueous solution. These properties are responsible for early penetration at the mucosal surface of the intestine and

thus facilitate the prompt symptoms observed in allergic patients (Stanley and Bannon 1999). The allergens are resistance to protease, heat, and denaturants, preventing them from degradation during food preparation and digestion (Metcalfe 1985). Teenagers, young adults, and asthmatics patients are at high risk of severe allergic reactions. Infants are more susceptible to absorption of food allergens because of ill-developed intestinal epithelium. Food allergy may be associated with other allergic conditions such as atopic dermatitis, eczema, hay fever, allergic bronchopulmonary aspergillosis, insect stings (bee venom), and eosinophilic gastrointestinal diseases. Atopy is a form of allergy in which there is exaggerated IgE response. Hereditary factor may appear to develop hypersensitivity reaction in response to allergens. IgE antibodies bind to IgE receptors of mast cells. This binding triggers the activation of mast cells. Degranulation of mast cells releases mediators, which are responsible for the clinical manifestations of food allergy (Colledge et al. 2010). The mediators are (1) histamine, which causes vasodilatation, increased capillary permeability, mucus secretion, chemotaxis, and bronchoconstriction; (2) prostaglandins (PGs)—PGD$_2$ and PGF$_{2\alpha}$—which are strong airway constrictors due to the contraction of smooth muscle (mainly bronchi) whereas PGE$_2$ is a vascular relaxant and causes mucus secretion; (3) thromboxanes, which are produced by platelets, and they cause constriction of airway (mainly bronchi) and promote platelet aggregation; (4) platelet activating factor, which is a bronchoconstrictor and promotes chemotaxis of white blood cells; and (5) tryptase, which activates complement C3. Food allergens may be present in wheat, soya, milk, egg, crustacean, shellfish (crab, crayfish, and lobster), peanuts, tree nuts (walnuts, almonds, and cashews), seafoods, etc. Cow's milk protein is one of the commonest causes of food allergy in the infants. Peanut/tree nut allergies are commonly found in adulthood (Iweala et al. 2018). An individual is more susceptible to food allergy due to eating in a restaurant/cafeteria/roadside stalls. Due to food allergy, the individual suffers immediately from dermatitis, urticaria, itchy rash, wheezing, erythema, sneezing, rhinorrhea, angioedema (non-pitting swelling of face), nausea, vomiting, etc. Related food allergy may result in chronic inflammation mediated by macrophage, basophil, and eosinophil. Severe food allergy may trigger fatal life-threatening anaphylaxis characterized by laryngeal edema, stridor, bronchospasm, dyspnea, and marked lowering of blood pressure (due to positive feedback), leading to anaphylactic shock. Cardiac patients having aspirin/nonsteroidal anti-inflammatory drugs may have aggravated anaphylactic shock by blocking the formation of vasoconstrictors (thromboxane, leukotrienes, PGD$_2$, and PGF$_{2\alpha}$).

2. Insect venom allergy: Multiple simultaneous venom bites trigger the release of IgE from mast cells, causing severe allergic reactions. It may be noted that venom sting bite locally causes non-IgE-mediated allergic reaction.

3. Food-borne parasites (Chatterjee and Chatterjee 2009)

    (a) *Ascaris lumbricoides* (round worm) is the largest intestinal nematode. The body fluids of round worm, containing antigen, when absorbed will cause allergic reactions such as urticaria, conjunctivitis, and irritation of the respiratory tract. Children are commonly affected.

(b) Echinococcus granulosa (dog tapeworm). The larva via the small intestine enters inside various organs such as liver, lungs, and brain and forms hydatid cyst. Rupture of cyst causes severe allergic reactions and even anaphylactic reaction, since the fluid of the cyst is toxic and contains antigen.

Food-borne parasites cause severe allergic reactions due to the activation of IgG classes associated with IgE.

6.2   Activation of T Cells

Food intolerance
1. Celiac disease (gluten-sensitive enteropathy) is due to severe reaction to gluten present in cereals (wheat, rye, barley, and oats). Gliadin (a type of prolamin having high proline amino acid content) is the main component of the gluten fraction of the wheat. Other prolamins are hordein of barley, secalin of rye and avenin of oats. Celiac disease is characterized by blisters on the skin surface, which are intensely pruritic ("Dermatitis herpetiformis"). The individual may have genetic predisposition towards developing celiac disease. The disease is associated with human leukocyte antigen (HLA) genes known as HLA-DQ2 and HLA-DQ8. Prolamins cause upregulation of proinflammatory cytokines, resulting in severe inflammation of the intestinal mucosa that hampers the absorption of nutrients from the intestine. The disease causes steatorrhea, flatulence, weight loss, stunted growth, and infertility. There may be a number of cases of undiagnosed celiac disease patients in the population.
2. Irritable bowel syndrome (IBS) is a chronic intestinal inflammation due to the activation of inflammatory cytokines. It is characterized by bloating, altered bowel habit, diarrhea alternating with constipation, feeling of incomplete defecation, colicky abdominal pain, dyspepsia, etc. (Crowe 2019). Individuals (more commonly young women) may be intolerant of specific dietary component. Food intolerance is common among IBS patients. The symptoms may be caused by abnormal intestinal muscular contraction due to visceral hypersensitivity or dismobility of the intestine. IBS may cause fibromyalgia. Fibromyalgia is characterized by aching pain in fibrous tissue of muscles without any inflammation. Note: Comparison of food allergy and food intolerance is shown in Table 6.1.

6.3   Non-IgE-Mediated Food Hypersensitivity

Non-IgE-Mediated Food Hypersensitivity (Connors et al. 2018) includes a variety of disorders including celiac disease, food protein-induced enteropathy (FPE), food protein-induced enterocolitis syndrome (FPIES), and allergic protocolitis (AP).

Table 6.1 Comparison between food allergy and food intolerance

Food allergy	Food intolerance
1. IgE-mediated food hypersensitivity	1. Non-IgE-mediated food hypersensitivity
2. Allergic reactions usually occur within minutes of ingestion of the food and provoke prompt clinical symptoms	2. Delayed allergic reactions. Symptoms usually occur from hours to even weeks after ingestion of the food
3. Food allergen may be present in wheat, fish, soya, crustacean shellfish, peanuts, tree nuts, seafoods, milk, egg, etc.	3. It may be caused by any food as mentioned under food allergy. The most common allergens are cow milk protein and soy (more common in infants and children). Some individuals may suffer due to lactose intolerance, galactosemia or due to hypersensitivity to triacylglycerol or fat present in the milk
4. Clinical features are dermatitis, urticaria, itchy rash, eczema, erythema, angioedema, nausea, vomiting, etc.	4. Clinical features are similar to food allergy
5. Small amount of food can cause severe reaction	5. Severity of symptoms depends on the amount of food ingested
6. It may trigger life-threatening anaphylaxia	6. It is usually not life-threatening

1. Celiac disease is described under Sect. 6.2.
2. FPE is characterized by small intestinal mucosa, which prevents absorption of nutrients. It is caused mainly by the ingestion of cow's milk and soy. Infants suffer from diarrhea and vomiting. T-helper 2 lymphocytes may be involved.
3. FPIES occurs in infants and is characterized by vomiting, diarrhea, colitis, dehydration, and hypotension. Stool contains blood. Like FPE it is caused mainly by the ingestion of cow's milk and soy as well as by the ingestion of other foods such as egg, chicken, fish, pea, oat, and barley. Food allergens may activate T cells in the intestinal mucosa, resulting in inflammation.
4. AP affects generally rectosigmoid. It is one of the common causes of rectal bleeding in infancy.

6.4 Nonallergic Food Hypersensitivity

1. Lactose intolerance is characterized by intolerance to milk sugar (alactasia). Certain individuals are not able to tolerate due to hypersensitivity of milk protein β-lactoglobulin. Susceptible elderly individuals may suffer from lactose intolerance due to reduced expression of the enzyme. Bacterial fermentation of undigested lactose produces CO_2, H_2, and methane, leading to flatulence, bloating, abdominal cramp, and diarrhea.
2. Galactosemia is inability to synthesize glucose from galactose due to congenital deficiency of the enzyme galactose-1-phosphate uridyltransferase, causing

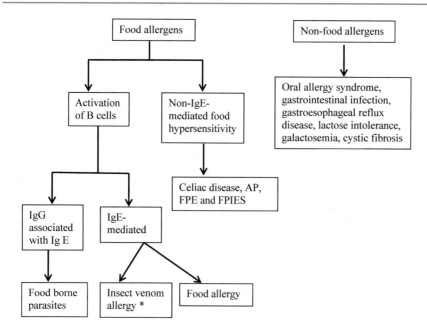

Fig. 6.1 Classification of allergic and nonallergic adverse food reactions. AP allergic proctocolitis, FPE food protein-induced enteropathy, FPIES food protein-induced enterocolitis. *Multiple simultaneous venom bites trigger the release of IgE from mast cells (local venom sting causes non-IgE-mediated allergic reaction)

accumulation of galactose in the blood. Excessive accumulation of galactose in the blood causes vomiting, diarrhea, enlargement of the liver, and jaundice.

3. Cystic fibrosis causes pancreatic insufficiency and destruction of acinar cells due to obstruction of pancreatic ductioles. As a result, steatorrhea occurs due to the loss of pancreatic lipase and proteolytic enzyme, leading to malabsorption syndrome. Food intolerance occurs due to difficulty in digesting food.

4. Gastrointestinal infections, gastroesophageal reflux disease, deficiency of digestive enzymes, etc. (Commins 2019) can cause adverse food reactions.

5. Oral allergy syndrome/pollen allergy syndrome indicates that contact with benzoic acid present in certain fruits (e.g., apple) may lead to allergic reactions in the mouth and throat such as urticaria, erythema, and angioedema (Sussman et al. 2010; Price et al. 2015).

6. Inflammatory bowel disease (IBD), which causes inflammation of the intestine include mainly Crohn's disease (CD) and ulcerative colitis (UC). IBD triggers the release of inflammatory mediators such as TNF and IL-2, which damage the intestinal tissue. (1) CD is due to long-standing inflammation of the alimentary tract, particularly ileum, and jejunum. This leads to pain in the abdomen, nausea, vomiting, and borborygmia. (2) UC involves rectum and sigmoid (proctosigmoiditis) and causes bloody diarrhea with mucus, tenesmus, abdominal pain, and even megacolon. Note: Classification of allergic and nonallergic reactions is shown in Fig. 6.1.

6.5   Vitamin D Hypothesis

An association between vitamin D and allergic diseases has been reported (Searing and Leung 2010; Abuzeid et al. 2012; Osborne et al. 2012). It has been reported that patients suffering from chronic spontaneous urticaria (CSU) had significantly lower serum vitamin D levels (Abdel-Rehim et al. 2014; Nasir-Kalmarzi et al. 2018; Tsai and Huang 2018; Tuchinda et al. 2018). Vitamin D can bind with mast cells, macrophages, T cells, and B cells (Deluca and Cantorna 2001; Yu et al. 2011). It causes reduction of IgE production by inhibiting B-cell function (Cheng et al. 2014; Di Filippo et al. 2015). Furthermore, atopic dermatitis (Heine et al. 2013), psoriasis (Bergler-Czop and Brzezinska-Wcislo 2016), vitiligo (Upala and Sanguankeo 2016), and acne (Lim et al. 2016) had low serum 25-hydroxycitamin D levels. Activation of immune mediators IgE and pro- and anti-inflammatory cytokines may mediate allergic reactions. Contradictory reports appeared about the association between vitamin D and allergic patients (Liu et al. 2011; Persson et al. 2013) and CSU patients (Wu et al. 2015).

6.6   Diagnosis

Identification of food allergens should be based on the observation of clinical manifestation after ingestion of the suspected food.

1. Allergens should be identified by skin prick tests. Antihistamines should be discontinued for at least 7 days prior to allergy testing. A drop of diluted allergen solution is placed on the forearm, and the skin is punctured through the drop of antigen with a sterile lancet. After 10–15 min appearance of wheal and flare ($\geq$2 mm) indicates positive result.
2. Absolute eosinophil count.
3. Radioallergosorbent test for the measurement of serum total IgE.
4. Measurement of mast cell tryptase.
5. Oral food challenge test (Ito and Urisu 2009): After 4–6 weeks of strict identified allergen avoidance, the suspected food is reintroduced to find out whether tolerance to the food has developed.
6. Measurement of hydrogen gas is a test for lactose intolerance.

6.7   Management

There is no permanent cure for food allergies. Life-long avoidance of identified allergen is necessary. Breastfeeding prevents allergy in infants (0–6 months). Allergic reactions can be treated by long-acting and nonsedative antihistamines such as cetirizine and chlorpheniramine. Omalizumab, which inhibits the binding of IgE to mast cells, may be effective in rectifying allergic reactions. Immediate intramuscular injection of epinephrine, antiallergic drug, or even intravenous injection of hydrocortisone (inhibition of proinflammatory cytokine) is to be administered within minutes under the supervision of a physician (medical emergency). Gluten-free diet with the correction of nutritional deficiencies is needed in celiac disease. Lactose-free diet is recommended for lactose intolerance. Curd contains bacterial lactase and is better tolerated than milk. One can take soya milk if the individual is not allergic. De-worming by albendazole/mebendazole/piperazine is needed to correct allergic reactions due to parasitic infections. Hydatid cysts should be removed by surgeon with great care to avoid spillage. The incidence of allergic reactions due to parasitic infections can be minimized by improving sanitation ("Hygiene hypothesis").

Oral immunotherapy (Wood 2017) is an experimental treatment in which individuals having food allergy are allowed to ingest gradually increasing quantities of offending food in order to induce desensitization (an increase in the threshold for reactivity). The initial phase is carried out with a very small dose for 1–2 days and then progressively to a larger intake of the culprit food. If well tolerated, the intake is increased gradually over a week to reach maintenance dose (dose-limiting symptoms). Sensitization is induced by initial exposure to aln allergen. During re-exposure of the sensitized individual, dendritic cells take up food allergen and digest the antigen (Poole et al. 2018).

Role of dietary fibers: A report by Zhang et al. (2016) showed that butyrate inhibited FCERI-dependent release of TNF and IL-6 from mouse bone marrow-derived mast cells. Furthermore, Wang et al. (2018) demonstrated that sodium butyrate pretreatment reduced the percentage of degranulated mast cells, decreased mast cell mediator content, and lowered mRNA expression of proinflammatory cytokines. A recent report indicates that supplementation of fiber and its metabolites prevent the development of immune disorders through the regulation of B- and T-cell activation (Folkerts et al. 2018).

6.8 Summary

Food allergy causes serious worldwide health hazards. Food allergy is hypersensitive immune reaction to food protein (allergen) mediated by IgE antibodies. Food allergens may be wheat, rye, milk, soya, crustacean, shellfish, egg, peanut, tree nut, and so on. IgE antibodies bind to IgE receptors of mast cells. Degranulation of mast cells release mediators, which causes serious immediate clinical manifestations such as dermatitis, urticaria, itchy rash, angioedema, sneezing, and rhinorrhea. It may trigger life-threatening anaphylaxis characterized by laryngeal edema and obstruction, bronchospasm, and marked lowering of blood pressure. Anaphylactic shock must be treated without delay. Food intolerance may be due to cell-mediated hypersensitivity to a food. Severe allergic reactions due to celiac disease, food-borne parasitic infestations, and non-IgE-mediated food hypersensitivity are discussed.

References

Abdel-Rehim AS et al (2014) Vitamin D level among Egyptian patients with chronic spontaneous urticaria and its relation to severity of the disease. Egypt J Immunol 21:85–90

Abuzeid WM et al (2012) Vitamin D and chronic rhinitis. Curr Opin Allergy Clin Immunol 12:13–17

Bergler-Czop B, Brzezinska-Wcislo L (2016) Serum vitamin D level—the effect on the clinical course of psoriasis. Postepy Dermatol Alergol 33:445–449

Chatterjee KD, Chatterjee D (2009) Parasitology (prootozoology and helminthology), 13th edn. CBS Publishers and Distributers Pvt Ltd, New Delhi

Cheng HM et al (2014) Low vitamin D levels are associated with atopic dermatitis, but not allergic rhinitis, asthma or IgE sensitization, in the adult Korean population. J Allergy Clin Immunol 133:1048–1055

Colledge NR et al (eds) (2010) Davidson's principles and practice of medicine, 21st edn. Churchill Livingstone, New York

Commins SP (2019) Food intolerance and food allergy in adults: an overview. https://www.upto-date.com/contents/food-intolerance-and-food-allergy-in-adults-an-overview

Connors L et al (2018) Non-IgE-mediated food hypersensitivity. Allergy Asthma Clin Immunol 14(Suppl 2):56

Crowe SE (2019) Food allergy vs food intolerance in patients with irritable bowel syndrome. Gastroenterol Hepatol (NY) 15(1):38–40

Deluca HF, Cantorna MT (2001) Vitamin D: its role in immunology. FASEB J 15:2579–2585

Di Filippo P et al (2015) Vitamin D supplementation modulates the immune system and improves atopic dermatitis in children. Int Arch Allergy Immunol 166:91–96

Folkerts J et al (2018) Effect of dietary fiber and metabolites on mast cell activation and mast cell-associated diseases. Front Immunol 9:1067

Heine G et al (2013) Association of vitamin D receptor gene polymorphisms with severe atopic dermatitis in adults. Br J Dermatol 168:855–858

Ito K, Urisu A (2009) Diagnosis of food allergy based on oral food challenge test. Allergol Int 58:467–474

Iweala OI et al (2018) Food allergy. Curr Gastroenterol Rep 20(5):17

Lim SK et al (2016) Comparison of vitamin D levels in patients with and without acne: a case-control study combined with a randomized controlled trial. PLoS One 11:0161162

Liu X et al (2011) Gene-vitamin D interactions on food sensitization: a prospective birth cohort study. Allergy 66:1442–1448

Metcalfe DD (1985) Food allergens. Clin Rev Allergy 3:331–349

Nasir-Kalmarzi R et al (2018) Evaluation of 1,25 dihydroxyvitamin D3 pathway in patients with chronic urticaria. QJM 111(3):161–169

Osborne NJ et al (2012) Prevalence of eczema and food allergy is associated with latitude in Australia. J Allergy Clin Immunol 129:865–867

Persson K et al (2013) Vitamin D deficiency at the arctic circle—a study in food-allergic adolescents and controls. Acta Paediatr 102:644–649

Poole A et al (2018) Cellular and molecular mechanisms of vitamin D in food allergy. J Cell Mol Med 22:3270–3277

Price A et al (2015) Oral allergy syndrome (pollen-food allergy syndrome). Dermatitis 26(2):78–88

Searing DA, Leung DY (2010) Vitamin D in atopic dermatitis, asthma and allergic diseases. Immunol Allergy Clin North Am 30:397–409

Stanley JS, Bannon GA (1999) Biochemistry of food allergens. Clin Rev Allergy Immunol 17:279–291

Sussman G et al (2010) Oral allergy syndrome. CMAJ 182(11):1210–1211

Tsai TY, Huang YC (2018) Vitamin D deficiency in patients with chronic and acute urticaria: a systematic review and meta-analysis. J Am Acad Dermatol 79:573–575

Tuchinda P et al (2018) Relationship between vitamin D and chronic spontaneous urticaria: a systematic review. Clin Transl Allergy 8:51

Upala S, Sanguankeo A (2016) Low 25-hydroxyvitamin D levels are associated with vitiligo: a systematic review and meta-analysis. Photodermatol Photoimmunol Photomed 32:181–190

Wang CC et al (2018) Sodium butyrate enhances intestinal integrity, inhibits mast cell activation, inflammatory mediator production and JNK signaling pathway in weaned pigs. Innate Immun 24(1):40–46

Wood RA (2017) Oral immunotherapy for food allergy. J Investig Allergol Clin Immunol 27(3):151–159

Wu CH et al (2015) Association between micronutrient levels and chronic spontaneous urticaria. Biomed Res Int 2015:926167

Yu C et al (2011) Vitamin D (3) signalling to mast cells: a new regulatory axis. Int J Biochem Cell Biol 43:41–46

Zhang H et al (2016) Butyrate suppresses murine mast cell proliferation and cytokine production through inhibiting histone deacetylase. J Nutr Biochem 27:299–306

Further Reading

Connors L et al (2018) Non-IgE-mediated food hypersensitivity. Allergy Asthma Clin Immunol 14(Suppl 2):56

Folkerts J et al (2018) Effect of dietary fiber and metabolites on mast cell activation and mast cell-associated diseases. Front Immunol 9:1067

Poole A et al (2018) Cellular and molecular mechanisms of vitamin D in food allergy. J Cell Mol Med 22:3270–3277

Tuchinda P et al (2018) Relationship between vitamin D and chronic spontaneous urticaria: a systematic review. Clin Transl Allergy 8:51

Wood RA (2017) Oral immunotherapy for food allergy. J Investig Allergol Clin Immunol 27(3):151–159

Food Groups, Balanced Diet, and Food Composition

7

Abstract

Food groups can be divided into six categories. Each food group is a prerequisite for the maintenance of vitality of health. Proteins of high biologic value are obtained from egg, fish, meat, poultry, milk, and milk products. Cereals, legumes, and pulses contain maximum amount of carbohydrate and dietary fibers and have low fat content. Vegetables including green leafy vegetables provide minerals, vitamins especially vitamin A, and less amount of fat. Fruits are rich sources of vitamins, containing low sodium and low fat. Balanced diet is one of the major determinants of health. It provides optimal energy and nutrition for optimal growth and development. Excess of any food group should be avoided. Dietary fibers, antioxidants, and foods of low glycemic index should be included. Balanced diet prevents age-related decline of physiological functions of the body, improves longevity and immunity, and slows the progression of degenerative diseases. While constructing a balanced diet, one should avoid excessive intake of sugar, sweet, chocolate, salt, saturated fat, trans fat, cholesterol-rich food, and processed food.

Keywords

Food group · Energy source · Proteins of high biologic value · Balanced diet · Dietary fibers · Antioxidants · Low glycemic index foods

7.1 Food Groups

1. Egg, fish, meat, poultry, and liver

   These are protein-rich foods of high biologic value containing less carbohydrate and are good sources of energy. Egg contains most of the B group vitamins and, in addition, contains vitamin A, vitamin D, vitamin K, and calcium. Meat contains vitamin B_1, vitamin B_{12}, niacin, iron, calcium, and zinc. Meat and egg

K. Chakrabarty, A. S. Chakrabarty, *Textbook of Nutrition in Health and Disease*,
https://doi.org/10.1007/978-981-15-0962-9_7

Table 7.1 Comparison of milk composition and milk products

	Carbohydrate	Protein	Fat
Buffalo milk	8.4	3.7	6.6
Cow, whole milk	4.9	3.3	4.5
Human milk	7.3	1.2[a]	4.6
Indian cottage cheese (paneer)	12.4	18.9	14.8
Reduced dried milk	16.5	16.3	20.6
Plain yoghurt (without any flavor)	10.8–13.0	4.5–5.3	2.1–3.0

Values are expressed as g/100 g
Note: Skimmed milk contains fat content of about 0.2 g/100 g
[a]Protein content in colostrum is higher compared to mature milk

Table 7.2 Cholesterol content (mg/100 g) and saturated and unsaturated fatty acid content (g/100 g): A comparison of milk composition between buffalo, cow, and human milk

	Cholesterol	Saturated fatty acid	MUFA	PUFA
Buffalo milk	19	4.6	1.8	0.1
Cow, whole milk	10	2.5	1.0	0.1
Human milk	14	1.8	1.6	0.5

are rich in fat and cholesterol. Fish contains vitamins (B_1, B_2, B_3, B_{12}, A, and D), magnesium and, in addition, contains EPA and DHA acids. Liver contains most of the B group vitamins and also contains vitamin A, vitamin D, vitamin K, iron, copper, and calcium.

2. Milk and milk products (other than butter)

   These are proteins of high biologic value, containing vitamins (B_1, B_2, B_{12}, and A) and calcium. Cheese contains more or less equal amounts of high protein and high fat. Curd is a rich source of vitamin B_{12} and calcium (Tables 7.1 and 7.2).

3. Cereals (wheat flour, rice, bajra, barley, maize, and jowar)

   These contain maximum amount of carbohydrate, less fat, protein of low biologic value, and dietary fiber and are inexpensive sources of energy.

4. Legumes and pulses

   These, similar to cereals and in addition, contain iron, calcium, and less abundant vitamins.

5. Vegetables including green leafy vegetables

   Provide vitamins, minerals (iron, copper, and magnesium), and dietary fibers. Fat content is very less.

6. Fruits

   Fruits are low in fat and sodium, are good sources of vitamins and calcium, and contain dietary fiber. Mango (ripe) and papaya (ripe) are rich sources of vitamin A. Citrus fruits and tomatoes are rich sources of vitamin C. Banana, apple, dates, and coconut water are rich sources of potassium.

7.2   Balanced Diet (Tables 7.3, 7.4 and 7.5)

A balanced diet provides enough optimal energy and nutrition for normal growth and development. A balanced diet should be designed to prevent diet-related diseases. Proper taste and availability of the diet should be considered. In order to achieve a balanced diet, essential foods from each of the food groups should be selected depending on the socioeconomic condition and caloric requirement. Caloric requirement depends on age, sex, body weight, height, and activities. Balanced diet should be constructed to meet the energy requirement of an individual as well as the need for all macro and micronutrients. Dietary fibers and antioxidants should be included. Food of low glycemic index should be selected, and excess of any food group should be avoided.

Main aim of a balanced diet is to
1. Prevent or slow age-related decline of physiological functions of the body. Maintain growth and development, and functions of brain and nervous system.
2. Slow the progression of degenerative diseases.
3. Improve longevity.
4. Improve immunity.
5. Maintain vitality of health.

Foods to be avoided while constructing a balanced diet:
1. Sugar, sweet, biscuit, jam, and chocolate.
2. Foods fried in unhealthy trans fat. Acrylamide found in a variety of fried foods is neurotoxic.
3. Cholesterol-rich food and food containing high saturated fat.
4. High amount of salt intake, processed foods, and refined flour.
5. Excessive intake of sugar in combination with a hypercaloric diet will increase fat deposition, particularly in the liver. It may induce fatty liver (Ma et al. 2016).

7.2.1   Beneficial Effects of Vegetarian Diet

Vegetarian people are at lower risk for developing colorectal cancer, coronary ischemic heart disease, obesity, gastrointestinal diseases, hypertension, and stroke. High intake of red and processed meat was found to be associated with higher risk of colorectal cancer (Schwingshackl et al. 2017) and stroke (Deng et al. 2018). Significant benefit of vegetarian diet for cognitive performance was reported (Cheung et al. 2014). The Mediterranean diet (Slavin and Lloyd 2012) constitutes vegetables, fruits, legumes, nuts, whole-grain cereals, and fish with moderate amount of dairy products and lean meat. The dietary approach to stop hypertension (DASH) diet (Slavin 2008) is more or less similar to the Mediterranean diet with restricted amount of sodium. Increased consumption of nuts, legumes, and whole grains can prevent metabolic disturbances and diseases compared to other food groups (Schwingshackl et al. 2018). Fruits are low in energy density and contain

Table 7.3 Balanced diet (g/day) at different age groups having sedentary life, and during pregnancy and lactation

| | School children | | Adolescents | | Adults | | Elderly | Pregnancy |
	Boys	Girls	Boys	Girls	Boys	Girls	person	and lactation
Cereals	150	140	340	300	340	300	250	340
Mixed pulses and legumes	60	50	70	60	70	60	50	70
Leafy green vegetables	50	50	50	50	50	50	50	50
Other vegetables including roots and tubers	100	100	120	120	120	120	120	120
Fruits	120	120	120	120	120	120	120	130
Skimmed milk	250	250	250	250	250	250	250	300
Milk products (except butter and ghee)/egg without yolk/fish/poultry without visible fat	30	30	40	40	50	40	40	50
Fats and oils	20	20	25	25	30	25	20	25
Sugar	20	20	20	20	20	20	15	20

Refer to Table 1.2 (Chap. 1) for weaning foods for infants (7–12 months). The amount of balanced diet increases proportionately with moderate and heavy activities

Table 7.4 Energy intake of balanced diet for a sedentary male (60 kg)

1. Proteins of high biologic value, for example, egg (without yolk), fish, milk, and lean meat	65 g/day (260 kcal/day) ~20% of total energy	
2. Carbohydrates (complex carbohydrates should be consumed more than simple carbohydrates)	300 g/day (1200 kcal/day) ~60% of total energy	Will provide 2000 kcal/day
3. Fats: Avoid saturated, trans fat, and cholesterol-rich diets. PUFA and MUFA should be consumed. Diet must have essential fatty acids. Saturated and trans fat <8% of energy. PUFAs $n = 6$ and $n = 3 < 8\%$ of energy. MUFA <6% of energy	60 g/day (540 kcal/day) ~20% of total energy	

Note: A sedentary female (50 kg) needs about 1800 kcal/day. Energy intake increases with activities. Pregnant women and lactating mothers need more energy intake (see Chap. 1)

dietary fibers (fructooligosaccharides) and antioxidants such as polyphenols and carotenoids (β-carotene, lycopene, lutein, and zeaxanthin) with small amount of fructose and sucrose (Fernandez and Marette 2017). Vitamin C, vitamin E, and selenium are strong antioxidants as they can inactivate oxygen free radicals. Berries and grapes are good sources of phenolic compounds. Tomatoes are good antioxidants as they contain vitamin C and lycopene. Flavinoids present in a variety of foods are antioxidant nutrients (Scalbert and Zamora-Ros 2015). Melatonin (present in vegetables, fruits, flowers, walnuts, and a variety of herbs) is reported to have the property of free radical scavenger (Reiter et al. 2000).

Table 7.5 Daily nutrients intake of balanced diet for adult male

1.	Carbohydrate	60% of total energy
2.	Protein	20% of total energy
3.	Fat	20% of total energy
(a)	Saturated fatty acid	<6% of total energy (about 5%)
(b)	$n = 6$ PUFAs	<6% of total energy (about 5%)
(c)	$n = 3$ PUFAs	<6% of total energy (about 5%)
4.	Cholesterol	<250 mg
5.	Fruits and vegetables	>400 g
6.	Vitamin A	600 µg
7.	Vitamin D	10 µg
8.	Vitamin E	10 mg
9.	Vitamin K	60 µg
10.	Vitamin B_1	1.5 mg
11.	Vitamin B_2	1.5 mg
12.	Vitamin B_3	16 mg
13.	Vitamin B_5	5 mg
14.	Vitamin B_6	2 mg
15.	Biotin	100 µg
16.	Vitamin B_{12}	1.5 µg
17.	Folate	100 µg
18.	Vitamin C	50 mg
19.	Calcium/phosphorus	800 mg
20.	Sodium chloride	<4 g
21.	Magnesium	400 mg
22.	Iron	20 mg
23.	Copper	1000 µg
24.	Iodine	150 µg
25.	Zinc	15 mg
26.	Fluoride	3 mg
27.	Manganese	2.5 mg
28.	Selenium	60 µg

See Chap. 4 regarding RDA/Dietary food intake at different age groups and during pregnancy and lactation

7.3 Food Pyramid

Food pyramid is shown in Fig. 7.1. One should eat a variety of foods according to the taste and liking. One should try to avoid processed food, which are rich in sugar, salts, and fats. One should take plenty of water (about 2.4 L/day). Daily exercise is essential. Consumption of fruits and vegetables should be greater than 400 g/day. Dietary fiber intake should be greater than 30 g/day.

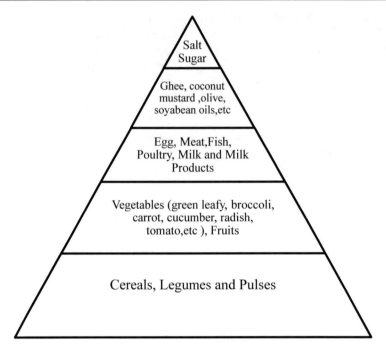

Fig. 7.1 Food pyramid

7.4   Food Composition

The approximate macronutrient content of various foods is shown in Table 7.6.

The approximate water-soluble vitamin content of various foods is shown in Table 7.7.

7.5   Summary

Excessive consumption of any food group should be avoided. Vegetarians generally have a well-balanced diet due to servings of cereal grains, pulses, nuts, flax, fruits, vegetables, dairy, or soya products. Vegetarian people are at a lower risk of cancer, CIHD, obesity, hypertension, and stroke. Without any pathological changes of the intestinal tract, they are lesser vulnerable to constipation. Significant benefits of vegetable and fruit consumption for cognitive performance is well documented. Excess sugar/sweet, salt, saturated fat, cholesterol, processed food, refined flour, and trans fat should be avoided while constructing a balanced diet. Excessive intake of sugar in combination with a hypercaloric diet will increase fat deposition, particularly in the liver. Vegetable soup, lemon/orange/tomato juice, cooked vegetables especially green leafy vegetables, fruits, and salads should be preferred. Balanced diet at different age groups, energy intake of balanced diet and daily nutrient intake of balanced diet have been described.

Table 7.6 Approximate macronutrient content of various foods (g/100 g edible portion)

	Protein	Fat	Carbohydrate
Bajra	11.0	5.4	61.8
Jowar	10.0	1.7	67.7
Barley	10.9	1.3	61.3
Parboiled rice	7.8	0.6	77.2
Wheat flour	10.6	1.5	64.2
Wheat flour, refined	10.4	0.8	74.3
Bengal gram	21.6	5.3	46.7
Black gram	23.1	1.7	51.0
Lentil	24.3	0.7	52.5
Dried peas	20.4	1.9	48.9
Soybean, white	37.8	19.4	10.2
Cabbage, green	1.4	0.1	3.3
Lettuce	1.5	0.3	3.0
Spinach	2.1	0.6	2.0
Brinjal	1.8	0.4	3.4
Cauliflower	2.1	0.4	2.0
Babycorn	2.7	1.3	11.7
Ripe tomato, hybrid	0.8	0.2	3.2
Broad beans	3.8	0.2	2.1
Green capsicum	1.1	0.3	1.8
Lady's finger	2.1	0.2	3.6
Egg poultry white, boiled	13.4	10.5	–
Chicken, boiled	19.4	12.0	–
Beef meat	20.6	14.6	–
Goat meat	20.0	11.8	–
Pork meat	17.4	18.8	–
Pomfret	19.0	5.0	–
Salmon	21.0	9.9	–
Tuna	24.5	1.4	–
Hilsa	21.5	18.5	–
Rohu	19.7	2.4	–
Lobster	18.5	0.8	–
Apple	0.3	0.6	13.1
Dried apricot	3.2	0.7	72.6
Banana, ripe	1.3	0.3	25.1
Blackberry	0.9	0.6	10.6
Red cherries	1.5	0.5	11.9
Seeded grapes, black	0.8	0.3	13.2
Dates, dry dark brown	2.4	0.4	72.7
Guava, white	1.4	0.3	5.1
Mango ripe, Himsagar	0.5	0.5	9.0
Orange pulp	0.7	0.1	7.9
Papaya, ripe	0.4	0.2	4.6
Pineapple	0.5	0.2	9.4

(continued)

Table 7.6 (continued)

	Protein	Fat	Carbohydrate
Pomegranate	1.3	0.2	11.6
Beetroot	1.9	0.1	6.2
Carrot, red	1.0	0.5	6.7
Potato, brown skin	1.5	0.2	14.9
Radish, elongate, white skin	0.8	0.1	6.6
Almond	18.4	58.5	3.0
Cashew nut	18.8	45.2	25.5
Coconut kernel, dry	7.3	63.3	8.0
Coconut kernel, fresh	3.8	41.4	6.3
Mustard seeds	19.5	40.2	16.8
Walnut	14.9	64.3	10.1
Mushroom oyster, dried	19.0	2.9	33.1
Cucumber, green, elongate	0.7	0.2	3.5
Coriander leaves	3.5	0.7	1.9
Onion	1.5	0.2	9.6

Reproduced with permission from T. Longvah et al. (2017)

Table 7.7 Approximate water-soluble vitamin content of various foods per 100 g edible portion

	B_1 (mg)	B_2 (mg)	B_3 (mg)	B_5 (mg)	B_6 (mg)	Biotin (µg)	Folate (µg)	Vit C (mg)
Bajra	0.25	0.20	0.86	0.50	0.27	0.64	36.11	–
Jowar	0.35	0.14	2.10	0.27	0.28	0.70	39.42	–
Barley	0.36	0.18	2.84	0.14	0.31	2.38	31.58	–
Parboiled rice	0.17	0.06	2.51	0.55	0.22	0.31	9.75	–
Wheat flour	0.42	0.15	2.37	0.87	0.25	0.76	29.22	–
Wheat flour, refined	0.15	0.06	0.77	0.72	0.08	0.58	16.25	–
Bengal gram	0.35	0.15	1.87	1.60	0.19	0.81	182	–
Black gram	0.21	0.09	1.76	2.95	0.22	0.81	88.75	–
Lentil	0.34	0.16	1.81	1.32	0.18	1.25	49.99	–
Dried peas	0.56	0.16	2.69	1.26	0.26	0.53	110	–
Soybean, white	0.61	0.23	2.28	1.97	0.45	0.77	288	–
Cabbage, green	0.03	0.05	0.24	0.24	0.13	1.41	48.36	33.25
Lettuce	0.05	0.09	0.17	0.11	0.08	2.15	30.69	11.91
Spinach	0.16	0.10	0.33	0.22	0.15	4.14	142	30.28
Brinjal	0.07	0.13	0.74	0.29	0.05	1.17	37.22	1.58
Cauliflower	0.04	0.07	0.31	0.62	0.13	2.47	45.95	47.14
Babycorn	0.15	0.07	0.53	0.94	0.16	0.79	45.53	8.59
Ripe tomato, hybrid	0.04	0.02	0.51	0.18	0.08	1.09	15.41	25.27
Broad beans	0.12	0.10	0.76	0.45	0.23	10.03	20.46	10.98
Green capsicum	0.05	0.03	0.56	0.21	0.15	4.59	51.85	123
Lady's finger	0.04	0.07	0.61	0.28	0.27	1.58	63.68	22.51
Egg poultry, white, boiled	0.02	0.18	0.01	0.18	–	4.37	4.10	–
Chicken	0.13	0.10	5.62	1.06	0.38	3.86	9.00	–
Beef meat	0.03	0.12	5.18	1.14	0.48	–	8.06	–

(continued)

Table 7.7 (continued)

	B_1 (mg)	B_2 (mg)	B_3 (mg)	B_5 (mg)	B_6 (mg)	Biotin (µg)	Folate (µg)	Vit C (mg)
Goat meat	0.07	0.17	5.14	1.07	0.26	–	2.08	–
Pork meat	0.18	0.10	4.22	0.86	0.41	–	6.70	–
Pomfret	0.05	0.03	1.38	1.11	0.13	–	9.61	–
Salmon	0.07	0.06	4.46	1.15	0.15	–	11.36	–
Tuna	0.06	0.07	4.73	1.34	0.07	–	13.74	–
Hilsa	0.01	0.04	2.85	2.33	0.12	–	28.75	–
Rohu	–	0.04	2.33	1.18	240	–	1263	
Lobster	0.01	0.02	1.87	1.25	0.16	–	19.97	–
Apple	0.03	0.01	0.25	0.09	0.04	0.34	3.04	3.57
Dried apricot	0.04	0.04	1.66	0.62	0.10	1.47	10.50	0.42
Banana, ripe	0.01	0.04	0.48	0.35	0.51	1.54	17.93	8.06
Blackberry	0.01	0.02	0.40	0.21	0.05	1.65	22.95	19.45
Red cherries	0.07	0.02	0.19	0.23	0.04	1.52	4.92	8.82
Seeded grapes, black	0.03	0.03	0.14	0.07	0.11	1.14	8.69	18.30
Dates, dry, dark brown	0.02	0.03	1.09	0.53	0.153	0.94	12.80	3.84
Guava, white	0.05	0.04	0.60	0.25	0.11	0.74	29.76	214
Mango, ripe, Himsagar	0.03	0.03	0.27	0.11	0.10	1.46	90.98	49.09
Orange pulp	0.07	0.02	0.28	0.20	0.04	2.88	19.46	42.72
Papaya, ripe	0.03	0.11	0.33	0.44	0.04	3.05	60.90	43.09
Pineapple	0.05	0.03	0.12	0.13	0.13	1.05	18.21	36.37
Pomegranate	0.06	0.01	0.20	0.42	0.29	0.60	38.64	12.69
Beetroot	0.01	0.01	0.21	0.26	0.07	2.56	97.37	5.26
Carrot, red	0.04	0.03	0.25	0.27	0.07	1.30	23.67	6.76
Potato, brown skin	0.06	0.01	1.04	0.38	0.10	1.35	15.51	23.15
Radish, elongate, white skin	0.02	0.02	0.30	0.15	0.07	2.48	29.75	19.91
Almond	0.15	0.26	3.71	0.73	0.09	2.39	36.46	0.74
Cashew nut	0.61	0.03	1.03	1.40	0.16	2.58	25.20	–
Coconut kernel, dry	0.04	0.04	0.71	0.21	0.15	1.01	24.27	–
Coconut kernel, fresh	0.03	0.08	0.03	0.21	0.10	0.63	25.41	0.80
Mustard seeds	0.55	0.33	3.80	0.48	0.24	1.45	94.88	–
Walnut	0.40	0.12	0.86	0.84	0.80	13.05	57.95	0.88
Mushroom oyster, dried	0.24	0.17	3.77	2.33	0.85	22.51	10.40	–
Cucumber, green, elongate	0.02	0.01	0.35	0.45	0.06	2.82	16.84	6.11
Coriander leaves	0.09	0.05	0.73	0.63	0.19	4.17	51.01	23.87
Onion	0.04	0.01	0.34	0.30	0.10	2.61	28.88	6.69
Milk whole buffalo	0.05	0.13	0.07	0.380	0.40	2.16	8.57	2.37
Milk whole cow	0.03	0.11	0.08	0.34	0.04	1.98	7.03	2.01
Paneer (cottage cheese)	0.02	0.10	0.13	0.49	0.04	21.04	93.31	–

Reproduced with permission from T. Longvah et al. (2017)

References

Cheung BH et al (2014) Current evidence on dietary pattern and cognitive function. Adv Food Nutr Res 71:137–163

Deng C et al (2018) Stroke and food groups: an overview of systematic reviews and meta-analyses. Public Health Nutr 21(4):766–776

Fernandez MA, Marette A (2017) Potential health benefits of combining yogurt and fruits based on their probiotic and prebiotic properties. Adv Nutr 8(Suppl):155S–164S

Longvah T, Ananthan R, Bhaskarachary K, Venkaiah K (2017) Indian food composition tables. National Institute of Nutrition (Indian Council of Medical Research), Hyderabad

Ma J et al (2016) Potential link between excess added sugar intake and ectopic fat: a systematic review of randomized controlled trials. Nutr Rev 74(1):18–32

Reiter RJ et al (2000) Actions of melatonin in the reduction of oxidative stress. A review. J Biomed Sci 7(6):444–458

Scalbert A, Zamora-Ros R (2015) Bridging evidence from observational and intervention studies to identify flavonoids most protective for human health. Am J Clin Nutr 101:897–898

Schwingshackl L et al (2017) Food groups and risk of colorectal cancer. Int J Cancer 142:1748–1758

Schwingshackl L et al (2018) Food groups and intermediate disease markers: a systematic review and network meta-analysis of randomized trials. Am J Clin Nutr 108:576–586

Slavin JL (2008) Position of the American dietetic association: health implication of dietary fiber. J Am Diet Assoc 108:1716–1731

Slavin JL, Lloyd B (2012) Health benefits of fruits and vegetables. Adv Nutr 3:506–516. https://www.viva.org.uk/white-lies/comparison-between-human-milk-and-cows-milk

Further Reading

Herforth A et al (2019) A global review of food-based dietary guidelines. Adv Nutr 10(4):590–605

Schulze MB et al (2018) Food based dietary patterns and chronic disease prevention. BMJ 361:k2396

Schwingshackl L et al (2017) Food groups and risk of all-cause mortality: a systematic review and network meta-analysis of prospective studies. Am J Clin Nutr 105:1462–1473

Schwingshackl L et al (2018) Food groups and intermediate disease markers: a systematic review and meta-analysis of randomized trials. Am J Clin Nutr 108:576–586

Nutritional Deficiencies and Disorders

8

Abstract

Diet history is an important parameter in nutrition science. A diet must provide optimal nutrition to prevent diet-related disorders (undernutrition or overnutrition). Balanced nutrition results in a healthy life. Humans may suffer from malnutrition due to starvation, malabsorption syndrome, maldigestion, protein energy malnutrition (PEM), and eating disorders (Fig. 8.1). Poverty causes health hazards. Low-income families due to lack of education, lack of family planning, and low consumption of nutrients are at high risk of PEM in developing countries. Infants and children of poor mothers die due to PEM. PEM is the most widespread nutritional problem in developing countries. Obesity increases mortality and reduces life span, especially in developed countries. The obese individuals should be educated about the serious health problems of obesity. The obese individuals should consume low-calorie food along with daily exercise. Adipocytes of adipose tissue secrete various biologically active cytokines termed as adipokines. Apart from regulation of adipose tissue mass and nutrients, adipokines take part in the regulation of glucose homeostasis, hemostasis, blood pressure, atherosclerosis, and inflammation. Exercise through the liberation of epinephrine, norepinephrine, and adipokines decreases body weight due to lipolysis. Bariatric surgery is necessary when BMI is more than 40 kg/m^2.

Keywords

Balanced nutrition · Malnutrition · PEM · Starvation · Malabsorption syndrome · Poverty · Obesity · Bariatric surgery · Adipokines

8.1   Body Mass Index and Autonomic Function

Measurement of body mass index (BMI) will determine whether the individual is suffering from malnutrition, undernutrition, and obesity. BMI is measured from weight in kg (W) and height in meters (H). BMI = W/H^2. BMI of a normal healthy individual is 18.5–24.9 kg/m^2. Overweight is considered when BMI is 25–30 kg/m^2. Obesity is considered when BMI is more than 30 kg/m^2. Undernutrition is considered if BMI is less than 18 kg/m^2. Severe obesity should be considered if BMI > 40 kg/m^2. Undernutrition can be considered also, if the body weight is less than 90% of ideal height. BMI is not an index of body fat mass. It does not indicate the mass of fat (Nuttall 2015). Malnourished children suffer from autonomic insufficiency, especially parasympathetic function (Bedi et al. 1999). Resting heart rate is higher in malnourished children. Pregnant women should not suffer from malnutrition. If pregnant women are underweight, infants may be born with low birth weight.

8.2   Protein Energy Malnutrition

PEM is the commonest nutritional disorder in many parts of the world. This is a disease of infants/children of poor mothers. Chronic poverty is the main cause (Bhutt et al. 2017). As many as one billion people suffer from various degrees of

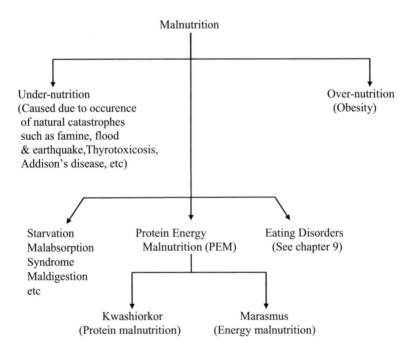

Fig. 8.1 Nutritional deficiencies and disorders

severity of PEM. It is the most widespread nutritional problem in developing countries, predominantly affecting infants and children. The prevalence rate varies from 20 to 50% in different areas, depending on socioeconomic status, level of education, and awareness. In developing countries, about 200 million infants and children under the age of 5 years suffer from PEM. In the developed countries, PEM is seen most frequently in hospital patients with chronic illness or in individuals who suffer from major trauma, severe infection, or the effects of major surgery.

Two forms of malnutrition are (1) kwashiorkor (protein malnutrition) and (2) marasmus (energy malnutrition).

8.2.1   Kwashiorkor

It is caused by inadequate intake of protein in the presence of adequate intake of calories. It is frequently seen in infants and children after weaning, i.e., about 1 year of age, when their diet consists of mainly carbohydrate. Protein deficiency occurs due to consumption of staple diets consisting of cereals and tubers only. The child is deprived of breast-feeding and is fed a starchy diet. Cicely Williams introduced the term kwashiorkor meaning "sickness the older child gets when the next child is born," after studies on the Ga tribe of Ghana.

Characteristic features
1. The hallmarks are hypoalbuminemia and edema (Fig. 8.2). The hypoalbuminemia reflects the inadequate supply of amino acids derived from the protein, thus impairing the synthesis of albumin and other proteins (e.g., transferrin which transports iron from the plasma to the bone marrow for the synthesis of hemoglobin). Plasma albumin is below 2.5 g/100 mL.

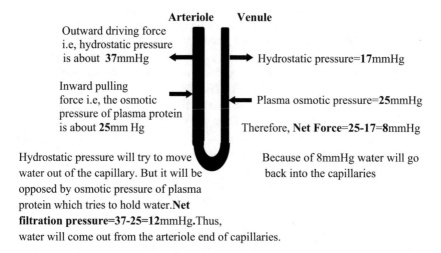

Fig. 8.2 Capillary showing Starling forces at the arteriole and venule end

2. Fatty liver and moderate enlargement of liver (hepatomegaly).
   Causes

   (a) Synthesis of plasma protein by the liver is decreased. This, in turn, impairs the export of triglycerides and other lipids from the liver, resulting in fatty liver.
   (b) Low level of epinephrine as epinephrine is synthesized from phenylalanine. Because of lower level of epinephrine, fat is not mobilized from the liver.
   (c) Impaired protein synthesis in the liver along with sufficient dietary carbohydrate ensures lipid synthesis, leading to the accumulation of triacylglycerides in the liver.

3. Antibodies are highly specific proteins. The immune system is impaired. Individuals are very susceptible to infections. Atrophy of thymus and lymphoid tissue with depressed T-cell function and B-cell antibody leads to impaired immune functions. Downregulation of immunity causes diarrhea, which will lead to dehydration and electrolyte imbalance. Apart from impaired immune system, carbohydrate malabsorption, particularly lactose malabsorption, can lead to osmotic diarrhea. In developing countries, diarrhea may be aggravated by enteric infection. As a result, carbohydrate malabsorption occurs due to "villus blunting" (Kvissberg et al. 2016). Intestinal infections hamper the absorption of both macronutrients and micronutrients, leading to impaired immune function as well as malnutrition (Farhadi and Ovchinnikov 2018). Downregulation of immunity is a cause of malnutrition, which, in turn, causes immune dysfunction (bidirectional interaction).

4. Swollen abdomen and swollen face (moon face) due to edema. Lower limb also shows edema.

5. Anorexia, i.e., loss of appetite, occurs due to fatty liver, which causes further restriction of food intake. Anorexia causes malnutrition, which aggravates anorexia (bidirectional interaction of anorexia and malnutrition).

6. Loss of emotions or apathy is the characteristic feature.

7. Growth retardation is present. It is due to deficiency of protein and associated deficiencies of zinc, phosphorus, and sulfur. Inflammation due to immune dysfunction causes reduced secretion of insulin-like growth factor (IGF-1), which in part may cause growth retardation (Bourke et al. 2016).

8. Muscles undergo wasting. Proteins are the major component of muscle. Milestones like crawling and walking are delayed. Impaired muscle strength and fatigue predispose to falls or accidents. Cough may enter the respiratory tract due to reduced strength of respiratory muscles, which is fatal. Wasting partly may be due to reduced expression of IGF-1.

9. Dermatitis causes skin cracks and leads to denuded areas of ulceration (Edwards and Bouchier 1991). Due to dermatitis, the skin resembles "crazy paving" (type of paving made up of irregular shaped slabs of stone or concrete).

10. Anemia is due to low hemoglobin and transferrin.

11. Note: Characteristic features of kwashiorkor are shown in Figs. 8.3 and 8.4.

Fig. 8.3   A child with kwashiorkor, showing edema of face, feet and hands, and skin lesions. Reproduced, with permission, from Elsevier, Book Title: Davidson's Principles and Practice of Medicine A textbook for Students and Doctors, 1991, Edited by Christopher R.W. Edwards and Ian A.D. Boucher (Chapter number 3, Title: Nutritional factors in disease, Fig 3.2, Pages 53 to 53) Copyright © Elsevier

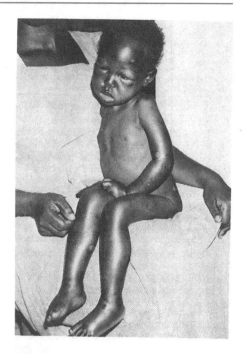

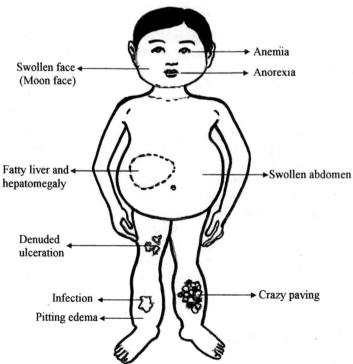

Fig. 8.4   Illustration of the features of kwashiorkor

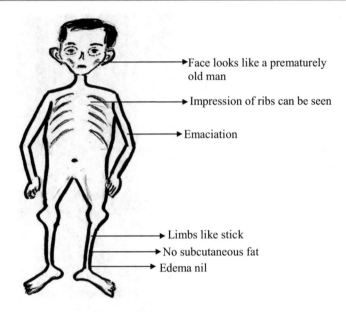

Face looks like a prematurely old man

Impression of ribs can be seen

Emaciation

Limbs like stick

No subcutaneous fat

Edema nil

Fig. 8.5 Illustration of the features of marasmus

8.2.2   Marasmus

Marasmus is a condition caused by generalized starvation. It is due to severe and prolonged restriction of all foods. Thus, it is a calorie-deficient malnutrition (due to energy-deficient diets). It occurs below 1 year of age. This is a disease of infants of poor mothers in developing countries.

Characteristic features
1. The infant is very thin with no subcutaneous fat (Fig. 8.5). Marked reduction of body weight compared with length. Weight is reduced below 60% of the standard (stunted growth).
2. The combination of low insulin and high cortisol greatly favors the catabolism of muscle. Thus, the muscle wasting is greater in marasmus than in kwashiorkor. Extreme muscle wasting known as emaciation is the characteristic feature of marasmus. The limbs look like sticks or broomsticks.
3. Lowering of BMR.
4. Poor hygiene leads to gastroenteritis (diarrhea and vomiting). Persistent diarrhea and vomiting will aggravate marasmus. Diarrhea may be due to lactose malabsorption.

8.2.3   Prevention of PEM

1. Family planning should be encouraged.
2. Awareness and education of parents is essential.

Table 8.1 Differences between kwashiorkor and marasmus

	Kwashiorkor	Marasmus
1. Age of onset	1–5 years	Below 1 year
2. Deficiency of	Protein	Calorie
3. Cause	Starchy diet after weaning	Early weaning
4. Edema	Present	Absent
5. Hypoalbuminemia	Severe	Mild
6. Fatty liver and hepatomegaly	Present	Absent
7. Level of insulin	Maintained	Low
8. Level of cortisol	Normal	High
9. Muscle wasting	Mild	Severe
10. Body weight as % of standard	60–80	<60

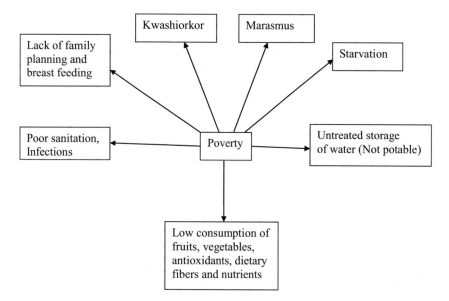

Fig. 8.6 Effects of poverty on health

3. Breast-feeding to marasmic child should be encouraged.
4. After weaning, dried milk powder mixed with some sugar and well-cooked cereal should be given four times a day.
5. Supplementation with multivitamin mixture is essential.
6. Oral hydration by dissolving the ingredients (sodium chloride 2.6 g, glucose anhydrous 13.5 g, potassium chloride 1.5 g, and trisodium citrate dehydrate 2.9 g) in 1 L of clean water should be started to correct electrolyte imbalance and thus prevent mortality from gastroenteritis.
7. Immunization should be started.
8. Hygiene must be improved to prevent infection.
9. Note: Differences between kwashiorkor and marasmus are shown in Table 8.1. Effects of poverty on health is shown in Fig. 8.6.

8.3   Starvation

Starvation is the inability to eat food for a prolonged period of time due to extreme poverty, carcinoma, burns, surgery, and so forth and also under natural catastrophes and famine. Anorexia nervosa and dementia of aging may lead to starvation.

Starvation due to nutrients deprivation is characterized by

1. Blood glucose is initially maintained because of glycogenolysis. When glycogen is completely depleted from liver after days of starvation, gluconeogenesis will try to maintain blood glucose.
2. Low insulin and high glucagon levels of blood due to starvation augment lipolysis with the inhibition of fatty acid synthesis. Catabolism of triglycerides stored in white adipocytes provides energy. High cortisol level augments protein catabolism. Glycerol liberated due to catabolism of triglycerides will try to maintain blood glucose because of glycolysis and gluconeogenesis.
3. Increased lipolysis → free fatty acids → glycerol → acetyl CoA → ketone bodies (see Chap. 2). The circulating ketone bodies are the main source of energy in fasting, permitting an individual to survive after prolonged fasting.
4. Anorexigenic leptin levels secreted by the white adipose tissue drops highly with starvation (Perry and Shulman 2018). In starvation, hypoleptinemia increases lipolysis of adipose tissue and tries to maintain blood glucose due to gluconeogenesis and glycolysis.
5. In prolonged starvation, most of the protein is burnt or exhausted, resulting in the prevention of gluconeogenesis. Prevention of gluconeogenesis causes hypoglycemia. Significant loss of both lean and fat body mass with reduction in total body water occurs due to increased catabolism, which will lead to extreme dehydration.

Death occurs due to severe hypoglycemia, depletion of body proteins, severe dehydration, and ketosis.

8.4   Malabsorption Syndrome

The digestion and absorption of food in the small intestine is essential for the maintenance of normal health. Macronutrients and micronutrients are absorbed from the intestine. Malnutrition occurs due to deficient nutrient absorption.

Nutrient absorption (arbitrarily upper 40% of small intestine is duodenum and jejunum and lower 60% is the ileum)

1. Iron, fat-soluble vitamins, water-soluble vitamins except vitamin B_{12}, fats (after hydrolysis), and calcium are absorbed mainly from upper small intestine.

2. Sugars and amino acids are mainly absorbed from mid small intestine.
3. Vitamin B_{12}, bile salts, and water are absorbed mainly from lower small intestine.
4. Water and electrolytes are absorbed from colon.

Common causes of malabsorption syndrome
1. Due to deficiency of bile salt, for example, cirrhosis.
2. Massive intestinal resection by surgery (more than 50% of small intestine) due to malignant tumor or bypass surgery causes inadequate absorptive surface. Resection decreases the surface area for absorption as well as decreases the brush border enzyme activity.
3. Maldigestion due to diseases of the pancreas.
4. Inflammatory disease of the small intestine (especially, Crohn's disease).
5. Mucosal diseases due to infections (giardiasis and tropical sprue), causing maldigestion.
6. Mutations in Sglt (sodium glucose cotransporter) genes cause glucose-galactose malabsorption (Wright et al. 2017).

Effects of Malabsorption Syndrome
1. Intestinal malabsorption of nutrients will cause malnutrition or deficiencies of iron, vitamin B_{12}, folate, etc.
2. Defective absorption of amino acids causes kwashiorkor.
3. Due to malabsorption, the amount of fat in the stool is increased and causes steatorrhea.

8.5   Overnutrition and Obesity

More than a million children and adults are obese worldwide. Obesity is a state of excessive deposition of adipose tissue mass in the body. Adipose tissue is composed of lipid-storing adipose cells. Excess of energy intake in food over energy expenditure is the fundamental cause of obesity. In other words, intake of excess amount of calorie-rich food and insufficient exercise causes obesity (Fig. 8.7). Hyperplasia (increase in number) and hypertrophy (increase in size) of adipocytes (Fig. 8.8) are responsible for the development of obesity. Relationship between height, weight, and BMI is shown in Fig. 8.9. For clinical purpose, the arbitrary threshold of 20% overweight according to height constitutes obesity. Obesity should be distinguished from overweight of muscular individuals. Obesity occurs at all ages, but is more prevalent in middle age. The fat content of adult women is more than that of adult men. Obesity in women begins at puberty and continues during pregnancy and at menopause.

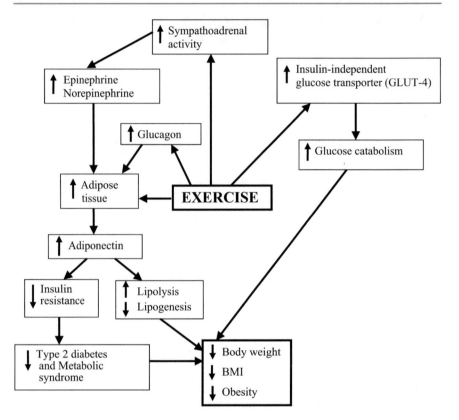

Fig. 8.7 Effects of exercise on overweight/obesity. ↑ indicates stimulation; ↓ indicates inhibition

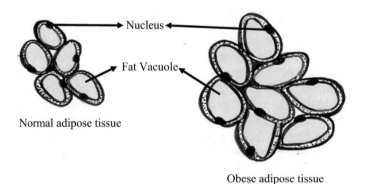

Fig. 8.8 Hyperplasia and hypertrophy of obese adipose tissue

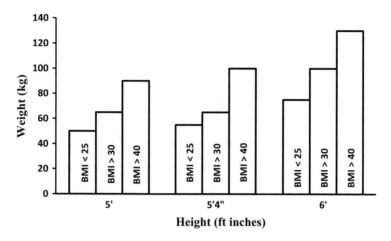

Fig. 8.9 Relationship between weight, height, and BMI

Adipose tissue is now considered as an active endocrine organ. Adipocytes are the main cell population of adipose tissue and secrete various biologically active cytokines termed as adipokines. Adipokines include adiponectin, leptin, plasminogen activator inhibitor-1 (PAI-1), IL-6, TNF-α, resistin, retinol-binding protein 4 (RBP4), angiotensin II, and various other mediators. Apart from the regulation of adipose tissue mass and nutrients, adipokines take part in the regulation of glucose homeostasis, hemostasis, blood pressure, atherosclerosis, and inflammation (Kwon and Pessin 2013). Adiponectin reduces adipose tissue mass and body weight due to inhibition of lipogenesis and stimulation of lipolysis. Furthermore, adiponectin inhibits hepatic gluconeogenesis and decreases insulin resistance (Rabe et al. 2008). Adiponectin level has been reported to be low in obese individuals. It takes part in reducing the incidence of type 2 diabetes mellitus and metabolic syndrome. Leptin is an appetite-inhibiting hormone and promotes lipolysis, resulting in loss of body weight. Lipoprotein lipase present in adipose tissue converts triglycerides into FFA and glycerol. Glycerol via glyceraldehyde 3-phosphate promotes glycolysis and gluconeogenesis. FFA is reesterified in adipocytes to form triglycerides. FFA and resistin promote insulin resistance and type 2 diabetes. RBP4 and TNF-α may induce insulin resistance. PAI-1 can cause fibrinolysis. Angiotensin II causes obesity-induced hypertension. Exercise through the liberation of epinephrine, norepinephrine, and adiponectin reduces fat tissue mass and body weight. All the effects of adipokines, as described above, are integrated (Fig. 8.10).

Causes
1. Hyperphagia due to pituitary tumor (hypothalamic obesity)
   Pituitary tumor will press and thus may damage or destroy the satiety center. Feeding center initiates hunger and appetite. If the satiety center is damaged

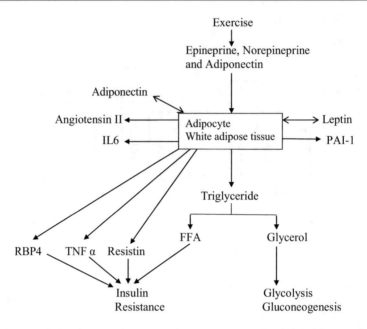

Fig. 8.10 Role of adipokines and lipoprotein lipase secreted by adipocytes in the regulation of fat mass, glucose homeostasis, hemostasis, blood pressure, atherosclerosis, and inflammation (see details in the text). PAI-1 = plasminogen activator inhibitor-1; IL-6 = interleukin-6; RBP4 = retinol binding protein 4; TNF-α = tumor necrosis factor α

Fig. 8.11 Illustration of hypothalamic obesity

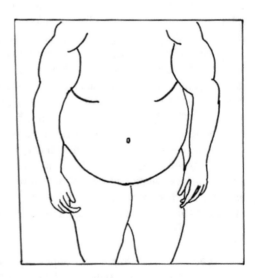

Fig. 8.12 Pendulous abdomen, moon face and buffalo hump (deposition of fat in the upper back not shown in the figure) of Cushing's syndrome (Illustration)

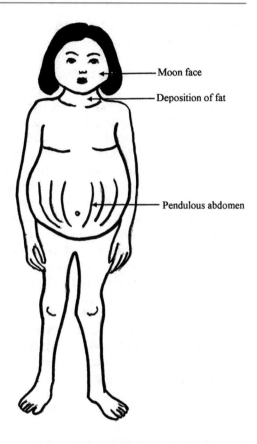

Moon face

Deposition of fat

Pendulous abdomen

due to pituitary tumor, satiety center is not able to inhibit the feeding center. There will be unchecked activity of feeding center, leading to hyperphagia and obesity (Fig. 8.11).

2. Depression causes overeating and ultimately overweight and obesity (Luppino et al. 2010).

3. Hypothyroidism

     Normally thyroid hormone increases the metabolic rate by increasing the catabolism of protein, carbohydrate, and fatty acids. Due to deficiency of thyroid hormones, weight is gained, and obesity results due to decreased catabolism of fatty acids.

4. Cushing's syndrome

     Tumor of adrenal cortex causes excessive secretion of glucocorticoids. Fat collects in the abdominal wall (pendulous abdomen), face (moon face), and upper back (buffalo hump) (Fig. 8.12).

5. Stress: It is well documented that stress causes cognitive dysfunction such as executive dysfunction and lack of self-control. As a result, overeating, consumption of high calorific food with sugar, decreased physical activity, and insomnia due to stress will lead to obesity (Tomiyama 2019).

6. Prefrontal cortex (PFC) is associated with control of appetite, craving for food, executive functioning, inhibition of impulsive behavior, and regulation of limbic reward or approach system. Self-stimulation of reward system such as medial forebrain bundle evokes motivation of feeding behavior. Neuroimaging studies have demonstrated lower activation in the PFC of obese individuals compared to lean individuals. A reciprocal relationship between obesity and PFC has been suggested (Gluck et al. 2017; Lowe et al. 2019).

7. Type 2 diabetes mellitus is associated with obesity. Increased body weight increases insulin resistance, resulting in type 2 diabetes mellitus. Adipokines (resistin, TNF-α and RBP4) released by the adipocytes increases insulin resistance and promotes type 2 diabetes mellitus. It is suggested that insulin resistance may be the cause of excess fat accumulation associated with type 2 diabetes mellitus (Malone and Hansen 2019).

8. Insensitivity to leptin or mutation of leptin gene

Leptin is an anorectic hormone produced by increased deposition of fat. It activates leptin receptors of the satiety center and decreases food intake. Obesity may be produced by mutation of leptin gene or due to insensitivity of leptin on the satiety center.

9. Insulinoma is an insulin-producing benign tumor of the beta cells of islets of Langerhans of the pancreas. High insulin level increases fatty acid synthesis and triglyceride deposition in the adipose tissue and causes obesity.

10. Male hypogonadism causes increased adipose tissue mass, resulting in increased body weight and obesity.

11. Genetic predisposition

If the parents are obese, there are chances for the children (50%) to become obese. Studies of twins indicate the genetic component. Similar BMIs are observed in identical twins. Mutation of Ob gene in genetically obese mouse causes obesity, insulin resistance, and hyperphagia. The Ob gene is present in humans. A few genetic disorders causing obesity are given below

(a) Laurence–Moon–Biedl syndrome is an autosomal recessive condition characterized by obesity, mental retardation, hypogonadism, and short stature.

(b) Mutation of melanocortin-4-receptor (MC4R) causes childhood obesity and early-onset diabetes.

(c) Mutation of leptin gene, pro-opiomelanocortin gene, and prohormone convertase 1 gene are associated with overweight.

(d) Prader–Willi syndrome is a rare congenital disorder (autosomal dominant trait) characterized by short stature, obesity, hypotonia, hypogonadism, and mental retardation. The increased appetite, decreased physical activity, and reduced energy expenditure due to marked hypotonia lead to obesity (Khan et al. 2018). Genes on chromosome 15 are deleted or unexpressed on the parental chromosome. In this syndrome, high level of ghrelin increases food intake, leading to obesity, and is not related to insulin levels (Purtell et al. 2011). Role of satiety hormones such as leptin and PYY is not yet explored.

(e) Down syndrome (Fig. 8.13) is due to chromosomal aberration of chromosome 21 and is characterized by mental retardation, short stature, hypotonia, and small oral cavity with a high prevalence of obesity (Bertapelli et al. 2016).

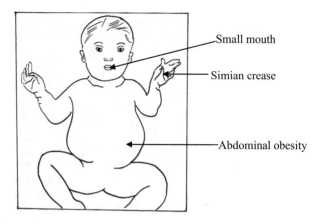

Small mouth

Simian crease

Abdominal obesity

Fig. 8.13 Illustration of Down syndrome

8.5.1 Complications

Obesity is a risk factor for increased mortality and reduces life span (Fontaine et al. 2003) due to the following reasons:

1. CIHD occurs due to atherosclerosis and hyperlipidemia leading to thrombosis.
2. Cerebral stroke occurs when the blood supply to a part of brain is prevented due to thrombosis. Depending on the area involved, there may be hemiplegia (paralysis of one half of the body) or paraplegia (paralysis of both legs).
3. Thrombosis at the vital center of the medulla may result in cardiac or respiratory failure.
4. Increases the prevalence of type 2 diabetes mellitus.
5. Heart enlarges and blood volume increases with increasing body weight, resulting in increased cardiac output and hypertension. Nevertheless, it has been reported that increased secretion of aldosterone due to the secretion of angiotensin II from abdominal adipose tissue mass may contribute to obesity-induced hypertension (Schütten et al. 2017). More recently, it has been suggested that obesity and increased sympathetic activity activate the RAAS, which induces hypertension (Roush 2019).
6. Aggravated osteoarthritis and flat foot may cause accidents. Osteoarthritis is partly due to the trauma of joints by overweight.
7. Dyspnea occurs due to exertion.
8. Sleep apnea is common in obese people and may result in death due to asphyxia. Pickwickian syndrome is characterized by obesity and sleep apnea.
9. Cholesterol gallstones are due to enhanced biliary secretion of cholesterol.
10. Steatosis (infiltration of fat in liver cells) may lead to nonalcoholic cirrhosis.
11. Varicose veins are characterized by distended and tortuous veins due to mechanical effect of weight.

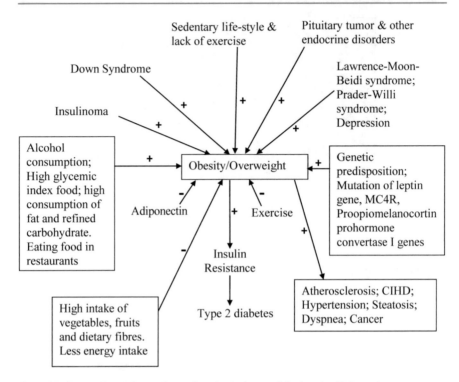

Fig. 8.14 Factors increasing or decreasing obesity/overweight (see details in text)

12. Cancer incidence increases with obesity (Stone et al. 2018). It has been reported that greater the body mass index, greater is the risk of cancer (e.g., stomach, liver, gall bladder, thyroid, and ovary).

13. Obese persons are more susceptible to respiratory tract infection (Hedge and Dhurandhar 2013).

14. Obese individuals are at great risk of asthma attack with exaggerated symptoms, frequent severe attack, and reduced response to asthma medication (Peters et al. 2018).

15. Obstructive sleep apnea is common in obese individuals. Increased fat deposition in the soft tissues of airways may reduce the size of airways. Moreover, superficial fat masses in the neck may compress the pharynx. Collapse of the pharynx is further enhanced by negative pressure during inspiration. Repetitive occurrence of intermittent hypoxia and hypercapnia due to apnea during sleep cycle may cause systemic hypertension, pulmonary hypertension, and myocardial ischemia in the obese individuals. Note: Factors increasing or decreasing obesity/overweight are shown in Fig. 8.14.

8.5.2 Prevention

The obese individual should be educated about the serious health hazards of obesity. The obese individual should be given the following advice:

1. Less calorie (energy) intake with high dietary fiber. The diet chart should be constructed in such a way that it provides energy of about 1000 kcal/day (approximately fat less than 40 g, protein 60 g, and carbohydrate 100 g). Protein because of thermogenic effect (higher SDA) has greater satiety effect compared with fat and carbohydrate.
2. Exercise will increase energy expenditure. Generally, obese individual leads sedentary life and avoids physical activities such as walking and swimming. The individual should be educated about the benefits of regular daily exercise. Brisk walking should be advised for about 1 h. It should not increase the cardiorespiratory capacity. The individual must take rest if there is difficulty in breathing. Exercise activates the hormone adiponectin released by the adipocytes. Adiponectin stimulates fatty acid oxidation, inhibits fatty acid synthesis, and prevents triglyceride deposition in the adipocytes. Exercise liberates epinephrine and norepinephrine from the adrenal medulla. Epinephrine and norepinephrine decrease body weight due to lipolysis. With low energy diet and regular daily exercise, weekly weight loss of 0.5–1 kg can be achieved.
3. Alcohol provides calorie without nutrients and stimulates appetite. The individual must avoid alcoholic drink.
4. Diet should contain plenty of fruits and green vegetables as they contain less calories. Recently it has been reported that high consumption of whole grains, vegetables, fruits, fish, low consumption of refined grains, red meat, and sugar-sweetened beverage prevents overweight and abdominal obesity (Schlesinger et al. 2019). The bulk of dietary fiber will fill the stomach. It may be noted that distension of stomach brings about a state of satiety. The vitamin and mineral contents of the diet should be sufficient to meet the body's requirement. High salt intake should be avoided. Dried salted fish should not be consumed. Nutritional interventions with antioxidants are essential for the prevention of cancer.
5. Drugs (anorectic drugs such as amphetamine or fenfluramine) must be avoided because of severe side effects. Pancreatic lipase inhibitor orlistat may be recommended. Orlistat, a saturated derivative of lipstatin and a potent inhibitor of pancreatic lipase, prevents the absorption of fat and is an anti-obesity drug (Fig. 8.15a). Daily multivitamin supplements are necessary as orlistat can interfere with the absorption of fat-soluble vitamins.
6. Surgery (Fig. 8.15b) may be required for individuals with severe obesity (BMI > 40 kg/m^2) to reduce the size of the stomach by band or by vertical banded gastroplasty (Colledge et al. 2010). Thus, the bulk of food will not be able to enter into the small intestine for digestion and absorption. By gastric bypass, stomach is divided into small upper pouch and much larger lower remnant pouch. The small intestine is connected to both.

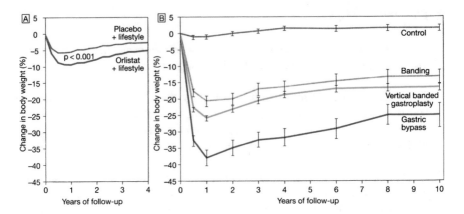

Fig. 8.15 Effects of orlistat and bariatric surgery on weight loss (**A**, **B**). Reproduced, with permission, from Elsevier, Book Title: Davidson's Principles and Practice of Medicine, 2010, Edited by Nicki R. Colledge, Brian R. Walker and Stuart H. Ralston (Chapter number 5, Title: Environmental and Nutritional factors in disease, Fig 5.13, Pages 128 to 128) Copyright © Elsevier

7. Hospital admission is recommended for severe obese individuals. Hospitalized obese is treated with only water, vitamins, minerals, and protein supplements (less quantity). Initially, weight loss of about 1 kg/day will be observed and afterwards will be stabilized at about 0.5 kg/day. This type of treatment cannot be continued for many days because of severe complications of starvation.

In conclusion, low glycemic index foods, regular exercise, less energy intake with high dietary fibers, and high consumption of fruits and vegetables prevent weight gain in the obese individuals. Conversely, sedentary lifestyle, eating food outside home, i.e., restaurants, and alcohol consumption promote weight gain in obese individuals.

8.5.3   Anthropometric Measurements

1. Measurement of BMI is previously explained (Sect. 8.1). If height cannot be measured, for example, in older people with kyphosis. A substitute measure is demispan and knee height (Colledge et al. 2010).
2. Obesity should be distinguished from overweight of muscular individuals (athletes). BMI can be increased by muscle mass. Muscle mass can be calculated by deducting triceps skinfold thickness (with the help of caliper) from mid-arm circumference.
3. Abdominal obesity due to increased intra-abdominal fat is strongly correlated with insulin resistance, type 2 diabetes mellitus, and metabolic syndrome. Abdominal obesity can be measured by the waist circumference at the level of umbilicus. Waist circumference is 90 cm for Asian men and 80 cm for Asian women. It is correlated with increased insulin resistance.

8.6 Summary

Both undernutrition and overnutrition have adverse effects on body functions. PEM is the commonest nutritional disorder in many parts of the world. This is a disease of infant/child of poor mother. Infections due to poor sanitation, diseases due to the use of untreated stored water (not potable), starvation, lack of family planning, and lack of breast-feeding increase the mortality and morbidity of infant/children of the low-income group families. Around the world, infants/children of low-income families are at the highest risk of nutritional deficiencies and inflammatory diseases due to low consumption of fruits, vegetables, antioxidants, dietary fibers, and nutrients. Natural catastrophes such as famine, flood, and earthquake lead to undernutrition, for example, subclinical PEM. Obesity is a risk factor for increased mortality, especially in developed countries. It reduces life span due to CIHD, cerebral stroke, and thrombosis of the vital center. Atherosclerosis is the main culprit for cardiovascular diseases, cerebral stroke, thrombosis of the cardiorespiratory center, and hypertension. Obesity is linked to increased prevalence of type 2 diabetes mellitus. It has been reported that greater the body mass index, greater is the risk of certain cancers. Obese individuals should be educated about the serious health hazards. Low glycemic index foods, regular exercise, less energy intake, and Mediterranean diet prevent weight gain in obese individuals. Conversely, sedentary lifestyle, eating junk food in restaurant/cafeteria, and alcohol consumption promote weight gain in obese individuals.

References

Bedi M et al (1999) Comparative study of autonomic nervous system activity in malnourished and normal children in India. Ann Trop Paediatr 19:185–189

Bertapelli F et al (2016) Overweight and obesity in children and adolescents with Down syndrome-prevalence, determinants, consequences, and interventions: a literature review. Res Dev Disabil 57:181–192

Bhutt ZA et al (2017) Severe childhood malnutrition. Nat Rev Dis Primers 3:17067

Bourke CD et al (2016) Immune dysfunction as a cause and consequence of malnutrition. Trends Immunol 37(6):386–398

Colledge NR et al (eds) (2010) Davidson's principles and practice of medicine, 21st edn. Churchill Livingstone, New York

Edwards CRW, Boucher IAD (eds) (1991) Chapter 3: Nutritional factors in disease. In: Davidson's principles and practice of medicine: a textbook for students and doctors. Churchill Livingstone, New York, p 53

Farhadi S, Ovchinnikov RS (2018) The relationship between nutrition and infectious diseases: a review. Biomed Biotechnol Res J 2:168–172

Fontaine KR et al (2003) Years of life lost due to obesity. JAMA 289(2):187–193

Gluck ME et al (2017) Obesity, appetite, and the prefrontal cortex. Curr Obes Rep 6(4):380–388

Hedge V, Dhurandhar NV (2013) Microbes and obesity-interrelationship between infection, adipose tissue and the immune system. Clin Microbiol Infect 19:314–320

Khan MJ et al (2018) Mechanism of obesity in Prader-Willi Syndrome. Pediatr Obes 13(1):3–13

Kvissberg MA et al (2016) Carbohydrate malabsorption in acutely malnourished children and infants: a systematic review. Nutr Rev 74(1):48–58

Kwon H, Pessin JE (2013) Adipokines mediate inflammation and insulin resistance. Front Endocrinol 4(71):1–13

Lowe CJ et al (2019) The prefrontal cortex and obesity: a health-neuroscience perspective. Trends Cogn Sci 23(4):349–361

Luppino FS et al (2010) Overweight, obesity, and depression: a systematic review and meta-analysis of longitudinal studies. Arch Gen Psychiatry 67(3):220–229

Malone JI, Hansen BC (2019) Does obesity cause type 2 diabetes mellitus (T2DM)? Or is it the opposite. Pediatr Diabetes 20(1):5–9

Nuttall FQ (2015) Body mass index: obesity, BMI, and health: a critical review. Nutr Today 50(3):117–128

Perry RJ, Shulman GI (2018) The role of leptin in maintaining plasma glucose during starvation. Postdoc J 6(3):3–19

Peters U et al (2018) Obesity and asthma. J Allergy Clin Immunol 141(4):1169–1179

Purtell L et al (2011) In adults with Prader-Willi syndrome, elevated ghrelin levels are more consistent with hyperphagia than high PYY and GLP-1 levels. Neuropeptides 45(4):301–307

Rabe K et al (2008) Adipokines and insulin resistance. Mol Med 14:741–751

Roush GC (2019) Obesity-induced hypertension: heavy on the accelerator. J Am Heart Assoc 8:e012334

Schlesinger S et al (2019) Food groups and risk of overweight, obesity, and weight gain: a systematic review and dose-response meta-analysis of prospective studies. Adv Nutr 10:205–218

Schütten MTJ et al (2017) The link between adipose tissue renin-angiotensin-aldosterone system signalling and obesity-associated hypertension. Physiology 32:197–209

Stone TW et al (2018) Obesity and cancer: existing and new hypotheses for a causal connection. EBioMedicine 30:14–28

Tomiyama AJ (2019) Stress and obesity. Annu Rev Psychol 70:703–718

Wright EM et al (2017) Novel and unexpected functions of SGLTs. Physiology 32:435–443

Further Reading

Black RE et al (2013) Maternal and child undernutrition and overweight in low-income and middle-income countries. Lancet 382:427–451

Schlesinger S et al (2019) Food groups and risk of overweight, obesity, and weight gain: a systematic review and dose-response meta-analysis of prospective studies. Adv Nutr 10:205–218

WHO Management of severe malnutrition (1999) A manual for physicians and other senior health services. WHO, Geneva. https://www.who.int/nutrition/publications/en/manage_severe_malnutrition_eng.pdf

Williams PCM, Berkely JA (2016) Severe acute malnutrition update. Current WHO guidelines and the WHO essential medicine list for children. http://www.int/selection_medicines/.../paed_antibiotics_appendix7_sam.pdf

Eating Disorders

9

Abstract

Eating disorders (EDs) are abnormal and pathological eating habit, leading to many psychiatric and somatic complications and thus constitute a major public health problem. There are many forms of EDs, which are described in the Diagnostic and Statistical Manual of Mental Disorders (DSM). Anorexia nervosa (AN) is common in adolescence, especially in girls. Due to intense fear of weight gain, an individual avoids high caloric diet and induces repeated vomiting, leading to emaciation, hypokalemia, alkalosis, and fluid and electrolyte imbalance. Bulimia nervosa (BN) is characterized by recurrent bouts of binge eating in a short period of time. Binge eating is followed by self-induced vomiting and use of laxative and purgative. Complications of the patients suffering from AN and BN occur due to hypokalemia, alkalosis, and electrolyte imbalance. Eating disorder not otherwise specified (EDNOS) is more or less identical to BN. Binge eating disorder (BED) is different from BN as episodes of binge eating are not followed by purging, fasting, and vigorous exercise. BED patients lose control over his or her eating and become obese due to hyperphagia. EDs are also common during childhood, pregnancy and in Type 1 diabetes mellitus.

Keywords

Anorexia nervosa · Bulimia nervosa · Eating disorder not otherwise specified · Binge eating disorder · Type 1 diabetes mellitus

9.1 Anorexia Nervosa

AN is commonly found during adolescence, especially in girls of the upper social classes. AN is rarely found after the age of 40 years. Due to morbid fear of obesity or intense fear of weight gain and a distorted body image, an individual avoids high caloric diet along with fasting and induced vomiting. This will result in extreme

© Springer Nature Singapore Pte Ltd. 2019
K. Chakrabarty, A. S. Chakrabarty, *Textbook of Nutrition in Health and Disease*,
https://doi.org/10.1007/978-981-15-0962-9_9

weight loss of at least 30–40% of original body weight and low BMI (<18 kg/m²) and, thus, leads to emaciation. Anorexia is a misnomer as the individual does not suffer from anorexia. Appetite remains normal. The individual has reduced appetite due to severe cachexia. Two forms of AN occur (Attia and Walsh 2018). (1) He/she avoids feeding without self-induced vomiting or use of laxative, purgative, and diuretic. The individual does not engage in binge eating (consumption of excessive amount of food in a short period of time) and will engage in frequent exercise. (2) Other individuals engage in recurrent bouts of binge eating followed by self-induced vomiting and misuse of purgatives and laxatives.

Complications: The patients may suffer from loss of sex drive, cognitive dysfunction, hypothermia, irritability, depression, and anxiety. Social isolation is common. The girls/women may have amenorrhea (without oral contraceptive), cardiomyopathy, low cardiac output, bradycardia, and hypotension. Mitral valve prolapse, pericardial effusion, osteoporosis, and gastroparesis are commonly present (Mehler et al. 2010). Complete food avoidance or frequent fasting leads to hypoglycemia characterized by agitation and impaired judgement. Induced vomiting and purging lead to hypokalemia, hypochloremia, alkalosis, and fluid and electrolyte imbalance. ECG/EKG finding shows prolonged PR interval, prominent U wave, inversion of T wave and occasionally, prolonged QT interval. EKG finding indicates hypokalemia. Hypokalemia can cause cardiac arrhythmia and renal damage. AN was associated with fibromyalgia (Udo and Grilo 2019). Physical findings include acrocyanosis of the digits (bluish-purple discoloration due to slow circulation of the blood through capillaries of the skin), lanugo hair (fine downy hair) covering the body, particularly on the back, forearms and cheeks, and slight hirsutism. Due to excessive vomiting, patients have eroded dental element and an inflamed esophagus. The patients may suffer from abdominal bloating, abdominal discomfort, and constipation.

Endocrine findings: (1) Low levels of FSH and LH inhibit ovulation, (2) Low levels of T_3 and T_4 cause cold intolerance, bradycardia, and hypotension, (3) Growth-hormone resistance results in muscle atrophy and stunted growth, (4) High level of cortisol results in loss of muscle mass.

Genetic predisposition: It probably exists as the concordance rate is high in identical twin and concordance ratio is lower in fraternal twins. Genetic factors may be involved (Steinhausen and Jenson 2015). Duncan et al. (2017) reported a locus on chromosome 12. However, no genetic risk factors have been conclusively identified (Mayhew et al. 2018).

Management: Weight restoration to normal is the primary goal for the treatment. Balanced diet with macro- and micronutrients should be given, so that weight gain is achieved of 1 kg/week. Caloric intake at the beginning should be 40 kcal/kg/day and gradually increased up to 70–100 kcal/kg/day (Rock 2010). Knowledge of cognitive neuroscience can guide effective treatment (Steinglass et al. 2019). Bullying and teasing regarding irregular eating habits from friends, colleagues, and even from relatives will aggravate EDs and may constitute a risk factor (Lie et al. 2019).

The patients must be encouraged to get rid of distress about body weight and shape. Selective serotonin reuptake inhibitor (SSRI) sertraline or mood stabilizing lithium may be tried under the direction of psychiatrist in order to reduce depression, anxiety, obsessive compulsive disorder (OCD), attention deficit hyperactivity disorder (ADHD), and social phobia. It was reported by American Psychiatric Association (2006) that dopamine antagonists increase gastric emptying and reduce abdominal distension and bloating. Despite the beneficial effect of dopamine antagonist on EDs, it may not be recommended as it may induce features of Parkinsonism. Cognitive behavioral therapy may be necessary to alter cognition of phobia, anxiety, and extreme sadness. Patients should be hospitalized if the body weight is less than 15% of body weight, systolic blood pressure falls below 90 mmHg, heart rate is below 50/min, core body temperature is less than 36 °C and if there is incidence of suicidal attempt (Akel et al. 2018; Hay et al. 2014). Fluid and electrolyte imbalance should be rectified. Nasogastric feeding may be required to restore body weight. Calcium and vitamin D should be prescribed for bony loss.

9.2 Bulimia Nervosa

The individual suffering from BN is extremely dissatisfied about body size and shape. BN is characterized by recurrent bouts of binge eating in a short period of time (consumption of excessive amount of food, especially junk food) without any self-control over eating. Binge eating is followed by self-induced vomiting and misuse of laxative, purgative and diuretic, leading to purgation and diuresis. The individual performs severe exercise after binges due to fear of weight gain. Dieting and fasting in between binge eating is also common. The binge eating and purging are done in secret. Especially the obese women suffer from shame, feeling of guilt, depression, impulsive behavior, negative thought, and sexual conflict. Frequent amenorrhea may be present. Cortico-striatal (Steinglass et al. 2019) and cortico-limbic circuits (Akel et al. 2018) abnormalities in BN may be responsible for cognitive dysfunction and behavioral disorders. Overeating may be due to inhibition of satiety center leading to unchecked activity of feeding center of the hypothalamus. Complications occur due to the effects of hypokalemia, hypochloremia, metabolic alkalosis, and fluid and electrolyte imbalance. Hypokalemia may cause cardiac arrhythmia and renal damage. Physical findings due to violent vomiting include Mallory–Weiss syndrome (longitudinal tear of the mucosa around the gastroesophageal junction), Russell's sign (calluses on the knuckles), and discoloration and surface irregularities of the teeth.

Management: Psychotherapy/counselling is essential. Cognitive behavioral therapy is required to alter cognition of phobia, anxiety, and adverse behavior. SSRI should be started if there is tendency of suicide. Fluid and electrolyte imbalance should be rectified. Note: Comparison of AN and BN is shown in Table 9.1.

Table 9.1 Comparison of anorexia nervosa and bulimia nervosa

Anorexia nervosa	Bulimia nervosa
1. Emaciation with extreme weight loss of 25–30% of original body weight. Low BMI (<18)	1. Absence of emaciation. Weight is maintained within normal limits. Commonly associated with obesity
2. Amenorrhea is present for about 3 months	2. Frequent amenorrhea may be present
3. Hypokalemia, hypochloremia, and alkalosis occur due to vomiting. Fluid and electrolyte balance should be maintained	3. Hypokalemia, hypochloremia, and alkalosis occur due to vomiting. Fluid and electrolyte balance should be maintained
4. Enlargement of parotid gland is not reported	4. Enlargement of painless noninflammatory parotid gland is reported
5. Low levels of FSH and LH of the anterior pituitary. As a result, ovulation is inhibited. Low level of T_3 and T_4 results in cold intolerance, bradycardia, and orthostatic hypotension. Growth hormone resistance results in muscle atrophy and stunted growth. High level of cortisol results in loss of muscle mass	5. Levels of FSH, LH, T_3, T_4, growth hormone, and cortisol are not reported (?)
6. Psychologist is required for counselling. Cognitive behavioral therapy may be essential	6. Psychologist is required for counselling. Cognitive behavioral therapy may be essential

9.3 Binge Eating Disorder

BED is characterized by episodes of binge eating without any control over eating. BED is not associated with purging or exercise and thus differs from BN. BED is more common in women. Because of social phobia, the individual eats alone and will not attend get-together. The individual is obese with BMI > 30 kg/m². Obesity is a risk factor for increased mortality and morbidity. It reduces life span due to CIHD, cerebral stroke, and thrombosis of the vital center of the medulla. It increases the prevalence of Type 2 diabetes/metabolic syndrome. The individual suffering from BED should be educated about the serious health hazards of obesity. Obesity, social isolation, and depression of BED individuals result in "poor health-related quality of life" (Singleton et al. 2019). Counselling by psychiatrist/psychologist is required. Cognitive behavioral therapy is essential to treat anxiety, social phobia, guilt complex, and suicidality. Psychotropic medicines such as antidepressants may not be recommended as they may further aggravate weight gain.

9.4 Eating Disorder Not Otherwise Specified

The individual of EDNOS maintains normal weight and normal BMI by consuming small amount of food. She/he repeatedly chews large amount of food and spits out without swallowing (unhealthy eating disorder). In contrast to anorexia nervosa, the women will have regular menstrual cycle without amenorrhea.

9.5   Avoidant-Restrictive Food Intake Disorder

ARFID is disorder of infancy or early childhood. It is characterized by anorexia and phobia related to the act of eating such as fear of choking. They generally reject food on the basis of color and texture. ARFID may be associated with gastrointestinal discomfort. They suffer from malnutrition (Kennedy et al. 2018).

9.6   Night Eating Syndrome

NES is an eating disorder due to abnormal circadian rhythm of food intake. Sleep pattern may be disturbed after heavy dinner. When the individual wakes up at night, he/she may start eating. Unlike BE, the individual will not consume heavy amount. It is different from sleep-related eating disorder (SRED). It is included in the other specified feeding or eating disorder (OSFED) category of the DSM-5.

9.7   Sleep-Related Eating Disorder

SRED is an eating disorder characterized by uncontrolled episode of eating while in a state of sleep. It is associated with high-calorie diet (rich in carbohydrate and fat), leading to weight gain and obesity. The DSM-5 classifies SRED under sleep-walking while ICSD (International classification of sleep disorders) classifies it as non-rapid eye movement (NREM/slow wave sleep)-related parasomnia. SRED usually occurs during NREM sleep in the early hours of the night and occurs during the transition period from NREM to arousal. The individuals who have a history of sleep-walking usually suffer from SRED. SRED is more common in women and starts in teenage years or early 20s. SRED should be differentiated from sleep-related parasomnia. (Parasomnia is a form of sleep disorder characterized by abnormal movements, emotions, dreams, etc. and occurs between rapid eye movement (REM) and wakefulness.)

9.8   Orthorexia Nervosa (McComb and Mills 2019; Saljoughian 2017)

ON is an emerging eating disorder characterized by excessive preoccupation or focus on eating healthy food, which may lead to obsession about foods (Plichta and Jezewska-Zychowicz 2019). Restriction of other macronutrients and micronutrients leads to health hazards. Dietary restriction intended to improve health causes aversion to other foods. ON is characterized by social withdrawal, avoidance of get-together, anxiety, and severe malnutrition.

9.9   Miscellaneous Eating Disorders

1. "Pervasive refusal syndrome" is an eating disorder during childhood. It occurs between the age of 7 and 15 years. Girls are more commonly affected than the boys. Social isolation, lack of hunger and appetite, lack of social communication, dysphagia, and food phobia are the common features of the syndrome. Psychotherapy, orogastric tube feeding, and nutritional intervention are essential (Mann and Truswell 2007).
2. EDs are often seen in infants of malnourished pregnant mothers. Infants have an obsession or aversion reaction to placing food in or near the mouth. Pregnant women suffering from EDs may deliver prematurely and may have a greater risk of miscarriage.
3. Endurance sports like distant running are at risk of EDs (Krebs et al. 2019). Females are more affected than males. Athletes deny exercise and restrictive diets recommended by the coach because of fear for thinness. Risk of EDs among athletes may be due to emphasis on thinness for improved performance. EDs among athletes may be due associated with lack of confidence and low self-esteem.
4. Type 1 diabetes or insulin-dependent diabetes mellitus with EDs: It occurs below the age of 40. The patient is usually underweight due to absolute deficiency of insulin. The patient is reluctant towards insulin therapy, because of fear of weight gain. He/she refuses to follow diabetic diet with poor glycemic control. As a result, the patient progresses rapidly to early onset of serious diabetic complications such as hyperlipidemia, hypercholesterolemia leading to CIHD, cerebral stroke, and diabetic retinopathy leading to blindness. This type of ED-associated Type 1 diabetes is described recently as "diabulimia" (Torjesen 2019).

In conclusion, overconcern about body weight and shape and overconcern to maintain slim figure causes EDs. EDs lead to many psychological/psychiatric and somatic complications. EDs are highly prevalent worldwide, especially in women and are a major public health problem. Prevalence of EDs in women may be due to mood and anxiety disorders (Ulfvebrand et al. 2015). The following are the types of EDs, which are described in DSM. In the most recent DSM-5 classification, EDs are AN, BN, and BED referred to "typical" EDs. Many other EDs are referred to as "atypical" types described by Galmiche et al. 2019, named as OSFEDs. These include NES, ARFID, EDNOS, and SRED. Typical EDs such as AN, BN, and BED under the DSM-5 category are all characterized by loss of interest in all activities and enjoyment. Social withdrawal, helplessness, and extreme dejection suggest that the patient might be suffering from major depression. Classification of EDs according to DSM criteria is described in Fig. 9.1.

9.10   Summary

Overconcern about body weight leads to EDs, especially in girls of high economic groups. Psychiatrists/psychologists are required for counselling. Cognitive behavioral therapy may be essential to treat anxiety and adverse behavior. Use of laxative

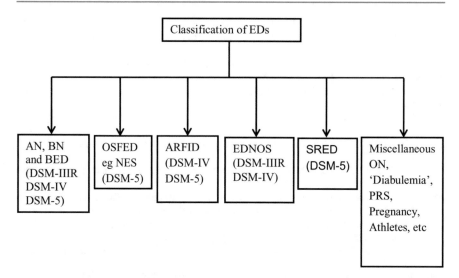

Fig. 9.1 Classification of EDs. AN anorexia nervosa, BN bulimia nervosa, BED binge eating disorder, OSFED other specified feeding or eating disorder, ARFID avoidant-restrictive food intake disorder, EDNOS eating disorders not otherwise specified, SRED sleep-related eating disorder, ON orthorexia nervosa, PRS pervasive refusal syndrome, NES night eating syndrome, DSM diagnostic and statistical manual of mental disorders

and purgative in AN and BN will have adverse effects on health. Hypokalemia, alkalosis, and fluid and electrolyte imbalance are detrimental to health. EDNOS is more or less similar to bulimia nervosa. BED patients are generally obese due to hyperphagia as they cannot control overeating. Death may occur due to suicide or due to complications. Eating disorders during childhood are often encountered between the age of 7 and 15 years. The child will have adverse response to feeding such as dysphagia, food phobia, and anorexia. Various types of eating disorders are compared and discussed. The individual with severe eating disorders should be hospitalized for tube feeding. Fluid and electrolyte imbalance must be rectified. Nutritional therapy having macronutrients and micronutrients is essential.

References

Akel M et al (2018) Eating and weight disorders: a mini review of anorexia and bulimia nervosa and binge eating disorder. IOSR J Pharm 8(4):12–16

American Psychiatric Association (2006) Treatment of patients with eating disorders, third edition. Am J Psychiatry 163(7 Suppl):4–54

Attia E, Walsh BT (2018) Anorexia nervosa. MSD manual professional version. https://www.msdmanuals.com/professional/psychiatric-disorders/eating-disorders/anorexia-nervosa

Duncan L et al (2017) Significant locus and metabolic genetic correlations revealed in genome-wide association study of anorexia nervosa. Am J Psychiatry 174(9):850–858

Galmiche M et al (2019) Prevalence of eating disorders over the 2000-2018 period: a systematic literature review. Am J Clin Nutr 109:1402–1413

Hay P et al (2014) Royal Australian and New Zealand College of Psychiatrists clinical practice guidelines for the treatment of eating disorders. Aust N Z J Psychiatry 48(11):977–1008

Kennedy GA et al (2018) Eating disorders in children: is avoidant-restrictive food intake disorder a feeding disorder or an eating disorder and what are the implications for treatment? F1000Res 7:88

Krebs PA et al (2019) Gender differences in eating disorder risk among NCAA division I cross country and track student-athletes. J Sports Med 5035871, 5 pages

Lie SØ et al (2019) Is bullying and teasing associated with eating disorders? A systematic review and meta-analysis. Int J Eat Disord 52(5):497–514

Mann J, Truswell SA (eds) (2007) Essentials of human nutrition, 3rd edn. Oxford University Press, Oxford

Mayhew AJ et al (2018) An evolutionary genetic perspective of eating disorders. Neuroendocrinology 106(3):292–306

McComb SE, Mills JS (2019) Orthorexia nervosa: a review of psychosocial risk factors. Appetite 140:50–75

Mehler PS et al (2010) Medical complications of eating disorders. In: The treatment of eating disorders: a clinical handbook. Guilford Press, New York, pp 66–80

Plichta M, Jezewska-Zychowicz M (2019) Eating behaviors, attitudes toward health and eating, and symptoms of orthorexia nervosa among students. Appetite 137:114–123

Rock CL (2010) Nutritional rehabilitation for anorexia nervosa. The treatment of eating disorders: a clinical handbook. Guilford Press, New York, p 187

Saljoughian M (2017) Orthorexia: an eating disorder emerges. US Pharm 42(12):9–10

Singleton C et al (2019) Depression partially mediates the association between binge eating disorder and health-related quality of life. Front Psychol 10:209

Steinglass JE et al (2019) Cognitive neuroscience of eating disorders. Psychiatr Clin N Am 42(1):75–91

Steinhausen HC, Jenson MC (2015) Time trends in lifetime incidence rates of first-time diagnosed anorexia nervosa and bulimia nervosa across 16 years in a Danish nationwide psychiatric registry study. Int J Eat Disord 48(7):845–850

Torjesen I (2019) Diabulimia: the world's most dangerous eating disorder. BMJ 364:1982

Udo T, Grilo MC (2019) Psychiatric and medical correlates of DSM-5 eating disorders in a nationally representative sample of adults in the United States. Int J Eat Disord 52:42–50

Ulfvebrand S et al (2015) Psychiatric comorbidity in women and men with eating disorders results from a large clinical database. Psychiatry Res 230(2):294–299

Website

www.mayoclinic.org/diseases-conditions/sleep-related-eating-disorder

Further Reading

Galmiche M et al (2019) Prevalence of eating disorders over the 2000-2018 period: a systematic literature review. Am J Clin Nutr 109:1402–1413

American Psychiatric Association (2013) Diagnostic and statistical manual of mental disorders. Fifth edition (DSM-5). American Psychiatric Publishing, Arlington

American Psychiatric Association (2006) Treatment of patients with eating disorders, third edition. Am J Psychiatry 163(7 Suppl):4–54

World Health Organization (2004) ICD-10: international statistical classification of diseases and related health problems: tenth revision, 2nd edn. World Health Organization, Geneva. https://apps.who.int/iris/handle/10665/42980

Nutritional Therapies, Exercise, and Diet for Mental Disorders

10

Abstract

Mental disorder is mainly due to any disturbance of emotion, i.e., disturbance of cognition, conation, and affect. Cognition is the study of any kind of mental operation by which knowledge is acquired. It includes learning, memory, language, perception, i.e., awareness of sensation, reasoning, execution of tasks, and so on. Cognitive behavioral therapy (in psychiatry) is directed to alter cognition in order to treat phobia, anxiety, and extreme sadness (depression). Conation, i.e., eager to take action, can sometimes lead to adverse reactions. Affect indicates the mental state at any particular moment associated with a particular idea. The hypothalamus and limbic system are mainly concerned with the genesis of emotion. Mental disorders may be due to mood, behavioral, biological, and psychological dysfunction. It may be associated with palpitation, sweating, fear, rage, nervousness, and lack of confidence. Mental disorders are caused by multifactorial pathogenesis, including genetics, oxidative damage, and neurotransmitter dysfunction. Nutritional therapies play an important role in reducing and even preventing the severity of a mental disorder.

Keywords

Nutritional therapies · Mental disorders · Emotion · Cognition · Conation · Behavioral therapy · Hypothalamus · Limbic system

10.1 Nutritional Therapies

Recommended macro- and micronutrients, balanced diet, and exercise can be prophylactic against mental disorders. Anxiety disorder (AD), major depressive disorder (MDD), bipolar disorder (BD), schizophrenia, obsessive compulsive disorder (OCD), and attention deficit hyperactivity disorder (ADHD) are the common mental disorders. These disorders are known to be due to deficiencies of neurotransmitters

© Springer Nature Singapore Pte Ltd. 2019
K. Chakrabarty, A. S. Chakrabarty, *Textbook of Nutrition in Health and Disease*,
https://doi.org/10.1007/978-981-15-0962-9_10

of the brain such as serotonin, norepinephrine (NE), dopamine (DA), gamma amino butyric acid (GABA), and N-acetylcysteine (NAC). The amino acids tryptophan, phenylalanine, tyrosine, glutamate, glycine, and methionine are converted to neurotransmitters, as shown below.

1. NE and DA are formed by hydroxylation and decarboxylation of the amino acid tyrosine.

$$\text{Phenylalanine} \rightarrow \text{Tyrosine} \rightarrow \text{Dopa} \rightarrow \text{DA} \rightarrow \text{NE}.$$

2. Serotonin is formed by hydroxylation and decarboxylation of tryptophan.

$$\text{Tryptophan} \rightarrow 5 \text{ hydroxytryptophan} \rightarrow \text{Serotonin} (5 \text{ hydroxytryptamine}).$$

3. GABA is formed by decarboxylation of glutamate.

$$\text{Glutamate} \rightarrow \text{GABA}.$$

4. Methionine is converted to S-adenosylmethionine (SAM).

$$\text{Methionine} + \text{ATP} \rightarrow \text{SAM} + p_i + pp_i$$

It has been suggested that SAM facilitates the synthesis of neurotransmitters of the brain. SAM provides methyl groups in the body.

5. Glycine (an inhibitory transmitter) can be synthesized from glutamate and choline. Choline is synthesized using methyl group donated by methionine.
6. Taurine is an end product of cysteine.
7. NAC is a sulfur-containing amino acid and is one of the most important nutritional therapies for many psychiatric conditions (Ooi et al. 2018). NAC facilitates dopamine receptor binding (Monti et al. 2016). NAC reduces the symptoms of schizophrenia (Chen et al. 2016) as well as improves psychopathology in schizophrenia (Zheng et al. 2018). Tourette syndrome (repetitive, stereotyped vocalization and involuntary movement), trichotillomania (hair pulling disorder), onychophagia (nail biting), and excoriation (skin pricking) can be treated by NAC as a glutamate-modulating agent (Oliver et al. 2015; Minarini et al. 2016). NAC may modulate the inflammatory pathways and ameliorates oxidative stress (Dean et al. 2011; Chen et al. 2016). NAC has been proposed as a treatment for OCD due to inhibition of the synaptic glutamate (Camfield et al. 2011). NAC significantly reduces depression, which can be minimized by antioxidants (Scapagnini et al. 2012).

Safety and side effects of NAC: NAC is well-tolerated oral therapy (about 2000 mg/day) without any considerable side effects (Oliver et al. 2015; Deepmala et al. 2015; Minarini et al. 2016). Higher dose may cause abdominal discomfort, heartburn, flatulence, nausea, vomiting, and diarrhea. NAC potentiates the action of nitrate vasodilator and causes hypotension. One should monitor blood pressure after therapy.

10.1.1 Role of Vitamins

Tryptophan takes part in the formation of active form of niacin. Deficiency of niacin causes dementia, delirium, and depressive psychosis. These symptoms of niacin deficiency may be due to reduced synthesis of serotonin from tryptophan. Vitamin B_{12} takes part in the formation of methionine from homocysteine. Supplementation with vitamin B_{12} improves cognitive functions and delays the onset of dementia in the elderly people. Folate consumption enhances the effectiveness of antidepressant drugs and its deficiency is associated with depressive symptoms. Vitamin B_6 is a cofactor for the conversion of DOPA to DA. Vitamin C is a cofactor for the conversion of DA to NE. Vitamin D regulates the conversion of tryptophan into serotonin. Vitamin D intake may prevent the severity of brain dysfunction. Low serotonin levels lead to impairment in executive function and sensory gating, for example, the ability of brain to filter out extraneous sensory inputs (Patrick and Ames 2015).

10.2 Anxiety Disorder and Nutrition

AD is characterized by tachycardia, sweating, tremor, restlessness, fearful facial expressions, dry mouth, dysfunction of cognition (poor concentration and distraction), affective symptoms like unpleasant fearfulness, sense of apprehension, irritability, insomnia, etc. It may lead to panic disorder, i.e., an acute intense episode of anxiety. AD may be due to the deficiency of GABA or due to the deficiencies of NE, serotonin, and DA (Liu et al. 2018). Administration of glutamate, phenylalanine/tyrosine, and tryptophan may alleviate anxiety disorder.

10.3 Major Depressive Disorder and Nutrition

MDD is characterized by melancholy, anxiety, anorexia, loss of interest in all activities and enjoyment, social withdrawal, helplessness, extreme dejection, loss of vigor and energy, and loss of motivation. MDD shares many similar symptoms with AD. MDD is the main cause of suicide worldwide. Glutamatergic modulator ketamine may act as an antidepressant and also, antisuicidal (Kadriu et al. 2019). Daily supplements of tryptophan, tyrosine, glutamate, and methionine may lead to recovery from major depression. It has been suggested that daily supplement of EPA can be helpful in treating patients suffering from major depression. Daily administration of vitamin B_{12}, folic acid, magnesium, zinc, and selenium are essential to elevate mood in depressed patients. Dysfunction of mono-aminergic neurotransmitters (serotonin, NE, and DA) is implicated in the pathogenesis of both AD and MDD (Liu et al. 2018).

10.4   Bipolar Disorder and Nutrition

BD is characterized by recurrent episodes of mania and depression. Severe mood elevation, talkativeness, flight of ideas, and grandiosity are the characteristic features of mania. There may be episodes of hypomania and depression. Genetic background and stressful life events are the etiology of BD. Several pathways underlying pathogenesis of BD have been reviewed (Sigitova et al. 2017). These include decreased brain-derived neurotrophic factor (BDNF), neurotransmitter deficits, oxidative damage leading to mitochondrial dysfunction, decreased tryptophan metabolism, accelerated apoptosis disrupting the structural and functional integrity of brain areas, disruptions of cortico-striatal-limbic circuits leading to emotional and cognitive dysfunction, etc. Mood stabilizing agent lithium is recommended for the prophylaxis and treatment of BD. High dose of lithium causes adverse side effects such as nausea, tremors of hand and foot, dulled personality, polydipsia, polyurea, weight gain, lethargy, loss of memory and emotions, mental confusion, and delirium. Prolonged use of lithium may damage the thyroid and kidneys. It is essential to get blood levels of lithium checked periodically in order to ensure that the level does not exceed the safe range. Vanadium is a trace element present in our diet. Elevated vanadium level of bipolar patients causes depression and mania. Lowering level of vanadium may improve BD. Lithium may antagonize and lower vanadium level. Vitamin C detoxifies vanadium and protects the body from toxic effect of vanadium. It has been suggested that a high dose (more than 1 g/day) of vitamin C decreases symptoms of bipolar patients (Lakhan and Vieira 2008). Studies indicate that excess acetylcholine receptors of bipolar patients contribute to the genesis of depression and mania. This is supported by the finding of reduced muscarinic acetylcholine receptor binding in subject with BD (Cannon et al. 2006). Glutamatergic dysfunction (Li et al. 2019) hyperdopaminergic function (Berk et al. 2007) and reduced expression of serotonin (Meyer et al. 2009) contribute to the genesis of depression and mania. It may be mentioned that attenuated BDNF contributes to the pathogenesis of BD (Lin 2009). Studies have also shown that taurine prevents the effects of excess acetylcholine. High intake of omega-3 fatty acids reduces depressive and manic symptoms. Deficiencies of amino acids (phenylalanine/tyrosine, tryptophan, methionine/choline, and cysteine/taurine) and vitamins (B_{12}, folate, B_6, and ascorbate) may aggravate BD. The combination of macronutrients and micronutrients, as mentioned above, omega-3 fatty acids and dietary intake of lithium will reduce the symptoms of bipolar patients.

10.5   Schizophrenia and Nutrition

Schizophrenia is characterized by auditory/visual hallucination, delusions, thought blocking, speech disorders, wrong perception of reality and paranoia, lack of relation between thoughts and feelings. An impaired synthesis of serotonin in the brain, glycine deficiency, and omega-3 fatty acid deficiency has been suggested in the pathogenesis of schizophrenia. Glutamate is an excitatory neurotransmitter in the human brain. Dysfunction of glutamatergic neurotransmission contributes to the development of schizophrenia Li et al. 2019. Dietary intake of vitamin D, folate, vitamin B_{12}, zinc, and antioxidants reduce symptoms of schizophrenia (Firth et al. 2017, 2018).

10.6 Obsessive Compulsive Disorder and Nutrition

OCD is characterized by obsession of contamination with dirt and excreta, etc. followed by compulsion, i.e., washing of hands or the whole body repeatedly many times a day. Multiple doubts regarding the locking of doors and counting of money, etc. Checking repeatedly about the doubts is the compulsion. Low serotonin level of the brain has been implicated for the cause of OCD. Normally serotonin released from the serotonergic neurons enters inside the neurons by an active reuptake mechanism. This reuptake decreases the serotonin level. Selective serotonin reuptake inhibitor (SSRI) inhibits neuronal uptake of the mid brain raphe nuclei and increases serotonin level and thus, can treat OCD. Dose more than 100 mg/day of serotonin produces side effects such as dry mouth, nausea, tremor, and dizziness. However, dietary intake of tryptophan increases serotonin level and reduces the symptoms of OCD. Glutamate is implicated in the development of OCD and addiction (Oliver et al. 2015). Increased level of glutamate in the cerebrospinal fluid was reported in patient with OCD.

10.7 Attention Deficit Hyperactivity Disorder and Nutrition

ADHD is characterized by aggressive behavior, quarrelsome behavior, impulsive behavior, lack of attention, and hyperactivity. It has been reported that in children and adolescents with ADHD, low serotonin level aggravated aggressive behavior. Tryptophan supplementation has been shown to decrease social anxiety, quarrelsome behavior, and impulsive behavior in irritable individuals. High consumption of refined sugar and saturated fat can increase the risk of ADHD, whereas high consumption of fruits and vegetables will reduce the symptoms of ADHD (Del-Ponte et al. 2019).

10.8 Oxytocin and Mental Disorders

Oxytocin is a hormone synthesized in the supraoptic and paraventricular nuclei of the hypothalamus, migrates down via the hypothalamo-hypophyseal tract and is stored in the posterior pituitary. It is released from the posterior pituitary for milk ejection and parturition. It has been reported that oxytocin may play an important role in many psychiatric disorders such as autism, schizophrenia, MDD, BD, and AD (Cochran et al. 2013). Further research is needed to confirm its therapeutic role due to conflicting reports.

10.9 Omega-3 Fatty Acids and Mental Disorders

EPA and DHA are two omega-3 fatty acids found in fish oil. EPA can be converted to prostaglandins and leukotrienes that are essential for fluidity and renewal of the membrane. The neural tissues have high levels of omega-3 fatty acids. They may maintain volume of the cerebral cortex and improve mental function. Brain cells require omega-3 fatty acids in order to stabilize mood and emotions (Cardosa et al. 2016). It has been shown that intake of a large amount of DHA reduces the risk of depression, BD, and even schizophrenia (Zarate et al. 2017). Lack of omega-3 fatty acids is observed in patients suffering from mental disorders (Lakhan and Vieira 2008). DHA is the main omega-3 fatty acid in the cortical gray matter. Furthermore, it has been suggested that EPA, DHA, arachidonic acid, and resolvins derived from DHA may play a crucial role for optimal brain function. Severity of ADHD symptoms was correlated with lower omega-3 fatty acid intake (Chang et al. 2016). Diet-induced decrease in brain DHA content causes lowering of serotonin and tryptophan hydroxylase (converts tryptophan to 5-hydroxytryptophan) in the CNS. Thus, diet-induced increase in the brain DHA content prevents various mental disorders. Postmortem study indicates that DHA content of PFC, entorhinal cortex, and amygdala which controls emotional behavior was found to be lower in patients with mental disorders (Schneider et al. 2017).

10.10 Oxidative Damage and Mental Disorders

Oxidative stress/damage and reduced intake of antioxidants contribute to the prevalence of many psychiatric disorders such as schizophrenia, depression, BD, and OCD (Dean et al. 2011; Alici et al. 2016). Increased levels of proinflammatory cytokines (IL-6, TNF-α, and C-reactive protein) are found in patients suffering from depression and other psychiatric disorders (Lavebratt et al. 2017; Fasick et al. 2015; Howren et al. 2009; Krysta et al. 2017). Meta-analyses revealed that increased proinflammatory cytokine levels in depressed individuals, compared to non-depressed individuals (Goldsmith et al. 2016; Dowlati et al. 2010). Oxidative damage, mitochondrial dysfunction and inflammation has been correlated with depressive symptoms (Kaplan et al. 2015). Consumption of anti-inflammatory diet may minimize depression (Tolkein et al. 2018). Consumption of whole grains inhibits proinflammatory cytokines, whereas lower whole grain intake increases proinflammatory cytokines (Hajihashemi et al. 2014; Goletzke et al. 2014; Gaskins et al. 2010; Oddy et al. 2018). Choline, a precursor of glycine, is a nutrient found in eggs, broccoli, cauliflower, and betaine. It is associated with lower proinflammatory cytokines (Detopoulou et al. 2008). Refined sugar can induce oxidative stress and increases the risk of depressive behavior (Sánchez-Villegas et al. 2018). Highly toxic ROS, i.e., free radical, induces oxidative damage. Activation of proinflammatory molecules (cytokines and chemokines) can inflict damage to nuclear DNA, resulting in mitogenesis (mutation). However, it is not yet examined that patients suffering from mental illness are associated with prevalence of cancer.

10.11 Depression in Elderly People

Elderly people generally like to live in isolation. There is a tendency for the elderly to stay inside the house and lead a sedentary life. Isolation syndrome is a disorder, which leads to depression, aggravates multiple forms of stress and is resistant to antidepressant drugs. Interactive and integrated association between depression and malnutrition is prevalent among geriatric population. Anorexia, poor dentition, difficulty in chewing and swallowing, impaired taste and smell sensation lead to malnutrition. Furthermore, atrophic gastroenteritis due to aging leads to anorexia, malabsorption, and consequently malnutrition. Malnutrition predisposes mental disturbances such as depression, agitation, and irritability (Al-Rasheed et al. 2018). Adequate carbohydrates, proteins, vitamins, antioxidants, zinc, calcium, and lithium may prevent depression. Depression among geriatric population is found to be more prevalent in females than males. Serotonin release from serotonergic fibers of the raphe nuclei of the midbrain is impaired (perhaps due to indoor living away from bright light), leading to depression, change in appetite, sleep disturbances, and cognitive dysfunction. Depression can be prevented by social interaction and meditation. Furthermore, nutritional therapies will be useful to prevent mental illness. Antioxidants are needed to decrease oxidative stress (Hedden and Gabrieli 2004).

10.12 Exercise and Mental Functions

Exercise increases BDNF, which stimulates the growth of new neurons and synapses. BDNF is active in the cerebral cortex and improves mental functions. Exercise prevents the decline of loss of volume of cerebral cortex and thus prevents the decline of intellectual and mental functions. It is well documented that exercise increases the liberation of norepinephrine, which will try to rectify anxiety disorder. Exercise increases β-endorphin level, which will stimulate elevation of mood. Exercise increases blood circulation of the brain and stimulates function of the limbic system, which improves motivation and mood. Exercise improves cognitive function and distracts mind that draws away from negative thought, anxiety, and depression. Physical activity benefits brain function and may prevent or delay neurodegeneration associated with mental health in humans (Duzel et al. 2016). Furthermore, exercise benefits memory, executive function and maintains neural gray and white matter volume, which improves motivation and mood (Duzel et al. 2016; Voss et al. 2013). Yoga, meditation, and pranayama will be useful to alleviate mood. Pranayama (deep inspiration through one nostril and hold breath for few seconds followed by deep expiration through another nostril) for about 15 min daily will provide oxygen to the brain and will support the functions of neurons as every cell is an oxygen sensor.

10.13 Diet and Mental Disorders (Fig. 10.1)

The Mediterranean and the DASH diets were found to be associated with lower risk of depression (Sánchez-Villegas et al. 2009; Jacka et al. 2010; Akbaraly et al. 2009; Torres and Nowson 2012; Khayyatzadeh et al. 2018; Adjibade et al. 2018). Indeed, a number of studies indicate that daily consumption of green and yellow vegetables (Tanaka and Hashimoto 2019), legumes, fruits, vegetables, and nuts (Daneshzad and Azadbakht 2018; Grases et al. 2019; Lassale et al. 2019; Molendijk et al. 2018) leads to lower risk of depressive symptoms. Conversely, consumption of excessive amount of sugar and sugar-sweetened beverages such as soft drinks and processed food is associated with depression risk and mental stress (Shi et al. 2010; Rucklidge and Kaplan 2016; Sánchez-Villegas et al. 2018; Pan et al. 2011). The patients suffering from mental illness desire to eat particularly proinflammatory-induced diets (for example, processed and junk food) and avoid anti-inflammatory food (fruits, vegetables, etc.), inviting inflammation (Papier et al. 2015). It may be noted that methionine is present in egg, meat, fish, sesame seed, cereal grains, etc. Diet containing methionine produces SAM, a substance which facilitates the synthesis of neurotransmitters of the brain by providing methyl group (Lakhan and Vieira 2008). Tyrosine-containing food (present in chicken, fish, milk, yoghurt, cottage cheese, and peanut) can augment arousal and alertness through the activation of reticular activating system. Legumes are rich sources of tryptophan and magnesium. Magnesium can help to support recovery in depressed patients (Lakhan and Vieira

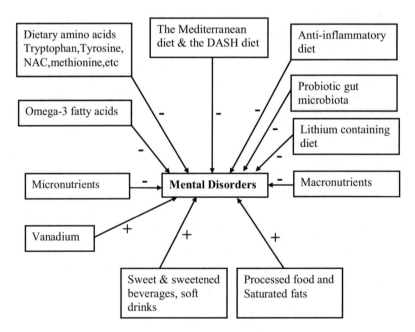

Fig. 10.1 Dietary intake facilitating or inhibiting symptoms of mental disorders. − indicates inhibition, + indicates facilitation, NAC *N*-acetylcysteine

2008). Ripe bananas contain serotonin, DA, and NE. Dietary intake of lithium may reduce depressive and manic symptoms of bipolar disorder. Foods rich in lithium are all kinds of dairy products, including natural milk, sugarcane, seaweed, eggs, tomatoes, cucumbers, potatoes, lemons, spices, crude salts, rock salts, and grains.

Probiotic gut microbiota and fermented food can reduce proinflammatory cytokines and oxidative stress and thus minimize depressive and mood disorders. Probiotic bacteria produce GABA from glutamate and utilize tryptophan for the synthesis of serotonin. Thus, probiotic gut microbiota has profound influence on mental health (Nasr 2018).

10.13.1   Role of Carbohydrates

Carbohydrates-rich foods release insulin, which facilitates the entry of glucose into the cells and increases energy and body activity. Also, carbohydrate consumption triggers tryptophan release, which is essential for the synthesis of serotonin. Low carbohydrate diet on the other hand is associated with high risk of depression (Rao and Asha 2008). If the carbohydrate consumption is less 100 g/day, cognitive function will be impaired due to ketosis.

10.13.2   Role of Proteins

Dietary protein provides amino acids, tyrosine and tryptophan. Tyrosine is required for the synthesis of dopamine, whereas tryptophan is required for the synthesis of serotonin. Both dopamine and serotonin are concerned with the pathogenesis of depression.

10.13.3   Role of Lipids

Brain lipids are essential for the localization and functioning of proteins in the cell membrane of neurons. Lipids (phospholipid fatty acids, sphingolipids, and endocannabinoids) represent preventive interventions for the psychiatric disorders, including drug addiction. Nutrition, mood, and psychological dysfunction can influence the lipid composition of the brain, which, in turn, controls behavioral activity and mental disorders. Lipids can prevent the severity of mental disorders by nutritional intervention (Schneider et al. 2017) (Fig. 10.1).

10.14 Effect of Bright Light on Mental Disorders

Exposure to bright light is very essential to reduce symptoms of mental illness. Impulse from the retina projects to the serotonergic fibers originated from the raphe nucleus of the midbrain and releases serotonin. Serotonin plays an important role in neurogenesis, neural repair, and mood (Vivar and van Praag 2017).

10.15 Summary

Nutritional supplements may reduce the severity of mental illness. Administration of glutamate, phenylalanine/tyrosine, and tryptophan will elevate mood in depressed patients and may alleviate anxiety disorder. Role of vitamin C to detoxify vanadium and thus to improve bipolar disorder is highlighted. Exercise has global effect on mental health. Serotonin plays an important role in preventing the severity of bipolar disorder, schizophrenia, anxiety disorder, OCD and ADHD. Serotonin plays an important role for neurogenesis and alleviation of mood. The role of GABA, glutamatergic neurotransmitter, and mono-aminergic neurotransmitters in pathophysiology of mental disorders has been discussed. The role of oxytocin in psychiatric disorders should be further examined.

References

Adjibade M et al (2018) Prospective association between adherence to the Mediterranean diet and risk of depressive symptoms in the French SU.VIMAX cohort. Eur J Nutr 57(3):1225–1235

Akbaraly TN et al (2009) Dietary pattern and depressive symptoms in middle age. Br J Psychiatry 195:408–413

Alici D et al (2016) Evaluation of oxidative metabolism and oxidative DNA damage in patients with obsessive-compulsive disorder. Psychiatry Clin Neurosci 70(2):109–115

Al-Rasheed R et al (2018) Malnutrition in elderly and its relation to depression. Int J Community Med Public Health 5:2156–2160

Berk M et al (2007) Dopamine dysregulation syndrome: implications for a dopamine hypothesis of bipolar disorder. Acta Psychiatr Scand Suppl 434:41–49

Camfield DA et al (2011) Nutraceuticals in the treatment of obsessive compulsive disorder (OCD): a review of mechanistic and clinical evidence. Prog Neuro-Psychopharmacol Biol Psychiatry 35(4):887–895

Cannon DM et al (2006) Reduced muscarinic type 2 receptor binding in subjects with bipolar disorder. Arch Gen Psychiatry 63:741–747

Cardosa C et al (2016) Dietary DHA and health: cognitive, function, ageing. Nutr Res Rev 29:281–294

Chang J et al (2016) Delay aversion, temporal processing and n-3 fatty acids intake in children with attention deficit hyperactivity disorder (ADHD). Clin Psychol Sci 4:1094–1103

Chen A-T et al (2016) Placebo-controlled augmentation trials of the antioxidant NAC in schizophrenia: a review. Ann Clin Psychiatry 28(3):190–196

Cochran D et al (2013) The role of oxytocin in psychiatric disorders. Harv Rev Psychiatry 21(5):219–247

Daneshzad E, Azadbakht L (2018) A quick review of DASH diet and its effect on mental disorders. JIMC 1(1):45–48

Dean O et al (2011) N-acetylcysteine in psychiatry: current therapeutic evidence and potential mechanisms of action. J Psychiatr Neurosci 36(2):78–86

Deepmala D et al (2015) Clinical trials of N-Acetylcysteine in psychiatry and neurology: a systematic review. Neurosci Behav Rev 55:294–321

Del-Ponte B et al (2019) Dietary patterns and attention deficit/hyperactivity disorder (ADHD): a systematic review and meta-analysis. J Affect Disord 252:160–173

Detopoulou P et al (2008) Dietary choline and betaine intakes in relation to concentrations of inflammatory markers in healthy adults: the ATTICA study. Am J Clin Nutr 87:424–430

Dowlati Y et al (2010) A meta-analysis of cytokines in major depression. Biol Psychiatry 67:446–457

Duzel E et al (2016) Can physical exercise in old age improve memory and hippocampal function? Brain 139:662–673

Fasick V et al (2015) The hippocampus and TNF:common links between chronic pain and depression. Neurosci Behav Rev 53:139–159

Firth J et al (2017) The effects of vitamin and mineral supplementation on symptoms of schizophrenia: a systematic review and meta-analysis. Psychol Med 47:1515–1527

Firth J et al (2018) Nutritional deficiencies and clinical correlates in first-episode psychosis: a systematic review and meta-analysis. Schizophr Bull 44(6):1275–1292

Gaskins AJ et al (2010) Whole grains are associated with serum concentrations of high sensitivity C-reactive protein among premenopausal women. J Nutr 140:1669–1676

Goldsmith DR et al (2016) A meta-analysis of blood cytokine network alterations in psychiatric patients: comparisons between schizophrenia, bipolar disorder and depression. Mol Psychiatry 21:1696–1709

Goletzke J et al (2014) Increased intake of carbohydrate from sources with a higher glycemic index and lower consumption of whole grains during puberty are prospectively associated with higher IL-6 concentrations in younger adulthood among healthy individuals. J Nutr 144:1586–1593

Grases G et al (2019) Possible relation between consumption of different food groups and depression. BMC Psychol 7:14

Hajihashemi P et al (2014) Whole-grain intake favourably affects markers of systematic inflammation in obese children: a randomized controlled crossover clinical trial. Mol Nutr Food Res 58:1301–1308

Hedden T, Gabrieli JD (2004) Insights into the ageing mind: a view from cognitive neuroscience. Nat Rev Neurosci 5(2):87–96

Howren MB et al (2009) Associations of depression with C-reactive protein, IL-1, and IL-6: a meta-analysis. Psychosom Med 71(2):171–186

Jacka FN et al (2010) Associations between diet quality and depressed mood in adolescents: results from the Australian healthy neighbourhood study. Aust N Z J Psychiatry 44:435–442

Kadriu B et al (2019) Glutamatergic neurotransmission: pathway to developing novel rapid-acting antidepressant treatment. Int J Neuropsychopharmacol 22(2):119–135

Kaplan BJ et al (2015) The emerging field of nutritional mental health: inflammation, the microbiome, oxidative stress, and mitochondrial function. Clin Psychol Sci 3:964–980

Khayyatzadeh SS et al (2018) Adherence to a DASH-style diet in relation to depression and aggression in adolescent girls. Psychiatry Res 259:104–109

Krysta K et al (2017) Sleep and inflammatory markers in different psychiatric disorders. J Neural Transm 124:179–186

Lakhan SE, Vieira KF (2008) Nutritional therapies for mental disorders (review). Nutr J 7:1–8

Lassale C et al (2019) Healthy dietary indices and risk of depressive outcomes: a systematic review and meta-analysis of observational studies. Mol Psychiatry 24(7):965–986. https://doi.org/10.1038/s41380-018-0237-8

Lavebratt C et al (2017) Interleukin-6 and depressive symptom severity in response to physical exercise. Psychiatry Res 252:270–276

Li C-T et al (2019) Glutamatergic dysfunction and glutamatergic compounds for major psychiatric disorders: evidence from clinical neuroimaging studies. Front Psychiatry 9:767

Lin PY (2009) State-dependent decrease in levels of brain-derived neurotrophic factor in bipolar disorder: a meta-analytic study. Neurosci Lett 466:139–143

Liu YI et al (2018) Emotional roles of mono-aminergic neurotransmitters in major depressive disorder and anxiety disorders. Front Psychol 9:2201. https://www.frontiersin.org/article/10.3389/fpsyg.2018.02201

Meyer JH et al (2009) Brain monoamine oxidase a binding in major depressive disorder: relationship to selective serotonin reuptake inhibitor treatment, recovery, and recurrence. Arch Gen Psychiatry 66:1304–1312

Minarini A et al (2016) N-acetylcysteine in the treatment of psychiatric disorders: current status and future prospects. Expert Opin Drug Metab Toxicol 13(3):279–292

Molendijk M et al (2018) Diet quality and depression risk: a systematic review and dose-response meta-analysis of prospective studies. J Affect Disord 226:346–354

Monti DA et al (2016) N-acetyl cysteine may support dopamine neurons in Parkinson's disease: preliminary clinical and cell line data. PLoS One 11(6):e0157602

Nasr NF (2018) Psychological impact of probiotics and fermented foods on mental health of human in integrated healthy lifestyle. Int J Curr Microbiol App Sci 7(8):2815–2822

Oddy WH et al (2018) Dietary patterns, body mass index and inflammation: pathways to depression and mental health problems in adolescents. Brain Behav Immun 69:428–439

Oliver G et al (2015) N-acetyl cysteine in the treatment of obsessive compulsive and related disorders: a systematic review. Clin Psychopharmacol Neurosci 13(1):12–24

Ooi SL et al (2018) N-Acetylcysteine for the treatment of psychiatric disorders: a review of current evidence. Biomed Res Int 2018:Article 2, 469, 486, 8p. https://doi.org/10.1155/2018/2469486

Pan X et al (2011) Soft drink and sweet food consumption and suicidal behaviours among Chinese adolescents. Acta Paediatr 100(11):e215–e222

Papier K et al (2015) Stress and dietary behaviour among first-year university students in Australia: sex differences. Nutrition 31:324–330

Patrick RP, Ames BN (2015) Vitamin D and the omega-3-fatty acids control serotonin synthesis and action, part 2: relevance for ADHD, bipolar, schizophrenia and impulsive behaviour. FASEB J 29(6):2207–2222

Rao TSS, Asha MR (2008) Understanding nutrition, depression and mental illnesses. Indian J Psychiatry 50(2):77–82

Rucklidge JJ, Kaplan BJ (2016) Nutrition and mental health. Clin Psychol Sci 4(6):1082–1084

Sánchez-Villegas A et al (2009) Association of the Mediterranean dietary pattern with the incidence of depression: the Seguimiento Universidada de Navarra/University of Navarra follow up (SUN) cohort. Arch Gen Psychiatry 66:1090–1098

Sánchez-Villegas A et al (2018) Added sugars and sugar-sweetened beverage consumption, dietary carbohydrate index and depression risk in the Seguimiento Universidada de Navarra (SUN) project. Br J Nutr 119(2):211–221

Scapagnini G et al (2012) Antioxidants as antidepressants fact or fiction? CNS Drugs 26(6):477–490

Schneider M et al (2017) Lipids in psychiatric disorders and preventive medicine. Neurosci Biobehav Rev 76(Pt B):336–362

Shi Z et al (2010) Soft drink consumption and mental health problems among adults in Australia. Public Health Nutr 13(7):1073–1079

Sigitova E et al (2017) Biological hypotheses and biomarkers of bipolar disorder. Psychiatry Clin Neurosci 71:77–103

Tanaka M, Hashimoto K (2019) Impact of consuming green and yellow vegetables on the depressive symptoms of junior and senior high school students in Japan. PLoS One 14(2):e0211323

Tolkein K et al (2018) An inflammatory diet as a potential intervention for depressive disorders: a systematic review and meta-analysis. Clin Nutr 38(5):2045–2052. https://doi.org/10.1016/j.clnu.2018.11.007

Torres SJ, Nowson CA (2012) A moderate-sodium DASH-type diet improves mood in postmenopausal women. Nutrition 28(9):896–900

Vivar C, van Praag H (2017) Running changes the brain: the long and the short of it. Physiology 32(6):410–424

Voss MW et al (2013) Bridging animal and human models of exercise-induced brain plasticity. Trends Cogn Sci 17:525–544

Zarate R et al (2017) Significance of long chain polyunsaturated fatty acids in human health. Clin Transl Med 6(1):25

Zheng W et al (2018) N-acetylcysteine for major mental disorders: a systematic review and meta-analysis of randomized controlled trials. Acute Psychiatr Scand 137(5):391–400

Further Reading

Korn LE (2016) Nutrition essentials for mental health: a complete guide to the food-mood connection, 1st edn. W.W. Norton, New York

Ooi SL et al (2018) N-Acetylcysteine for the treatment of psychiatric disorders: a review of current evidence. Biomed Res Int 2018:Article 2, 469, 486, 8p. https://doi.org/10.1155/2018/2469486

Sigitova E et al (2017) Biological hypotheses and biomarkers of bipolar disorder. Psychiatry Clin Neurosci 71:77–103

Lifestyle-Related Diseases and Disorders

Abstract

Diabetes mellitus, hypertension, and atherosclerosis are the biggest global health problems and are the leading causes of morbidity and mortality in different population at different age groups, and during pregnancy. According to the American Diabetes Association (ADA), diabetes mellitus is classified into type 1 (due to absolute deficiency of insulin), type 2 (due to insulin resistance), gestational diabetes, and secondary diabetes (ADA, Diabetes Care 41:S13–S27, 2018). Hyperglycemia along with abnormalities of fat and protein metabolism cause various symptoms such as polyphagia, polyuria, polydipsia, weight loss, kussmaul breathing, and drowsiness. Uncontrolled diabetes increases risk of various complications such as nephropathy, neuropathy, retinopathy, hypovolemia, CIHD, and peripheral vascular diseases. Diet plans along with regular daily exercise will slow or prevent the progression to the development of complications of diabetes. Persistent hypertension is one of the risk factors for myocardial infarction due to CIHD, cerebral stroke, retinal hemorrhage, papilloedema, progressive renal failure, and left ventricular failure. DASH diet and regular aerobic exercise are the preventive measures to slow or even prevent the development of hypertension. Weight reduction (BMI < 25 kg/m^2), total abstinence from alcohol and smoking, and meditation to minimize stress are other preventive measures. Atherosclerosis is due to infiltration of cholesterol into macrophages of connective tissue of the arterial walls leading to the formation of foam cells. Foam cells release growth factors and inflammatory cytokines, which cause proliferation of smooth muscle, thickening of elastin, and ultimately narrowing of the arterial wall. Atherosclerosis predisposes to myocardial infarction, cerebral stroke, and hypertension. Diets containing low cholesterol, low saturated fat and avoidance of simple sugar and trans fat are the measures to prevent atherosclerosis. Total abstinence from smoking and alcohol is necessary. Daily exercise is advisable.

© Springer Nature Singapore Pte Ltd. 2019
K. Chakrabarty, A. S. Chakrabarty, *Textbook of Nutrition in Health and Disease*,
https://doi.org/10.1007/978-981-15-0962-9_11

Keywords
Diabetes mellitus · Persistent hypertension · Atherosclerosis · Dietary plan ·
Exercise

11.1 Diabetes Mellitus and Nutrition

Before discussion of diabetes mellitus, the physio-biochemical actions of insulin
are summarized in brief.

1. Carbohydrate Metabolism.
   (a) Insulin facilitates the entry of glucose into the cells by increasing the number
       of glucose transporters of the cell membrane and thereby facilitates the uti-
       lization of glucose by the cells.
   (b) Insulin enhances glycolysis by activating several enzymes of glycolytic
       pathways, for example, glucokinase, phosphofructokinase, pyruvate kinase,
       and enhances the breakdown of glucose into CO_2 and H_2O via citric acid
       cycle and electron transport chain. Moreover, insulin inhibits glucose-6-
       phosphatase present only in the liver (normally glucose-6-PO_4 is converted
       to glucose by glucose-6-phosphatase), resulting in the retention of glucose
       inside the liver. By activating glycolysis and by inhibiting glucose-6-
       phosphatase, insulin decreases release of glucose into the plasma.
   (c) Insulin enhances the synthesis of glycogen in the liver and muscles by acti-
       vating glycogen synthase as well as by inhibiting the breakdown of glycogen
       by inhibiting phosphorylase. It enhances glycogen storage.
   (d) Insulin inhibits gluconeogenesis by inhibiting phosphoenolpyruvate
       carboxykinase.
2. Fat Metabolism
       Insulin promotes lipogenesis by inhibiting hormone-sensitive lipase and by
   activating lipoprotein lipase and ultimately causes increased deposition of triglycerides in the adipose tissues with the inhibition of ketogenesis.
3. Protein Metabolism
       It enhances protein synthesis and inhibits protein catabolism.
       Therefore, insulin is a storage hormone of carbohydrate, fat, and protein
   metabolism (Fig. 11.1).

11.1.1 Diabetes Mellitus

Adverse consequences of insulin deficiency cause diabetes mellitus (Fig. 11.2).
Effects of insulin deficiency are aggravated by unchecked activity of glucagon
secreted by A or α cells of the islets of pancreas. Normally insulin inhibits glucagon
secretion. Glucagon is a hyperglycemic hormone, whereas insulin is a hypoglycemic hormone. Glucagon increases gluconeogenesis and glycogenolysis. In addition,
glucagon is lipolytic and ketogenic.

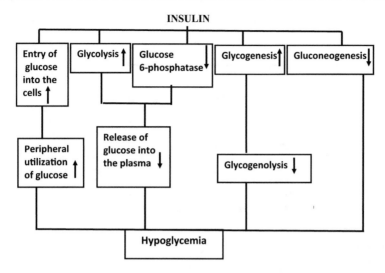

Fig. 11.1 Hypoglycemic effect of insulin. ↑ Indicates facilitation and ↓ indicates inhibition

11.1.2 Characteristic Features of Uncontrolled Diabetes Mellitus

1. The hallmark of uncontrolled diabetes mellitus is hyperglycemia caused by
   (a) Reduced entry of glucose into the cells.
   (b) Reduced peripheral utilization of glucose.
   (c) Reduced glycolysis.
   (d) Reduced glycogenesis.
   (e) Increased glycogenolysis.
   (f) Increased gluconeogenesis.
   (g) Increased release of glucose into the plasma.
      Hyperglycemia occurs, but glucose cannot be utilized, leading to excess extracellular glucose and loss of intracellular glucose ("starvation in the midst of plenty") (Ganong 2003). Hyperglycemia causes glycosuria (presence of glucose in urine). Normally 100% glucose is reabsorbed by proximal tubules of the kidneys. When plasma glucose (venous blood) exceeds the renal threshold of glucose (180 mg/100 mL), i.e., exceeds the transport maximum of glucose (Tmg), glucose cannot be absorbed anymore and appears in the urine. Glycosuria causes polyuria (frequent and excessive urination). Glucose holds water and is excreted, causing osmotic diuresis. Osmotic diuresis causes dehydration, leading to increased osmolality surrounding the thirst center of hypothalamus. Increased osmolality of ECF and dryness of mouth due to dehydration stimulate thirst center, resulting in polydipsia (excessive thirst). Polyphagia is excessive food intake due to excessive hunger. Glucostats of the satiety center (ventromedial nucleus of hypothalamus) are inhibited due to incomplete glucose utilization, leading to unchecked activity of the feeding center (lateral hypothalamus). Increased activity of the feeding center leads to excessive hunger.

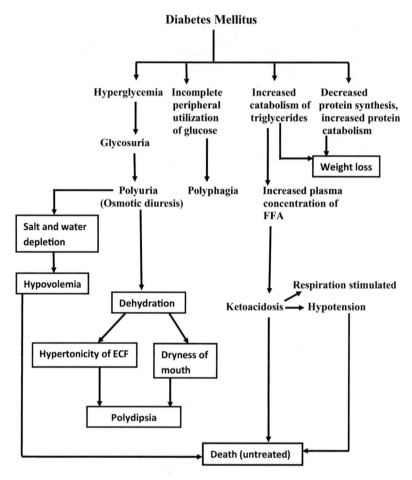

Fig. 11.2 Effects of untreated diabetes mellitus on carbohydrate, fat, and protein metabolism, showing a typical example of metabolic disease

2. The next hallmark is abnormalities in fat metabolism.
   (a) Decreased lipogenesis and increased lipolysis.
   (b) Decreased deposition of triglycerides in the adipose tissues.
   (c) Increased triglyceride catabolism.
       These lead to an increase in plasma concentration of triglycerides and FFA.
3. The next hallmark is abnormalities in protein metabolism characterized by decreased protein synthesis, increased plasma amino acids, increased protein catabolism, and negative nitrogen balance.
4. Weight loss in spite of polyphagia is due to enhanced protein and fat catabolism. Loss of calories due to glycosuria also contributes to loss of body weight.
5. Hypovolemia occurs due to loss of water as well as due to increased excretion of Na$^+$ and K$^+$ by the kidneys.

6. The final and fatal hallmark is ketosis (metabolic acidosis): Excess acetyl-CoA due to the oxidation of fatty acids causes excessive formation of ketone bodies, resulting in ketosis. Metabolic acidosis may be further aggravated by lactic acidosis if the tissues are hypoxic. Acidosis causes fall in BP due to vasodilatation and also causes kussmaul breathing (slow deep breathing).

11.1.3 Complications

Before the development of ketoacidosis, death may occur due to the following complications.

1. Excessive formation of acetyl-CoA increases the synthesis of cholesterol. Hypercholesterolemia occurs also due to increased plasma LDL. This may lead to atherosclerosis in vital organs (heart, cerebrum, etc.). The patient may die due to myocardial infarction or cerebral stroke. GLP-1RAs (glucagon-like peptide-1 receptor agonists) and SGLT-2 (sodium-glucose cotransporter-2) inhibitors may benefit patients from cardiovascular disease (Schmidt 2019; Pradhan et al. 2019).
2. Diabetic nephropathy, resulting in renal failure and uremia.
3. Diabetic retinopathy leads to retinal hemorrhage and blindness, diabetic polyneuropathy (burning sensation and pain in the legs), necrosis, infection, and gangrene of the feet (amputation may be indicated). Vitamin D deficiency is common in patients with diabetes. An inverse association between vitamin D level and cardiovascular disease (Wang et al. 2008) as well as retinopathy (Parveen et al. 2017) was reported. It has been suggested that vitamin D supplementation may prevent the risk of diabetic complications. Vascular dysfunction and impaired angiogenesis due to endothelial dysfunction was reported in diabetes (Sawada and Arany 2017).

11.1.4 Types of Diabetes Mellitus (Fig. 11.3; Table 11.1) (ADA 2018)

1. Type 1 diabetes or IDDM (Insulin-dependent diabetes mellitus) or juvenile diabetes. It occurs below the age of 40. It develops rapidly. The patient is usually underweight. Absolute deficiency of insulin is due to autoimmune destruction of the B cells. It may be triggered by viral infection or genetic factor, allowing the B cells to recognize as "non-self." It is due to T-lymphocyte infiltration and T-lymphocyte-mediated disease, leading to destruction of B cells. Main abnormality is on chromosome 6. It is treated by insulin therapy. In the absence of insulin therapy, the patient develops ketoacidosis, coma, and death.
2. Type 2 diabetes or NIDDM (Non-insulin-dependent diabetes mellitus). It occurs after the age of 40. It develops gradually. Genetic predisposition is stronger than

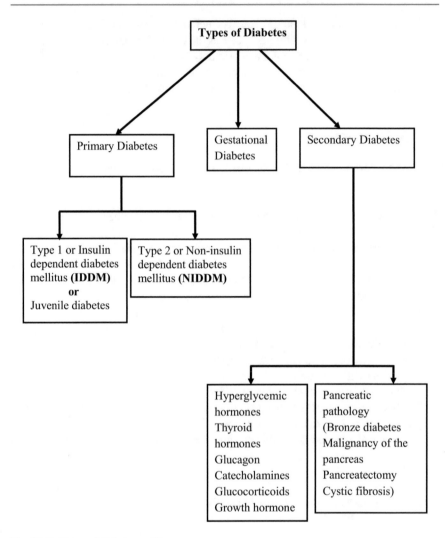

Fig. 11.3 Types of diabetes mellitus

Type 1 diabetes as the concordance rate is very high in identical twins. It is due to insulin resistance. It is associated with normal B cell morphology, hyperglycemia, and normal or even high plasma insulin level. Hyperglycemia stimulates B cells to secrete insulin until the B cell reserve is exhausted, leading to progressive loss of B cell insulin secretion. Insulin resistance may be due to genetic deficiency of glucokinase or due to abnormal insulin receptor or due to defect of glucose transporters. It is associated with obesity (BMI > 30) as obesity increases insulin resistance. High intake of saturated fats and sugar, low intake of dietary fibers, and high glycemic index foods cause increased risk of developing NIDDM. Excessive consumption of saturated fat and sugar may cause high

Table 11.1 Contrast between the two types of diabetes mellitus

	Type 1 (IDDM)	Type 2 (NIDDM)
1. Age of onset	During childhood and less than 40 years age. Juvenile onset	After 40 years age (increase in frequency with age)
2. Body weight	Normal or may be underweight	Obese
3. Morphology of beta cells	Destroyed or partial morphology	Normal morphology
4. Genetic factor component	Moderate	Strong
5. Family history of diabetes	Uncommon	Common
6. Plasma insulin	Very low or even absent	Normal to high
7. Polyphagia, polyuria, and polydipsia	Severe	Moderate
8. Incidence of ketoacidosis	High	Rare
9. Ketonuria	Present	Absent
10. Insulin therapy	Always essential	Not required

energy intake. High energy intake and positive nitrogen balance will contribute to the development of obesity and insulin resistance, which may be associated with an increased risk of NIDDM. Insulin resistance decreases with the decrease of body weight. Obesity may be a promoter of type 2 diabetes mellitus and may not be the cause of type 2 diabetes. Insulin resistance may lead to abdominal obesity, leading to hyperlipidemia (Malone and Hansen 2019). NIDDM of young people (under 25 years) is designated as Maturity Onset Diabetes (MOD). Unhealthy eating habits lead to progression of type 2 diabetes mellitus (Pot et al. 2019). The Mediterranean diet (high in vegetables, fruits, whole grains, legumes, nuts, and moderate amount of yoghurt) with the avoidance of processed red meat, refined grains, and sugar-sweetened drinks may be recommended for prevention and treatment of type 2 diabetes mellitus (Forouhi et al. 2018). Hypothyroidism may be associated with type 2 diabetes (Biondi et al. 2019). MOD is a monogenic form of diabetes characterized by autosomal dominant mode of inheritance. MOD patients exhibit mild diabetic symptoms. MOD can be managed by dietary interventions or by minimal dose of oral antidiabetic drug. Diagnosis is based on the screening of three marker genes like hepatocyte nuclear factor 1 alpha (HNF1α), hepatocyte nuclear factor 4 alpha (HNF4α), and glucokinase (Firdous et al. 2018).

Metabolic Syndrome or Syndrome X

In this syndrome, obesity (fat deposition, especially around the waist) is associated with hyperinsulinemia due to insulin resistance and hypertriglyceridemia due to low HDL and high LDL, leading to the development of atherosclerosis.

3. Gestational diabetes

It refers to hyperglycemia of pregnant women, having an inherited predisposition towards developing diabetes. It is common in obese women. Insulin resistance mainly occurs, especially in second half of pregnancy. Renal threshold for glycosuria is reduced. A family history of diabetes is common.

Gestational diabetes leads to ultimately type 2 diabetes mellitus in the mother. WHO report indicates that all pregnant women having impaired glucose tolerance after OGTT should be considered as having gestational diabetes. It has been suggested that vitamin D deficiency is a risk factor for gestational diabetes. Diabetic pregnant women are associated with stillbirths, neonatal deaths, and premature delivery. Congenital malformation is common. Maternal glucose crosses the placenta and stimulates fetal insulin production and therefore stimulates fetal growth resulting in abnormally large size of fetus (macrosomia). This may complicate delivery, resulting in cesarean delivery.

4. Secondary diabetes due to hyperglycemic hormones (antagonistic hypoglycemic insulin hormone).
   (a) Glucagon secreted by α cell of islets of the pancreas elevates blood glucose due to glycogenolysis and gluconeogenesis.
   (b) Catecholamines (epinephrine and norepinephrine) of adrenal medulla produce hyperglycemia as it causes incomplete peripheral glucose utilization, glycogenolysis as well as increased output of glucose from the liver.
   (c) Thyroid hormones increase absorption of glucose from the intestine and potentiate the actions of catecholamines on blood glucose.
   (d) Growth hormone of anterior pituitary produces hyperglycemia by increasing output of glucose from the liver and by incomplete peripheral utilization of glucose.
   (e) Glucocorticoids of the adrenal cortex produce hyperglycemia by increasing gluconeogenesis and by decreasing peripheral utilization of glucose.
5. Bronze diabetes (see Chap. 4; iron toxicity).

11.1.5 Genetic Factors of Type 1 and Type 2 Diabetes Mellitus

11.1.5.1   Type 1 Diabetes Mellitus
IDDM1 locus contains HLA genes on chromosome 6 and is linked with the autoimmune destruction of beta cells, whereas IDDM2 locus contains insulin gene located on chromosome 11 and is implicated with risk of developing Type 1 diabetes mellitus.

11.1.5.2   Type 2 Diabetes Mellitus
Long arm of chromosome 2 having a gene is correlated with increased incidence of type 2 diabetes mellitus. Type 2 involves multiple gene interactions. A genetic variant of CAPN10 (calpain 10) is implicated in causing Type 2 diabetes mellitus (Turner et al. 2005). Calpain 10 is a calcium-activated enzyme.

11.1.6 Diagnosis

1. Urine examination for the presence of glucose in the urine.
2. Oral glucose tolerance test (Fig. 11.4).
3. Glycosylated hemoglobin A_{1C} (HbA$_{1C}$). Glucose is attached to the valine in each β chain of hemoglobin. Normal level is about 3–5%. In diabetes, the value is increased to 6–15%.

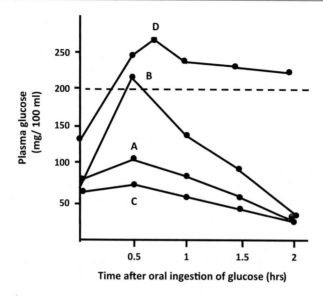

Fig. 11.4 Oral glucose tolerance test (OGTT): The subject must take rest and relax for at least half an hour before the test. He should miss the morning exercise and should avoid smoking. A sample of venous blood is taken to measure the fasting plasma glucose level. 75 g glucose dissolved in 300 mL of water is given to drink orally. Thereafter, samples of blood are collected at half hourly interval for 2–2.5 h and their glucose content is measured. *A = Normal*: Fasting plasma level is between 70 and 110 mg/100 mL. Following the glucose intake the level reaches a peak within 1 h. (below 140 mg) and comes back to the normal by 2–2.5 h. No value is greater than 200 mg. *B = Alimentary glycosuria*: Glycosuria occurs due to rapid gastric emptying after gastrectomy and also due to increased absorption of glucose in hyperthyroidism. After glucose intake plasma glucose level rises more than 200 mg, but comes back to normal within 2 h. *C = Subnormal type*: It is found due to malabsorption or hypothyroidism. Fasting glucose level and peak level are much lower than normal. *D = Diagnostic criteria for diabetes mellitus*: Fasting plasma glucose level is more than 140 mg/100 mL. Peak value and 2 h value are more than 200 mg/100 mL

4. Determination of plasma immunoreactive insulin. Insulin level is very low in Type 1 diabetes, whereas it is normal or even high in Type 2 diabetes.

Note: (1) False type of glucose tolerance test may occur due to increased secretion of epinephrine. Stress before the test or early morning exercise increases the secretion of epinephrine, leading to hyperglycemia. (2) Impaired glucose tolerance is diagnosed when the glucose level of hyperglycemic hormones may impair glucose tolerance, for example, excessive secretion of glucocorticoids (Cushing's syndrome) or due to gestational diabetes.

Glycemic Index (GI)

$$GI = \frac{\text{Area under the 2 h glucose tolerance curve (GTC) following the ingestion of 50 g test meal}}{\text{Area under GTC following ingestion of 50 g of glucose (reference meal)}}$$

This formula is multiplied by 100. Diet for diabetic patients is prepared from GI values.

Carbohydrates that can be digested to glucose [almost all monosaccharides and disaccharides, maltodextrins (oligosaccharides), and polysaccharides (digestible starch)] are referred to as glycemic carbohydrates. Resistant starch and oligosaccharides except maltodextrins are referred to as non-glycemic carbohydrates. Non-glycemic carbohydrates pass to the large intestine where they are fermented.

11.1.7 GI Range of Certain Food Items

11.1.7.1   Low GI 49
Most fruits except bananas and pineapples, all vegetables, especially green leafy vegetables, fructose, milk, curd, beans, legumes, pulses, barley, soya beans, lentils, and kidney beans.

11.1.7.2   Medium GI 50–69
Bananas, pineapples, parboiled brown rice, whole wheat products, lactose, brown bread, tomatoes, bran, green peas, orange, and pear.

11.1.7.3   High GI 70 and Above
Corn flakes, white rice (polished), white bread, watermelon, potatoes including chips, mango, sweet corn, ice cream, sweet potato, popcorn, honey, glucose, and sucrose.

GI of starch (complex carbohydrate) is lesser than glucose (simple sugar) as it takes more time for starch to be digested and converted to glucose. Similarly sucrose will have lesser GI compared with glucose. Fructose will have low GI as it is changed to glucose in the liver and used in the body.

The presence of dietary fibers will slow the gastric emptying time and will lower GI. One can take vegetables as desired as GI of vegetables is low due to low carbohydrate content with high dietary fibers.

11.1.8 Prevention

1. Diabetic diet.
    (a) Before preparing any dietary regimen, energy requirement depending on age, sex, weight, and activity should be decided. Energy intake of overweight middle aged person should not exceed 1500 kcal/day. BMI should be maintained within the range of 20–25.
    (b) Low energy weight reducing diet should be prescribed.
    (c) Frequent small meals should be taken. The patient should take fixed quantity of meal at fixed times.
    (d) Complex carbohydrates such as starch should be taken as they are not rapidly absorbed. Simple sugars, for example, glucose, should be avoided as they are rapidly absorbed to produce sudden high rise of blood glucose. Simple sugars have higher glycemic index compared to complex sugars. The

patient should be advised to take foods having low glycemic index, for example, vegetables, legumes, milk, peas, soya beans, kidney beans, etc. Parboiled brown rice should be more preferred than polished white rice. The intake of carbohydrate should be about 200–250 g/day or should not be more than 50–60% of total energy, depending on the severity of diabetes. Intake of carbohydrate should not be less than 150 g/day. Otherwise deficiency of carbohydrate will lead to catabolism of fat and protein (protein-sparing action). Increased catabolism of fat may lead to ketosis.

(e) Dietary fiber (>25 g/day) having negligible calories must be taken to prevent absorption of glucose from the intestine. Cereal-based foods should be whole grain.

(f) Protein intake should be about 60 g/day (~20% of total energy). Protein intake will lower blood glucose as amino acids stimulate the secretion of insulin.

(g) Fat intake should not be more than 50 g/day or <30% of total energy. Diabetic patients are more prone to develop atherosclerosis. Saturated fat like ghee, butter, animal fat, and cholesterol-rich diet must be reduced (saturated and trans fat <8% of energy, PUFAs $n = 3$ <8% of energy, MUFAs <8% of energy, cholesterol <300 mg). Dietary fibers decrease plasma cholesterol level by preventing absorption of cholesterol from the intestine. Food prepared by grilling, baking, and steaming should be more preferred than fried food. Processed food, refined flour, and trans fat present in fried food should be avoided.

(h) Salt intake should be reduced in hypertensive diabetic patients (less than 3 g/day).

(i) Omega-3 fatty acids should be preferred.

(j) Smoking must be avoided. Antioxidants should be taken.

(k) Vegetables, fruits, legumes, and cereal-derived food should be consumed.

(l) Alcohol consumption should be avoided, since it may cause hypoglycemia and may cause serious consequences, particularly in patients taking insulin or antidiabetic drugs.

An example of diabetic diet which will provide <1800 kcal/day is given below.

One cup of morning green tea without sugar and one biscuit.

Breakfast: One slice of toasted brown bread with one boiled egg (without yolk) and a glass of double toned milk.

Midmorning: One cup of green tea without sugar and one biscuit.

Lunch: Tomato soup, 30 g of rice or one chappati (Indian flat bread), pulses 20 g and vegetables 100 g (pulses and vegetables to be cooked in poly or monounsaturated fatty acid), curd 250 mL, one fruit (for example, orange/apple), and salads (carrot, cucumber, radish, etc.).

Evening: One cup of tea or coffee without sugar and one biscuit or one slice toasted brown bread.

Dinner: Orange juice or lemon juice, one chappati, pulses 20 g and vegetables 100 g, curd 250 mL, one fruit, fish/chicken/roasted paneer 30 g, and salads.

Bed time: One glass of double toned milk without sugar.

2. Exercise

Regular daily exercise will slow or prevent or minimize the progression to the development of diabetes mellitus and also will reduce the requirement of insulin and oral hypoglycemic drugs. Exercise increases the number of insulin-independent glucose transporters of the cell membrane and thereby increases the entry of glucose inside the cells in the absence of insulin. Exercise because of anaerobiosis increases the permeability of cell membrane for the entry of glucose inside the cell in the absence of insulin. It improves blood lipid profile. Severe exercise should be avoided to prevent lactic acidosis as well as to prevent the occurrence of myocardial infarction.

11.1.9　Ketogenic Diets

Very low carbohydrate diets are ketogenic diets, usually obtained from vegetables. Very low intake of carbohydrate will lead to the catabolism of fat and protein. Increased catabolism of fat leads to ketosis. Ketone bodies are the main source of energy alternative to glucose for the brain and other parts of the body. The possible role of ketogenic diet in the management of diabetes is debatable. Although ketogenic diets cause marked reduction of body weight with anorexia, it may not be advisable to recommend this diet for diabetic patients, especially type 2 diabetes with obesity. Protein is burnt or exhausted resulting in the prevention of gluconeogenesis. Prevention of gluconeogenesis causes severe hypoglycemia, which is detrimental. Ketogenic diet may have adverse effects (Bolla et al. 2019). Note: Features of diabetes and diabetic ketosis are shown in Tables 11.2 and 11.3.

Table 11.2 Characteristic features of diabetes mellitus

1. Polyphagia (predilection for sweet food)
2. Polyuria
3. Polydipsia
4. Weight loss despite polyphagia
5. Hyperglycemia
6. Glycosuria
7. Increased plasma concentration of triglycerides, FFA, and cholesterol
8. Ketoacidosis followed by coma

Table 11.3 Clinical signs of diabetic ketosis

1. Kussmaul breathing
2. Acetone odor of expiration
3. Hypotension
4. Low body temperature
5. Confusion
6. Drowsiness
7. Coma

11.2 Hypertension and Nutrition

Blood pressure is classified as systolic and diastolic blood pressure. Cardiac cycle consists of a period of contraction called systole followed by a period of relaxation called diastole. Blood pressure is the pressure exerted by the blood on the walls of the blood vessels. Systolic pressure is the maximum pressure during systole, i.e., during a heart beat, whereas diastolic pressure is the lowest pressure during diastole, i.e., between heart beats. When heart pumps blood into the arterial system, it expands because of elasticity. After systole, it recoils back and maintains diastolic pressure. Diastolic pressure also depends upon the peripheral resistance, i.e., resistance to the flow of blood through the peripheral vessels. Diastolic pressure is, therefore, determined by (1) elastic recoil of the vessels and (2) peripheral resistance. Peripheral resistance lies mainly in the arterioles. It depends on the tone of arterioles. Systolic pressure depends on cardiac output and elasticity. Whenever cardiac output is more or elasticity is less, there will be more pressure during systole.

How high is high blood pressure? (Chobanian et al. 2003)

	Systolic pressure (mmHg)	Diastolic pressure (mmHg)
Normal (under complete physical and mental rest)	90–120	60–80
Prehypertension	120–140	80–90
Mild hypertension (stage 1)	140–160	90–100
Moderate hypertension (stage 2)	160–180	100–110
Severe hypertension	More than 180	More than 110

Hypertension is classified as (1) essential or primary hypertension, (2) secondary hypertension, and (3) resistant hypertension.

11.2.1 Essential Hypertension

It is the common type of hypertension, affecting about 90–95% of hypertensive patients. Cause is not known. The following factors may contribute to the development of essential hypertension: (1) smoking, (2) sedentary lifestyle, (3) stress, (4) obesity, (5) hypersensitivity to sodium, (6) hypokalemia, (7) hyperactivity of sympathetic nervous system, leading to the liberation of norepinephrine, (8) aging process, (9) family history (which may be due to inherited genetic mutation), (10) metabolic syndrome or syndrome X, and (11) resetting of arterial baroreceptor (carotid sinus) at a higher level.

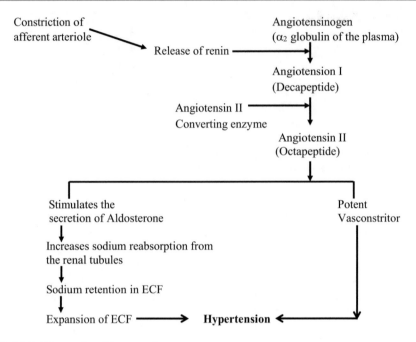

Fig. 11.5 Causes of renal hypertension

11.2.2 Secondary Hypertension

Cause is known. It affects 5–10% of hypertensive patients.

1. Renal hypertension (Fig. 11.5)

    Renin is a protease and is secreted by juxtaglomerular cells present in the afferent arterioles before entering the glomeruli. Narrowing of the afferent arterioles due to atherosclerosis or due to renal disease increases the secretion of renin. Furthermore, "dysfunctional adipose tissues" contribute to secretion of the RAAS (Schütten et al. 2017). Hypertension produces ROS, which causes oxidative damage through the activation of proinflammatory molecules, for example, cytokines. Oxidative damage can cause constriction of renal blood vessels, interstitial fibrosis, and glomerular injury (Norlander et al. 2018).

2. Pheochromocytomas

    Secretion of epinephrine and norepinephrine due to tumor of adrenal medulla causes sustained hypertension, palpitation, and headache (Fig. 11.6).

3. Primary hyperaldosteronism (Conn's syndrome)

    It is due to tumor of zona glomerulosa (which secretes aldosterone) of the adrenal cortex, causing excessive secretion of aldosterone. As discussed before, aldosterone increases the volume of ECF due to accelerated reabsorption of sodium from the renal tubules, resulting in hypertension.

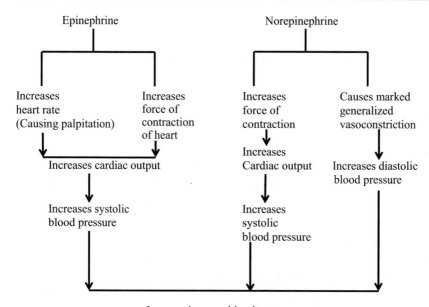

Increase in mean blood pressure

Fig. 11.6 Causes of hypertension due to pheochromocytomas

4. Cushing's syndrome

   It is due to excessive secretion of glucocorticoids due to tumor of glucocorticoid-secreting adrenal cortex. It causes hypertension. The cause of hypertension is not certain. Glucocorticoids may potentiate the action of cate-cholamines (epinephrine and norepinephrine) on heart and blood vessels or may cause increased production of angiotensinogen.

5. Diabetes mellitus causes hypertension. Hypertension is a risk factor for CIHD, stroke, heart failure, peripheral arterial disease, etc. (de Boer et al. 2017).

6. Pregnancy is sometimes associated with toxic stage, i.e., pre-eclampsia charac-terized by convulsions (not due to epilepsy or brain hemorrhage), hypertension, and oliguria. Pre-eclampsia may progress to eclampsia, leading to severe con-vulsions followed by coma. Hypertension of pre-eclampsia and eclampsia is the leading cause of maternal and perinatal morbidity and mortality (Braunthal and Brateanu 2019). It may be associated with vitamin D deficiency, which may be due to the hyperactivity of the RAAS (Garach et al. 2019).

7. Coarctation of the aorta

   Congenital narrowing of a segment of thoracic aorta increases the resistance to flow, resulting in hypertension in the upper part of the body and hypotension in the legs.

8. Oral contraceptives containing estrogens

   Prolonged use may cause hypertension. Estrogens may increase the production of angiotensinogen and may cause salt and water retention, resulting in hypertension.

9. Obesity leads to hypertension due to increased cardiac output as heart enlarges and blood volume increases with increased body weight.
10. Vitamin D deficiency leads to hypertension, which probably may be due to the activation of RAAS.

11.2.3 Resistant Hypertension

Resistant hypertension is the blood pressure of a hypertensive patient that remains high "above goal" despite repeated use of antihypertensive drugs, angiotensin receptor blocker, and a long-acting calcium channel blocker [(A scientific statement from the American Heart Association) (Carey et al. 2018)].

11.2.4 Symptoms

Mild hypertension is without any symptom (silent killer). Severe hypertension is characterized by headache, nausea, vomiting, drowsiness, and confusion. Symptoms are due to the constriction of small arterioles of the brain. Papilloedema (swelling of optic disc) may be present.

11.2.5 Complications

Persistent hypertension may lead to myocardial infarction and cerebral stroke due to constriction of coronary artery/small arterioles, leading to ischemia. It may cause brain hemorrhage due to the rupture of arterioles or capillaries, leading to paralysis (hemiplegia or paraplegia depending on the site of hemorrhage). Retinal hemorrhage will lead to blindness. Renal damage followed by renal failure may occur due to the hypertensive damage of the renal blood vessels. Heart has to pump blood against high pressure, resulting in left ventricular hypertrophy followed by left ventricular failure. Malignant hypertension is the accelerated phase, leading to myocardial infarction, cerebral stroke, progressive renal failure, and papilloedema.

11.2.6 Prevention

1. Regular brisk walking causes dilatation of peripheral blood vessels due to the accumulation of metabolites. Exercise will decrease peripheral resistance and therefore will decrease blood pressure.
2. Weight must be reduced.
3. Reduce salt intake (<3 g/day) with potassium in the diet. Calcium influx through voltage-gated calcium channels (due to depolarization) causes contraction of vascular smooth muscles. Increased potassium efflux increases the membrane potential and closes the voltage-gated calcium channels, causing relaxation of

Table 11.4 Factors which increase and decrease blood pressure

Factors increasing blood pressure	Factors decreasing blood pressure
1. Stress, leading to hyperactive sympathetic nervous system	1. Regular aerobic exercise (walking)
2. Smoking	2. Niacin causes hypocholesterolemia and hypolipidemia
3. Alcohol	3. Meditation
4. Obesity/metabolic syndrome, especially abdominal obesity	4. More omega-3 fatty acids
5. Excessive high caloric intake and sugar	5. High intake of dietary fiber
6. Hypersensitivity to sodium	6. Low cholesterol diet
7. Aging process	7. High intake of vegetables and fruits
8. Sedentary lifestyle	
9. Vitamin D deficiency	
10. Inherited genetic mutation	

vascular smooth muscles. Unopposed calcium influx due to low potassium maintains the contraction of vascular smooth muscle, leading to high blood pressure. One should not take excessive amount of potassium (not more than 5 g/day) as high potassium may result in tachycardia, ventricular fibrillation, and consequently cardiac arrest.

4. DASH diet should be followed. DASH diet was originally designed by US National Institute of Health (Sacks et al. 2001). Diets should be rich in fruits, vegetables, and dietary fibers. DASH diet contains low sodium (about 2 g/day) and more potassium (about 3 g/day). DASH diet should not include excess calcium as calcium may lead to high blood pressure due to the contraction of vascular smooth muscles. Low fat or fat free diet should be taken. Simple sugar should be avoided. Include more omega-3 fatty acids in the diet. Cholesterol-rich diet must be avoided. The DASH diet along with lifestyle factors such as exercise significantly lowered systolic and diastolic blood pressure (Ndanuko et al. 2016).

5. Smoking must be discontinued as it causes constriction of vascular smooth muscles, resulting in hypertension.

6. Heavy alcohol consumption should be discontinued. It causes obesity, which causes thromboembolism and atherosclerosis.

7. Meditation causes relaxation of mind and reduces stress.

8. Niacin supplementation.

9. Beetroot juice intake was reported to reduce blood pressure through the nitric oxide pathway (Ocampo et al. 2018) (Table 11.4).

11.3 Atherosclerosis and Nutrition

Cholesterol is the precursor of vitamin D, adrenocortical hormones, sex hormones, and bile acids. It is a major constituent of plasma membrane. Adverse effects of cholesterol are atherosclerosis and cholelithiasis. Cholesterol is synthesized from acetate. All the carbon atoms in cholesterol are provided by acetate (Fig. 11.7; Tables 11.5 and 11.6).

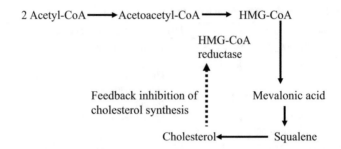

Fig. 11.7 High plasma cholesterol due to large intake of cholesterol-rich diet inhibits hepatic cholesterol synthesis by the above feedback mechanism and vice versa. However, their feedback mechanism is not fully operative as high plasma cholesterol level predisposes atherosclerosis. HMG-CoA = 3-hydroxy-3-methylglutaryl CoA. The "statin" drugs, for example, lovastatin, atorvastatin, etc., inhibit HMG-CoA reductase and decrease the cholesterol level

Table 11.5 Factors which increase and decrease plasma cholesterol level

Hypercholesterolemia	Hypocholesterolemia
1. Familial due to mutation of LDL receptor gene and loss of feedback control	1. Niacin supplementation
2. Excessive alcohol consumption	2. Dietary fiber intake
3. Obesity (especially abdominal obesity)	3. Diets rich in MUFA and PUFA
4. Type 2 diabetes mellitus	4. Exercise due to increased HDL level
5. Hypothyroidism	5. Hyperthyroidism
6. Obstructive jaundice	6. Statin drugs, for example, lovastatin
7. Nephrotic syndrome	7. Estrogens by increasing HDL level
8. Diets rich in saturated fats and cholesterol	8. Intake of omega fatty acids
9. Excessive intake of simple sugar and trans fat	
10. Smoking	

Atherosclerosis refers to the loss of the elasticity of the arteries with thickening of elastin and is characterized by infiltration of cholesterol into the macrophages of connective tissues of the arterial walls. This forms foam cells (lipid-laden macrophages). Foam cells release growth factors and inflammatory mediators (cytokines), which stimulate proliferation of smooth muscle and ultimately narrowing of the arterial wall. Cytokines such as interleukin-1, platelet-derived growth factors, and tumor necrosis factor cause proliferation of smooth muscles resulting in narrowing of the arterial wall. Narrowing of the arterial wall and rupture of atherosclerotic plaques may cause formation of clot (thrombosis) due to decreased blood flow. Clot will occlude the lumen of arterial wall (for example, coronary and cerebral, etc.) and will lead to ischemia followed by infarction of vital organs. Atherosclerosis predisposes to myocardial infarction, cerebral thrombosis, and hypertension.

Lipoproteins are classified according to the ratio of lipids (low density) to proteins (high density): (1) Very low density lipoprotein (VLDL), (2) Low density lipoprotein (LDL), and (3) High density lipoprotein (HDL). Triglyceride-rich glycoproteins are the mixture of VLDL and chylomicrons (Sandesara et al. 2019).

Table 11.6 Approximate cholesterol content expressed as mg per 100 g of food. Cholesterol is absent in plant foods (cereals, fruits, nuts, and vegetables)

1. Egg boiled	450
2. Egg yolk	260
3. Egg omelette fried in unsaturated fat	430
4. Butter	260
5. Ghee	300
6. Brown bread	1
7. Lard	95
8. Margarine	0–4
9. Crab	65
10. Fried fish fingers	50
11. Lobster	70
12. Salmon	90
13. Sole/mackerel	55
14. Beef cooked	80
15. Mutton cooked	80
16. Chicken cooked	70
17. Pork cooked	100
18. Cheese cottage	15
19. Cheese spread	70
20. Ice cream	50
21. Cream	100
22. Liver	300
23. Chocolate milk	90
24. Rice boiled/puffed	Nil
25. Wheat flakes	Nil
26. Honey	Nil
27. Sugar	Nil
28. Potato	Nil
29. Cornflakes	Nil
30. Milk	See Chap. 7
31. Plain yoghurt without any flavor	5
32. Cheese pizza	20

VLDL transports triglycerides for storage and can be converted to LDL. LDL enters cells by receptor-mediated endocytosis using the LDL receptor, whereas HDL transports excess cholesterol to the liver for excretion by bile. LDL delivers cholesterol to peripheral tissues including arterial wall and predisposes atherosclerosis, whereas HDL picks up cholesterol from peripheral tissues and transports it to the liver and lowers plasma cholesterol, preventing atherosclerosis. That is why LDL cholesterol is known as bad cholesterol and HDL cholesterol is known as good cholesterol (Fig. 11.8). Fat content of food is shown in Table 11.7.

Excessive and prolonged intake of saturated fatty acids increases LDL level and predisposes atherosclerosis, whereas intake of omega-3 fatty acids will have the opposite effect. High levels of trans fat increase LDL and decrease HDL and thereby increase plasma cholesterol level. Females, compared with males, are less susceptible to atherosclerosis and have a lower incidence of myocardial infarction

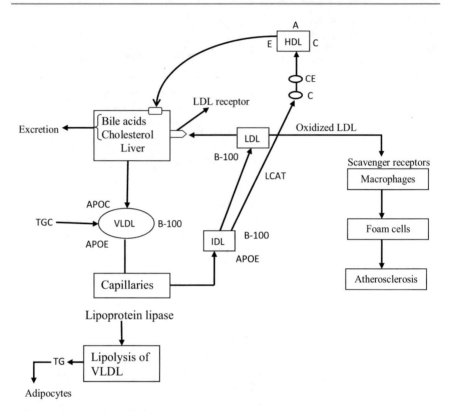

Fig. 11.8 Lipoprotein families for transporting lipids. VLDL very low density lipoproteins, IDL intermediate density lipoprotein, LDL low density lipoprotein, HDL high density lipoprotein, LCAT lecithin-cholesterol acyl transferase, TG triglycerides, C cholesterol, CE cholesteryl esters, APO apolipoprotein. VLDL containing TG, CE, and C is secreted from the liver and converted to LDL via IDL. IDL transports cholesterol to cells and tissues including arterial wall and predisposes atherosclerosis by forming foam cells. HDL transports cholesterol from peripheral tissues to the liver. Cholesterol from the liver is excreted in the bile. Thus, HDL lowers plasma cholesterol and prevents atherosclerosis

Table 11.7 Approximate total fat content of food (g/100 g)

High content >30	Medium content <30	Low content <5
Walnut, coconut (kernel, dry), almond, cashew nut, groundnut, mustard seed	Soya bean white, paneer (Indian cottage cheese), egg boiled (poultry), egg boiled (hen), egg boiled (duck), chicken, goat meat, beef meat, pork meat, hilsa fish, pomfret white, salmon, sardine	Fruits; vegetables; salads (cucumber green, ripe tomato, red carrot, radish, coriander leaves, onion, lettuce, etc.); Whole milk (cow and human); skimmed milk; yoghurt

Reproduced with permission from Longvah et al. (2017)

as estrogen increases HDL level and decreases plasma cholesterol. Lp(a) is a lipoprotein composed of a lipid core and a protein component. The protein component consists of apoB-100 and apoA. Lp(a) can inhibit activation of plasminogen, and binds to fibrin inhibiting clot breakdown (Lubitz and Mukerji 2018).

Skin xanthomas are tumor-like masses under the skin (especially, knee and elbow joints). These types of palpable growth masses are due to aggregation of cholesterol and cholesterol esters in the macrophages of subepithelial connective tissues. Xanthomas indicate hypercholesterolemia/hyperlipidemia (Colledge et al. 2010).

11.3.1 Prevention of Atherosclerosis

1. Minimize cholesterol-rich diet (<300 mg/day).
2. Minimize saturated fatty acids (<10% of total energy) as saturated fatty acids are converted to cholesterol and increase LDL level.
3. Diets rich in PUFA and MUFA should be consumed.
4. Nutraceuticals are functional foods with health benefits beyond their basic nutritional values (diet rich in fruits, vegetables, fish, cereal grains, or olive oil). Nutraceuticals with antioxidants may reduce the prevalence of cardiovascular diseases (Moss and Ramji 2016).
5. Daily exercise decreases cholesterol level due to increased HDL levels.
6. Avoid smoking as nicotine increases LDL levels and also increases fatty acid oxidation. Nicotine → Oxidation of fatty acid → Acetyl-CoA → Synthesis of cholesterol.
7. Avoid excess intake of simple sugar. Glucose through glycolysis and citric acid cycle increases the formation of acetyl-CoA and increases the synthesis of cholesterol.
8. Avoid trans fat as they increase plasma cholesterol level.
9. Avoid alcohol consumption due to formation of acetyl-CoA.

11.4 Summary

Diabetes mellitus is a metabolic disorder due to the absolute deficiency of insulin secreted by β cells (type 1 diabetes mellitus) or due to insulin resistance (type 2 diabetes mellitus), leading to severe abnormalities of carbohydrate, fat, and protein metabolism. It also comprises gestational diabetes and secondary diabetes. Adverse consequence of diabetes mellitus is aggravated by unchecked activity of glucagon secreted by α cells. Glucagon is a hyperglycemic hormone in contrast to the hypoglycemic action of insulin. The hallmark of uncontrolled diabetes mellitus is hyperglycemia, decreased lipogenesis, increased lipolysis, hypertriglyceridemia, hypercholesterolemia, increased protein catabolism, and ketosis. Signs and symptoms are characterized by polyphagia, polyuria, polydipsia, weight loss, kussmaul breathing, and drowsiness due to ketoacidosis. Uncontrolled diabetes mellitus increases risk of various complications such as nephropathy, neuropathy, retinopathy, hypovolemia, CIHD, and peripheral vascular disease. Diet plan along with daily regular exercise is essential to prevent the progression towards development of diabetes. Hypertension is classified as essential or primary hypertension and secondary hypertension. Primary hypertension is the common type affecting about 90–95%, whereas secondary hypertension affects 5–10% of hypertensive patients. Persistent hypertension is one of the risk factors for myocardial infarction, cerebral stroke, blindness, progressive renal failure, and left ventricular failure. The DASH diet and regular aerobic exercise are the preventive measures. Atherosclerosis is characterized by infiltration of cholesterol into the macrophages of connective tissues of the arterial walls. Foam cells (lipid-laden macrophages) release growth factors and cytokines, which stimulate proliferation of smooth muscle and ultimately cause narrowing of the arterial walls. Atherosclerosis predisposes to myocardial infarction, cerebral stroke, and hypertension.

References

American Diabetes Association (2018) 2. Classification and diagnosis of diabetes: standards of medical care in diabetes. Diabetes Care 41(Suppl 1):S13–S27

Biondi B et al (2019) Thyroid dysfunction and diabetes mellitus: two closely associated disorders. Endocr Rev 40(3):789–824

Bolla AM et al (2019) Low-carb and ketogenic diets in type 1 and type 2 diabetes. Nutrients 11(5):962

Braunthal S, Brateanu A (2019) Hypertension in pregnancy: pathophysiology and treatment. SAGE Open Med 7:2050312119843700. https://doi.org/10.1177/2050312119843700

Carey RM et al (2018) Resistant hypertension: detection, evaluation, and management: a scientific statement from the American Heart Association. Hypertension 72(5):e53–e90

Chobanian AV et al (2003) Seventh report of the Joint National Committee on Prevention, Detection, Evaluation and Treatment of High Blood Pressure. Hypertension 42:1206–1252

Colledge NR et al (eds) (2010) Davidson's principles and practice of medicine, 21st edn. Churchill Livingstone, London

de Boer IH et al (2017) Diabetes and hypertension: a position statement by the American Diabetes Association. Diabetes Care 40(9):1273–1284

Firdous P et al (2018) Genetic testing of maturity-onset diabetes of the young current status and future perspectives. Front Endocrinol 9:253

Forouhi NG et al (2018) Dietary and nutritional approaches for prevention and management of type 2 diabetes. BMJ 361:k2234

Ganong WF (2003) Review of medical physiology, 21st edn. McGraw-Hill, New York

Garach AM et al (2019) Vitamin D status, calcium intake and risk of developing type 2 diabetes: an unresolved issue. Nutrients 11:642

Longvah T, Ananthan R, Bhaskarachary K, Venkaiah K (2017) Indian food composition tables. National Institute of Nutrition, Indian Council of Medical Research, Hyderabad

Lubitz M, Mukerji V (2018) Lipoprotein(a) and atherosclerosis: a case report and literature review. Int Arch Cardiovasc Dis 2:006

Malone JI, Hansen BC (2019) Does obesity cause type 2 diabetes mellitus (T2DM)? Or is it the opposite. Pediatr Diabetes 20(1):5–9

Moss JWE, Ramji DP (2016) Nutraceutical therapies for atherosclerosis. Nat Rev Cardiol 13(9):513–532

Ndanuko RN et al (2016) Dietary patterns and blood pressure in adults: a systematic review and meta-analysis of randomized controlled trials. Adv Nutr 7(1):76–89

Norlander AE et al (2018) The immunology of hypertension. J Exp Med 215(1):21–33

Ocampo DAB et al (2018) Dietary nitrate from beetroot juice for hypertension: a systematic review. Biomol Ther 8:134

Parveen R et al (2017) Association of Vitamin D and diabetic retinopathy in patients with type 2 diabetes mellitus. Ann Natl Acad Med Sci (India) 53(3):156–165

Pot GK et al (2019) Nutrition and lifestyle intervention in type 2 diabetes: pilot study in the Netherlands showing improved glucose control and reduction in glucose lowering medication. BMJ Nutr Prev Health 0:1–8

Pradhan A et al (2019) Review on sodium-glucose cotransporter 2 inhibitor (SGLT2i) in diabetes mellitus and heart failure. J Family Med Prim Care 8:1855–1862

Sacks FM et al (2001) Effects on blood pressure of reduced dietary sodium and the dietary approaches to stop hypertension (DASH) diet. New Engl J Med 344(1):3–10

Sandesara PB et al (2019) The forgotten lipids: triglycerides, remnant cholesterol, and atherosclerotic cardiovascular disease risk. Endocr Rev 40(2):537–557

Sawada N, Arany Z (2017) Metabolic regulation of angiogenesis in diabetes and aging. Physiology 32:290–307

Schmidt AM (2019) Diabetes mellitus and cardiovascular disease: emerging therapeutic approaches. Arterioscler Thromb Vasc Biol 39:558–568

Schütten MT et al (2017) The link between adipose tissue renin-angiotensin-aldosterone system signalling and obesity-associated hypertension. Physiology 32:197–209

Turner MD et al (2005) Calpain-10: from genome search to function. Diabetes Metab Res Rev 21(6):505–514

Wang TJ et al (2008) Vitamin D deficiency and risk of cardiovascular disease. Circulation 117(4):503–511

Further Reading

Kitt J et al (2019) New approaches in hypertension management: a review of current and developing technologies and their potential impact on hypertension care. Curr Hypertens Rep 21(6):44

Punthakee Z et al (2018) Definition, classification and diagnosis of diabetes, prediabetes and metabolic syndrome. Can J Diabetes 42:S10–S15

Sami W et al (2017) Effect of diet on type 2 diabetes mellitus: a review. Int J Health Sci 11(2):65–71

Addiction-Related Health Problems

12

Abstract

Alcoholism, tobacco smoking, and drug abuse/drug dependence are the major social problems throughout the world. Sense of failure in life, problems in personal life, and mental tension drive a person to take shelter under alcohol, smoking, and drugs for relief. Cultural or family pressure and peer influences are other factors that are responsible for addiction. Consumption of alcohol, tobacco smoking, and illegal drugs are deleterious to health and cause adverse effect on health on most societies of the world. Severe alcoholism causes fatty liver followed by cirrhosis. Cirrhosis will result in complete dysfunction of the liver and ultimately will result in hepatic coma followed by death. The adverse ill effects of tobacco smoking on health are horrifying. In addition to many toxic chemicals such as nicotine, CO, and thiocyanate, many chemicals have been identified as carcinogens such as hydrogen cyanide, benzopyrene, polycyclic aromatic hydrocarbons, aldehydes, and nicotine-derived nitrosamine ketones. Carcinogens travel throughout the body, causing cancer at various sites of the body due to their mutagenic effects. Euphoria, elevated mood, sedation, etc. are some of the effects of illegal drugs, which tempt an individual to adopt these drugs. The temporary pleasurable experience of drugs is followed by depression, anxiety, personality disorder like paranoid behavior and intense craving of the drug, which creates social problems, including violence. High dose induces convulsion and cardiorespiratory failure.

Keywords

Alcohol · Cirrhosis · Tobacco smoking · Cancer · Drug abuse

K. Chakrabarty, A. S. Chakrabarty, *Textbook of Nutrition in Health and Disease*,
https://doi.org/10.1007/978-981-15-0962-9_12

12.1 Alcoholism

Consumption of alcohol causes detrimental effects in most societies of the world. The following are the major problems caused by alcohol

1. Social and behavioral problems: Paranoid personality disorder, especially sus-piciousness of the partner having sexual relationship with a neighbor/colleague may lead to violence and even murder of the partner. Suicide is common in chronic alcoholics due to extreme depression. Driving due to intoxication may be dangerous. Financial problem due to unemployment disrupts the peace of the family. Self-importance and hostility are other behavioral disorders. Alcohol intoxication causes aggressiveness and increased excitability (Edwards and Bouchier 1991).

2. Fatty liver followed by cirrhosis (Fig. 12.1): Alcohol dehydrogenase present in the cytoplasm of liver cell oxidizes alcohol into acetaldehyde, which is further oxidized by mitochondrial aldehyde dehydrogenase into acetate. Acetate is eas-ily converted to acetyl-CoA. Acetyl-CoA increases the synthesis of fatty acids and triglyceride in the liver as well as increased formation of ketone bodies. Simultaneously, oxidation of alcohol decreases fatty acid oxidation. Increased synthesis of fatty acids and triglycerides with inhibition of fatty acid oxidation will cause accumulation of fat in the liver (fatty liver), causing the loss of func-tion of liver cells. Alcohol is also converted to acetaldehyde by the microsomes of the smooth endoplasmic reticulum. Poor diet deficient in proteins and vita-mins will aggravate fatty liver. Acetaldehyde has direct toxic effect on liver cells and will cause necrosis of already damaged cells. Stellate cells are present in the space between hepatocyte and endothelial cells (Fig. 12.2). These are activated by liver injury and cause hepatic fibrosis with the formation of nod-ules in response to proinflammatory cytokines produced by Kupffer cells and hepatocytes (Colledge et al. 2010). Fibrosis of the liver is called cirrhosis. Cirrhosis leads to accumulation of fluid in the peritoneal cavity and causes ascites (Fig. 12.3). Many factors cause cirrhosis. Multifactorial causes of asci-tes are shown in Fig. 12.4. Cirrhosis will result in complete dysfunction of liver (liver failure) and ultimately will result in hepatic coma followed by death. Coma is aggravated by lactic acidosis and ketosis. High NADH + H$^+$ convert pyruvate to lactate, causing lactic acidosis.

3. Effect on carbohydrate metabolism: Due to liver dysfunction, gluconeogenesis is inhibited. In severe cases, glycogen of the liver is depleted. Glucose is not derived from glycogen (inhibition of glycogenolysis). Inhibition of gluconeogenesis and glycogenolysis results in hypoglycemia and glucose intol-erance. Hypoglycemia can cause agitation and impaired judgement.

4. Effect on brain: Alcohol inhibits neuronal activity as it is a CNS depressant. It augments GABA receptors with inhibition of NMDA receptors. Alcohol causes direct toxic effects by increasing fluidity of the biologic membranes. Excessive consumption of alcohol causes amnesia (alcoholic blackouts) for events that occurred during intoxication, impairment of speech, drunken gait, and ataxia.

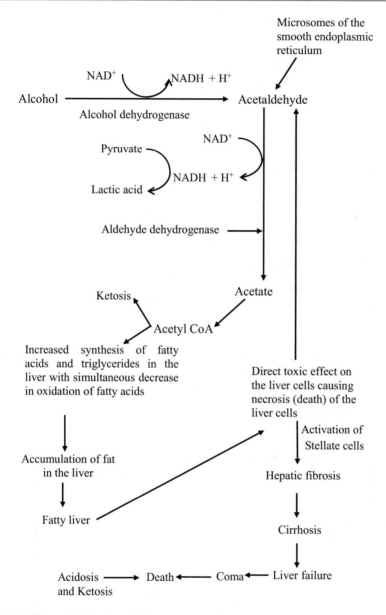

Fig. 12.1 Effects of alcohol on the liver

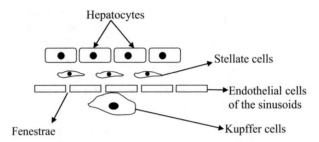

Fig. 12.2 Stellate cells present in the space between hepatocytes and endothelial cells

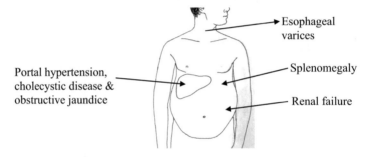

Fig. 12.3 Ascites (abdominal swelling) due to accumulation of fluid in the peritoneal cavity. Various complications due to loss of liver function are shown in the figure

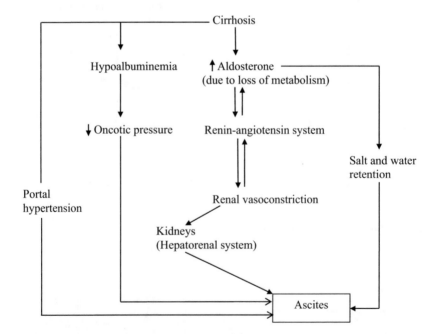

Fig. 12.4 Multifactorial causes of ascites

Chronic alcoholics develop tremor due to muscular incoordination. Atrophy of certain areas of brain after several years of drinking will even cause dementia. Loss of brain volume is most noticeable in two areas. (1) Cerebral cortex, which is concerned with higher intellectual mental functions. (2) Cerebellum, which is concerned with normal posture, balance, and muscular coordination. Chronic heavy alcoholics may develop brain-damage syndrome known as Wernicke–Korsakoff syndrome. Due to this disorder, alcoholics are not able to remember new information for more than few seconds. Confusion, hypotension, nystagmus, and ophthalmoplegia are the other features of Wernicke–Korsakoff syndrome.

5. Vitamin deficiencies: Alcoholism causes deficiencies of vitamins B_1 and B_6. Alcoholism causes deficiency of vitamin B_6 as acetaldehyde causes hydrolysis of phosphate of the coenzyme. Thiamin is an obligatory part of a coenzyme for the conversion of pyruvate to acetyl-CoA. In thiamin deficiency, the above reaction is slowed and pyruvate accumulates, leading to lactic acidosis. Alcohol interferes with the absorption of folic acid, niacin, and vitamin A and decreases their storage in the liver. Glucose intake leads to further accumulation of pyruvic acid and aggravates lactic acidosis, leading to a fatal state.

6. Effect on stomach: Excessive intake of alcohol causes gastritis followed by epigastric distress, gastric ulcer, and even bleeding from the stomach.

7. Effect on heart and skeletal muscle: Chronic alcoholics suffer from cardiac and skeletal muscle myopathy (weakness of muscle). Cardiomyopathy may lead to arrhythmias.

8. Peripheral neuropathy: Pain, tingling, numbness, and paresthesia of the extremities of limbs.

9. Pancreas: Chronic alcoholics suffer from acute pancreatitis, which is characterized by severe pain in the epigastrium and upper back.

10. Obesity-induced hypertension: Alcohol stimulates appetite, provides calories and may cause obesity. Increased secretion of angiotensin II from abdominal tissue mass may lead to hypertension (Toffolo et al. 2012).

11. Deficiency of phenylalanine hydroxylase: In the liver (see Chap. 1) as a result of cirrhosis can cause brain damage and epilepsy.

12. Abnormalities of the bone: Chronic alcoholics may lower bone density, lower growth in the epiphyses and osteonecrosis of the bone with an increased risk for fractures.

13. Spider naevi indicates dilated capillaries on the skin, which radiate like spider and are prominent in liver cirrhosis and hepatitis.

14. Mallory–Weiss syndrome is characterized by a longitudinal tear at the mucosa around the gastroesophageal junction due to violent vomiting caused by heavy drinking.

15. Cognitive dysfunction including loss of recent memory is common.

16. Cancer: Alcohol enhances the risk for breast, bowel, prostate, and liver cancer (Scheideler and Klein 2018). Postmenopausal women due to heavy drinking may suffer from breast cancer (McDonald et al. 2013).

17. Hematological parameters: Heavy drinking causes an increase in MCV, hypersegmented neutrophils, reticulocytopenia, and thrombocytopenia. Sideroblastic changes may be present.
18. Sexual dysfunctions: Although heavy drinking increases sexual drive, it inhibits erectile capacity in men. Chronic alcoholics suffer from testicular atrophy with shrinkage of the seminiferous tubules and lower sperm count. Excessive intake of alcohol by women can cause amenorrhea, a decrease in ovarian size, infertility, and abortion. Sexual dysfunction leads to marital tension and even divorce.
19. Hormonal changes: Increase in glucocorticoids and decrease in ADH secretion are associated with drinking. Decrease in ADH secretion causes excessive urination after consumption of alcohol. Alcoholics may have low serum T_3 and T_4.
20. "Fetal alcohol syndrome": Heavy drinking during pregnancy causes transfer of acetaldehyde through placenta, resulting in abnormal fetal development characterized by ventricular septal defect, microcephaly (small head in relation to the size of the rest of body due to underdeveloped brain), and mental retardation.

Note: Alcohol causes peripheral vasodilation, resulting in a mild decrease in the blood pressure. It may decrease the incidence of CIHD, probably by increasing coronary dilatation as well as by increasing HDL. It may minimize cognitive dysfunction of Alzheimer's disease. However, an individual should not consume more than two drinks per day.

12.1.1 Withdrawal Symptoms

Sudden cessation of drinking alcohol causes delirium tremens characterized by agitation, terrifying visual hallucination of animals and insects, bad dreams, anxiety, tremor, sweating, insomnia, head injury, etc. Without treatment chronic alcoholism can be fatal.

12.1.2 Prevention

1. Detrimental effects of alcoholism must be told to alcoholics. Life span of chronic alcoholics is shortened by many years. They must be motivated for total abstinence.
2. Counselling and psychotherapy will be crucial for alcoholics to avoid alcohol.
3. In serious cases, they should be admitted at specialized centers (alcoholic rehabilitation center or de-addiction center). Medications are essential for the treatment of withdrawal symptoms under the supervision of a physician.
4. High protein diet will be detrimental and must be avoided in patients suffering from liver diseases.

12.2 Tobacco Smoking

Smoking is deleterious to health all over the world. It is the most worldwide public health problem. Most of the smokers die at early stage. There are more than 4000 chemicals in cigarettes which are toxic and destructive. Many chemicals have been identified as carcinogens such as arsenic, benzene, hydrazine, N-Nitrosonornicotine, N-Nitrosopyrrolidine, N-Nitrososarcosine, polycyclic aromatic hydrocarbons, vinyl chloride, and many others (US Department of Health and Human Services). Apart from carcinogenic chemicals, many toxic substances are present in cigarette smoke, for example, nicotine, carbon monoxide, thiocyanate, hydrogen cyanide, nitric oxide, and many others. Tobacco smoke also produces ROS that induces oxidative stress and chronic inflammation. ROS can inflict damage to nuclear DNA, resulting in mitogenesis and cancer of various cells. Cancer of oral cavity is common in Asia due to addiction to smokeless tobacco (chewing tobacco, khaini, zarda, and gutkha). One must quit smoking if there is presence of leukoplakia (precancerous condition).

Adverse harmful effects of smoking are shown below:

1. Oxidative damage due to ROS activates proinflammatory molecules (cytokines), resulting in chronic inflammation of the airways and lungs.
2. Carcinogens travel throughout the body causing cancer at any site of the body such as oral cavity, larynx, esophagus, lungs, kidneys, urinary bladder, pancreas, and stomach (Fig. 12.5).
3. Smoking causes airway inflammation due to loss of function of cilia and thus causes chronic bronchitis.
4. Despite relaxation of smooth muscle of the airways, NO aggravates airway resistance by increasing mucus secretion of the submucosal glands due to increased blood flow (Chakrabarty and Chakrabarty 2006). Smoking causes airway inflammation due to loss of ciliary escalator. The inflammation of the airways along with increased mucus secretion causes chronic bronchitis.
5. Smoking inhibits the activity of macrophages inside the lungs. Inhibition of immune system causes inflammation of the lungs.
6. Smoking causes loss of elastic tissue leading to emphysema. Emphysema may lead to COPD, which ultimately causes cor pulmonale (Ganong 2003).
7. Nicotine causes lipolysis and increases acetyl-CoA, resulting in the synthesis of cholesterol. Hypercholesterolemia associated with constriction of coronary arteries may lead to myocardial infarction.
8. Atherosclerosis and constriction of arteries will cause cessation of blood flow of peripheral arteries, which may lead to gangrene of the legs.
9. Nicotine increases HCl secretion, which may cause ulcers of stomach and duodenum.
10. Smokers are susceptible to stroke because of atherosclerosis and reduced blood flow to the cerebrum.
11. If smokers constantly smoke in a closed room without ventilation, they may suffer from carbon monoxide poisoning and severe hypoxia, leading to death.

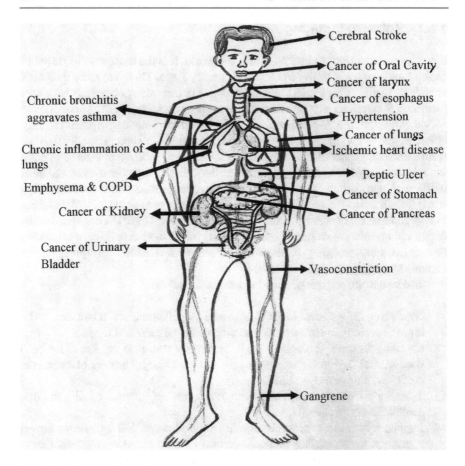

Fig. 12.5 Adverse effects of tobacco smoking

12. Nicotine can cross placental barrier and may cause miscarriage and fetal death. Smoking in women may cause developmental defect in the fetus and anencephaly. Maternal smoking may even cause sudden infant death syndrome (crib's death).
13. Smoking causes hypertension due to vasoconstriction.
14. Chronic smokers may suffer from hypothyroidism as tobacco contains thiocyanate. Thiocyanate inhibits iodine transport.
15. Passive smoking or environmental smoke causes bronchitis, cardio-related diseases and even lung cancer to non-smoker.

The smokers must be motivated to quit smoking. Counselling by family members and friends will be useful. They must be made aware of the detrimental effects of smoking. Nicotine gum/logenze may be tried. After the abstinence of smoking, they may be given antidepressant drug under the supervision of the physician.

12.3 Drug Abuse/Drug Dependence

Dependence on illegal drugs is a major problem throughout the world. Addicts of opiates, cannabis, cocaine, and hallucinogenic drugs are prevalent throughout the world. Cultural pressures, peer influences, and availability of drugs are some of the factors of drug abuse.

12.3.1 Opiates

Opiates include morphine, codeine, and heroin. Morphine and codeine are obtained from the milky juice of the poppy (*Papaver somniferum*). Heroin (diacetyl morphine) is a white powder derived from morphine. Opiates, when taken orally or by injection, give an immediate pleasurable experience, diminished pain perception, and sedation. Dependence occurs after a few weeks of regular dose. The addicts will try to obtain the drug by any means and at any cost. Death may occur due to overdose. Death occurs due to cardiorespiratory arrest. If they are not able to obtain the drug, within 12–24 h they suffer from palpitation, perspiration, shivering, abdominal cramps, flushing of face, etc. Intense craving for the drug may create social problems, including violence.

12.3.2 Cocaine (Longo et al. 2011)

Cocaine is obtained from the leaves of coca plant (erythroxylon coca). It is taken by sniffing the powder into the nostrils with the help of a tube. It is absorbed by the nasal mucus. Mental disorders due to cocaine toxicity may be related to increased synaptic concentration of dopamine, norepinephrine, and serotonin. Elevated mood is the characteristic feature. Psychosis develops after prolonged use. Tactile hallucination or formication (a sensation resembling the feeling of ants crawling over the skin) may develop. Other features are delirium, tremor, and hyperventilation. Cocaine is a potent vasoconstrictor and may cause hypertension and cerebral stroke. Excessive inhalation of cocaine may cause convulsion and even cardiac or respiratory failure. In women, cocaine addiction may cause amenorrhea and infertility. Congenital malformations in the fetus may occur in pregnant women due to cocaine abuse.

12.3.3 Amphetamine

Amphetamine causes insomnia and elevation of mood and confusion. Withdrawal of the drug after prolonged use causes depression, anxiety, and even personality disorder like paranoid behavior. It causes loss of hunger and appetite as it stimulates the release of norepinephrine in the hypothalamus. MDMA is a derivative of methamphetamine and is known as ecstasy. It produces euphoria and a sense of well-being followed by depression. MDMA causes depletion of brain dopamine as well as serotonin.

12.3.4 Hallucinogenic Drugs (Ganong 2003)

1. LSD-serotonin agonist.
2. Psilocin, a substance found in magic mushrooms.
3. DMT.

All these hallucinogenic agents are derivatives of tryptamine and act by blocking $5\text{-}HT_2$ receptor. Within an hour they produce euphoria, heightened visual sensation of objects, distorted images of objects (in shape or size), distorted sensation of sounds, and other transient hallucinations.

12.3.5 Cannabis Compounds (Zehra et al. 2018)

Cannabis compounds are derived from the plant Cannabis sativa, which include marijuana, hashish, ganja, charas, and bhang. They are prepared from the leaves and flowering top of the female plants. Hashish (smack/brown sugar), charas, and ganja are inhaled by smoking, whereas bhang is taken orally. Cannabis is smoked with tobacco (marijuana cigarettes) and is absorbed quickly from the lungs into the blood and tissues. Cannabinoid receptors are present in the spinal cord, cerebral cortex, basal ganglia, and hippocampus. It produces immediately a perception of relaxation, euphoria, and well-being. High dose resembles alcohol intoxication and may induce heart attack due to coronary insufficiency. Withdrawal effects include sweating, nausea, vomiting, anorexia, tremor, and insomnia.

12.3.6 Phencyclidine

PCP is a derivative of a cyclohexylamine and binds to NMDA receptors in the nervous system. Oral intake by smoking or by intravenous injection causes excitement, impaired motor incoordination, analgesia, hyperacusis, etc. Acute psychosis may develop.

12.3.7 Barbiturates and Sedatives

Barbiturates and sedatives cause relaxation, pain-relief, and sedation. Prolonged use may cause liver damage.

12.4 Summary

Alcoholism, tobacco smoking, and illegal drug abuse cause serious health and social problems in most societies of the world. Chronic alcoholics disrupt the peace of family due to behavioral disorders such as paranoid personality, self-importance, hostility, and aggressiveness. Marital tension due to suspiciousness and delusion of sexual infidelity leads to divorce and even murder of partner. Apart from behavioral problems, consumption of alcohol for a long period of time causes adverse and deleterious effects on health. Chronic alcoholics suffer from malnutrition due to anorexia and malabsorption syndrome. They suffer from deficiencies of vitamin B_1 and B_6, hypoglycemia, and glucose intolerance. Alcohol dehydrogenase present in the cytoplasm of hepatic cells oxidizes alcohol into acetaldehyde. Acetaldehyde has direct toxic effect on liver cells and causes cirrhosis. Cirrhosis will result in liver failure followed by hepatic coma and death. The morbidity and mortality are extremely high due to tobacco smoking. The adverse ill effects of tobacco smoking are mediated by the presence of toxic chemicals, including carcinogens. Tobacco smoke generates ROS, which induces oxidative damage. Oxidative damage activates proinflammatory cytokines, resulting in inflammation. ROS inflicts damage to nuclear DNA, resulting in mitogenesis and cancer of various cells of the body. Nicotine-induced hypercholesterolemia leads to CIHD, hypertension, cerebral stroke, and gangrene of the legs. Smoking causes chronic bronchitis, emphysema, COPD, and ultimately cor pulmonale. Passive smoking causes bronchitis, cardiac-related diseases, and even lung cancer. Addicts of opiates, cannabis compounds, ecstasy, and hallucinogenic drugs such as LSD are prevalent worldwide. The temporary pleasurable experience of drugs is followed by depression, anxiety, paranoid behavior, and intense craving for the drug which will create social problem, including violence. High dose will induce convulsion followed by cardiorespiratory failure and death.

References

Chakrabarty AS, Chakrabarty K (2006) Fundamentals of respiratory physiology, 1st edn. I.K. International Publishing House Pvt Ltd, New Delhi

Colledge NR et al (eds) (2010) Davidson's principles and practice of medicine, 21st edn. Churchill Livingstone, London

Edwards CRW, Bouchier IAD (eds) (1991) Davidson's principles and practice of medicine, 16th edn. Churchill Livingstone, London

Ganong WF (2003) Review of medical physiology, 21st edn. McGraw-Hill, New York

Longo DL et al (eds) (2011) Harrison's principles of internal medicine, 18th edn. McGraw-Hill, London

McDonald JA et al (2013) Alcohol intake and breast cancer risk: weighing the overall evidence. Curr Breast Cancer Rep 5(3). https://doi.org/10.1007/s12609-013-0114-z

Scheideler JK, Klein WMP (2018) Awareness of the link between alcohol consumption and cancer across the world: a review. Cancer Epidemiol Biomark Prev 27(4):429–437

Toffolo MCF et al (2012) Alcohol: effects on nutritional status, lipid profile and blood pressure. J Endocrinol Metab 2(6):205–211
Zehra A et al (2018) Cannabis addiction and the brain: a review. J Neuroimmune Pharmacol 13(4):438–452

Website

http://en.wikipedia.org/wiki/List_of_cigarette_smoke_carcinogens
11th Report on Carcinogens by the U.S. Department of Health and Human Services. http://ntp. niehs.nih.gov/ntp/roc/toc11.html

Further Reading

Kwok A et al (2019) Effect of alcohol consumption on food energy intake: a systematic review and meta-analysis. Br J Nutr 121(5):481–495
Peacock et al (2018) Global statistics on alcohol, tobacco and illicit drug use: 2017 status report. Addiction 113(10):1905–1926

Nutritional Management of Diseases

13

Abstract

Anorexia, malabsorption syndrome, impaired immune function, and increased susceptibility to infection will cause malnutrition in patients suffering from cancer. Malnutrition is a risk factor for cardiovascular diseases. Atherosclerosis is the main factor for the genesis of cardiovascular diseases. Liver dysfunction or failure causes malnutrition due to anorexia, impaired digestion, malabsorption syndrome, pancreatic insufficiency, and hepatorenal syndrome. Renal hypertension, hyperkalemia, metabolic bone disease, anemia, and uremia are the main features of renal diseases related to nutrition. Malnutrition due to gastrointestinal diseases is due to cancer, celiac disease, Crohn's disease, and cystic fibrosis. Malabsorption syndrome leading to malnutrition is one of the common features of patients infected with HIV. Proper nutrition will minimize the side effects of antiretroviral drugs and will slow progression to the final stage of AIDS.

Keywords

Cancer · Cardiovascular diseases · Liver diseases · Renal diseases · Gastrointestinal diseases · AIDS

13.1 Cancer and Nutrition

Anorexia, loss of body weight, malabsorption syndrome, impaired immune function, and increased susceptibility to infection will cause malnutrition in patients suffering from carcinoma. Carcinogens cause DNA damage and mutation of normal gene, i.e., proto-oncogene. Cancer cells damage cellular DNA by covalent binding and disrupt helical structure and block Watson–Crick base pairing. As a result, cancer arises from abnormal uncontrolled cell division. Cancer cells do not obey apoptosis, i.e., programmed cell death, which is essential for normal cell to maintain the "structural and functional integrity" after cell division (Mann and Truswell 2007).

© Springer Nature Singapore Pte Ltd. 2019
K. Chakrabarty, A. S. Chakrabarty, *Textbook of Nutrition in Health and Disease*,
https://doi.org/10.1007/978-981-15-0962-9_13

Cancer cells also inactivate tumor suppressor genes and thus promote inflammation. High levels of interleukin-6, C-reactive protein, and TNF receptor superfamily member 1β in the plasma induced by diet (red meat, possibly high fat diet) contribute to the gastrointestinal inflammation by the bacterial flora, especially *Helicobacter pylori* and *Fusobacterium nucleatum*. Bacterial flora in the colon may form carcinogenic N-nitrosocompounds by reacting with red meat. This effect is not caused by fish or poultry. Polycyclic aromatic hydrocarbons generated by barbequing or grilling meat may cause cancer. Apart from bacterial flora, several viruses are implicated such as: (1) Cervix cancer may be caused by human papilloma virus (Crosbie et al. 2013) and (2) Nasopharynx cancer may be caused by Epstein-Barr virus (Wu et al. 2018; Farrell 2019). Cancer cells may cause upregulation of proinflammatory genes (for example, cytokines), resulting in chronic inflammation (Grivennikov and Karin 2011). Chronic inflammation is one of the main culprits for the development of cancer (Fig. 13.1).

13.1.1 Prevention

1. Tobacco smoking must be completely avoided. More than 1000 carcinogenic chemicals present in cigarette smoke travel throughout the body and cause cancer at various sites of the body. Tobacco smoking produces ROS. ROS induces oxidative stress and chronic inflammation. ROS causes upregulation of proinflammatory genes (cytokines), resulting in chronic inflammation.
2. Obesity is a great risk factor for cancer for any part of the body, particularly endometrium and breast in postmenopausal women. Less energy intake and more energy expenditure (regular daily exercise) will reduce obesity. BMI should be less than 25 kg/mm^2. Weight gain must be avoided by regular physical activity.
3. Alcohol consumption should be minimized as it shows a positive association with cancer of the mouth, pharynx, and larynx.
4. One should protect from air pollution by using anti-pollution mask as polyaromatic hydrocarbons are mutagenic and carcinogenic. Automobile exhaust fumes, (especially diesel fumes), benzopyrene emitted from gasoline, fumes liberated from industrial plants and forest fire are at great risk for the skin, lungs, and urinary bladder cancer. Ionizing radiation (Υ-radiation and electromagnetic radiation) and non-ionizing radiation (UV radiation) lead to cancers, especially of parts exposed to solar radiation.
5. One should avoid consuming post-harvest spoilage of stored food, since growth of fungus *Aspergillus flavus* and *A. parasiticus* produces carcinogenic aflatoxin.
6. One should minimize the consumption of preserved food, for example, nitrites used as preservative in food may be converted to carcinogenic nitrosamines.
7. One should take moderate amount of red meat and probably saturated fat as they may form carcinogenic N-nitrosocompounds in the intestine.

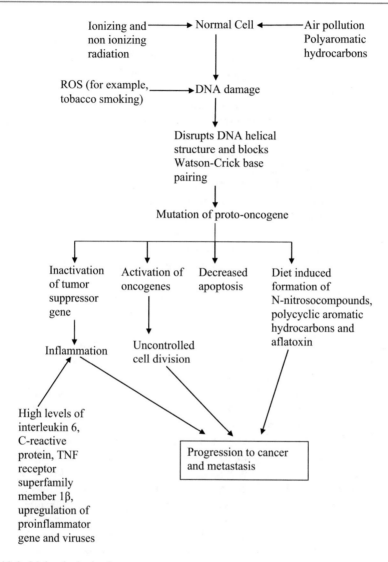

Fig. 13.1 Molecular basis of cancer

13.1.2 Nutritional Interventions

1. A well-balanced diet having more protein should be consumed to minimize malnutrition. Adequate nutrition (both macro- and micronutrients) will reduce the side effects of chemotherapy and radiotherapy.
2. Vitamin C, vitamin E, carotenoids, flavonoids, and selenium are radical-trapping antioxidants, and prevent oxidative damage by inactivating oxygen free radicals (Eliassen et al. 2015). Carotenoids contain β-carotene, β-cryptoxanthin, lutein,

zeaxanthin, and lycopene. They are present in tomatoes, dark green vegetables, oranges, carrots, and capsicum. Flavonoids present in variety of foods, for example, onions, berries, apples, etc., are antioxidant nutrients. Selenium supplementation lowers incidence of prostate cancer. Vitamin C may prevent the formation of carcinogenic nitrosamines.

3. Dietary fibers reduce the contact of gastrointestinal epithelium with carcinogenic chemicals and prevent the absorption of carcinogens. Thus, dietary fibers reduce the incidence of cancer of the gastrointestinal tract, particularly colonic cancer. Diet must be supplemented with dietary fibers (about 25 g/day) to prevent cancer of the gastrointestinal tract.

4. Diet should consist of high amounts of fruit and vegetables. WHO (2003) recommended intake of minimum 400 g/day of fruits and vegetables in order to decrease the risk of cancers. Seeds and leaves of vegetables contain cellulose and hemicellulose. Outer covering of fruits contains pectin.

5. It has been reported that omega-3 fatty acids may reduce cancer risk by decreasing inflammation. Thus consumption of fish may decrease incidence of cancer.

13.2 Cardiovascular Diseases and Nutrition

Cardiovascular diseases occur due to angina pectoris, myocardial infarction, systemic hypertension, and cardiac failure. Malnutrition is a risk factor for cardiovascular diseases. Excessive and prolonged consumption of saturated fat, and trans fat will increase LDL and decrease HDL levels, and will predispose to atherosclerosis. Intake of omega-3 fatty acids will have the opposite effect. Females compared with males are less susceptible to infarction as estrogens increase HDL level, causing hypocholesterolemia. Dyspnea and orthopnea are the common symptoms of cardiac diseases (Chakrabarty and Chakrabarty 2006). Angina is characterized by choking chest pain due to exertion. Pain is relieved by rest. Myocardial infarction causes irreversible severe pain, which persists at rest. Pain is felt behind the sternum and may radiate to the left arm, neck, jaw, and at the back. Accompanied by vomiting. Syncope is due to sudden fall in cardiac output and vasodilatation. Vasodilatation occurs due to the accumulation of metabolites and the activation of β-adrenergic receptors. Decreased cardiac output and vasodilatation causes decreased blood supply to the brain, resulting in loss of consciousness. Systemic hypertension leads to myocardial infarction, whereas pulmonary hypertension causes right ventricular failure.

Before discussion of myocardial infarction, the anatomic considerations of coronary arteries are summarized (Ganong 2003). The right coronary artery arises from the right sinus behind the aortic valve and supplies the right ventricle, right atrium, and the inferior part of the left ventricle. The left coronary artery arises from the left sinus and divides into (a) an anterior descending artery which supplies the anterior and apex of the heart and (b) a circumflex artery which supplies the lateral and posterior surface of the heart. Interventricular septum is supplied by both coronary arteries. Increased anastomosis between the coronary arteries is developed in the person due to aging as well as due to ischemia of the heart. The venous system drains into the coronary sinus and then to the right atrium. Pressure of the left ventricle is slightly more than the aorta and consequently blood flow is possible in the arteries of subendocardium of the left ventricle only during diastole. Ischemia of the left ventricle occurs due to tachycardia as duration of diastole becomes shorter during tachycardia. Furthermore, blood flow of the left ventricle markedly decreases due to aortic stenosis, because the coronary arteries are compressed due to high left ventricular pressure. The left ventricle is the main site of myocardial infarction. The right ventricle maintains the coronary blood flow during systole, since the right ventricular pressure is below the aortic pressure and less susceptible to infarction. Foam cells produced by infiltration of cholesterol into the macrophages of the arterial wall stimulate proliferation of smooth muscle with the loss of elasticity and cause narrowing of the arterial wall. Narrowing of the arterial wall and rupture of atherosclerotic plaque lead to the formation of clot (thrombosis). Clot leads to ischemia followed by infarction. Dyspnea (Fig. 13.2) and orthopnea (Fig. 13.3) are the common symptoms of myocardial infarction. Cerebral stroke caused by hemorrhage or thrombosis is commonly due to: (1) Rupture of an aneurysm at the circle of Willis (an anastomosis like a ring at the base of the brain between the internal carotid and basilar arteries). Rupture of middle and anterior cerebral arteries (branches of internal carotid artery) causes contralateral hemiparesis and hemianesthesia. (2) Rupture of lateral striate branch of middle cerebral arteries. Lateral striate branch ascends over the lower part of the lenticular nucleus, bends medially and traverses the internal capsule. Because of bending, it is a common site of hemorrhage due to high resistance to blood flow (termed as Charcot's artery of cerebral hemorrhage). Pyramidal tract, extra pyramidal tract, somatic sensory pathway, and visual pathway pass through the internal capsule (Fig. 13.4). Hemorrhage or thrombosis causes contralateral hemiplegia, contralateral hemianesthesia, and homonymous hemianopia. Thrombosis at the vital center of the medulla will result in cardiac or respiratory failure.

Fig. 13.2 Causes of cardiac dyspnea and Cheyne–Stokes respiration. Breathing is uncomfortable when PV is more than four times. Strenuous exercise causes dyspnea due to marked increase in PV. Congestive heart failure patients complain of suffocation or air hunger. PV pulmonary ventilation, VC vital capacity, PaCO$_2$ partial pressure of CO$_2$ in arterial blood

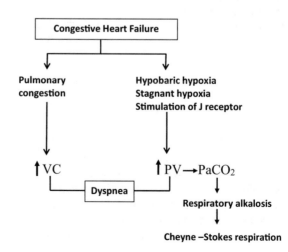

Fig. 13.3 Pulmonary
hypertension leads to cor
pulmonale followed by
right ventricular failure.
PaO₂ partial pressure of O₂
in arterial blood, PaCO₂
partial pressure of CO₂ in
arterial blood

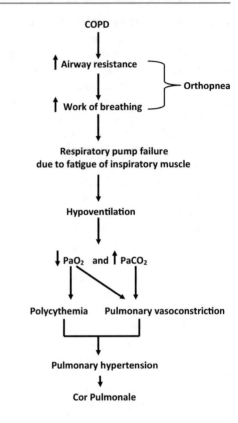

Fig. 13.4 Horizontal
section through the base of
the brain. CN caudate
nucleus, TH thalamus,
GLP globus pallidus,
PUTAM putamen, PYR
pyramidal tract

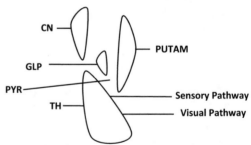

13.2.1 Nutritional Therapies

1. Minimize cholesterol-rich diet, saturated fatty acids, and trans fat.
2. Diets rich in PUFA and MUFA should be consumed. Fatty fish containing omega-3 fatty acids should be preferred.
3. Daily exercise increases HDL levels and thus decreases cholesterol level.
4. Smoking must be forbidden.
5. Avoid excess intake of sugar.
6. Alcohol, if consumed, should not exceed one unit/day.

7. DASH diet containing low sodium intake about 2 g/day and potassium in diet about 3 g/day should be consumed in order to prevent hypertension. One should not take excessive amount of potassium as high potassium may result in tachycardia, ventricular fibrillation, and even cardiac arrest.
8. Adequate nutrients, especially niacin and folic acid should be supplemented to prevent cardiovascular diseases.

13.3 Liver Diseases and Nutrition

Liver dysfunction or failure causes severe malnutrition (Marco et al. 2015). Anorexia, impaired digestion, and impaired intestinal absorption of nutrients due to malabsorption syndrome, and pancreatic insufficiency lead to malnutrition. Liver dysfunction occurs due to acute and chronic hepatitis, viral hepatitis, and cirrhosis. Cirrhosis of the liver will cause portal hypertension, esophageal varices, hematemesis, splenomegaly, ascites, hepatic encephalopathy, and renal failure. Complete dysfunction of the liver due to cirrhosis results in coma followed by death. Acidosis due to high $NADH^+ + H^+$ and ketosis due to high acetyl CoA aggravate hepatic coma. High level of NH_3 (derived from metabolism of amino acids) due to liver dysfunction also contributes to hepatic coma. Inhibition of glycogenolysis and gluconeogenesis due to liver failure causes severe hypoglycemia. Liver dysfunction may lead to liver cancer.

Chronic alcoholics suffer from deficiencies of vitamins B_1 and B_6. Alcohol interferes with the intestinal absorption of thiamin as well as the synthesis of thiamin pyrophosphate. Alcohol causes vitamin B_6 deficiency due to the hydrolysis of phosphate. Alcohol also interferes with absorption of folic acid, niacin, and vitamin A. Alcohol dehydrogenase present in the cytoplasm of liver cells oxidizes alcohol into acetaldehyde, which is further oxidized into acetate. Acetate is converted to acetyl CoA and thus increases synthesis of fatty acids and triglycerides in the liver. Increased triglycerides and inhibition of fatty acid oxidation causes fatty liver. Fatty liver ultimately causes necrosis of the liver cells followed by replacement of fibrous tissue resulting in cirrhosis (Colledge et al. 2010). Cirrhosis causes abdominal swelling due to accumulation of fluid in the peritoneal cavity (ascites). Complete dysfunction of the liver results in hepatic coma followed by death. Acidosis due to high $NADH + H^+$ and ketosis aggravate coma. Multifactorial causes of ascites are shown in Fig. 12.4 of Chap. 12. NH_3 derived from metabolism of amino acids is not converted to urea and may rise to the toxic level due to liver dysfunction. High level of NH_3 will contribute to hepatic coma. Liver failure causes renal failure due to decreased renal blood flow (hepatorenal syndrome). Liver dysfunction causes accumulation of phenylalanine due to deficiency of phenylalanine hydroxylase in the liver, resulting in brain damage and epilepsy. Glycogen of the liver is depleted due to liver dysfunction. Inhibition of glycogenolysis and gluconeogenesis causes hypoglycemia. Patients with hypoglycemia suffer from agitation, delirium, and impaired judgement. Liver dysfunction may lead to cancer of the liver (hepatoma). Acetaldehyde crosses the placenta during pregnancy and causes "fetal alcohol syndrome."

Cholestasis, acute and chronic hepatitis, and cirrhosis cause obstructive jaundice. Hepatitis (A, B, and E) viruses cause severe viral hepatitis, resulting in liver failure. Immuno-compromised individuals, particularly neonates, infants, pregnant women, and elderly persons are vulnerable to viral replication. Pregnant women are adversely affected. Virus can be transmitted from infected mother to fetus during pregnancy. Viral infection in pregnancy causes abortion.

13.3.1 Nutritional Interventions

1. Alcohol consumption must be completely forbidden.
2. Protein diet should be reduced below 20 g/day.
3. Carbohydrate should be the main diet (about 250–300 g/day). In case of severe liver dysfunction, glucose may be given orally or parenterally. Lipids should not be more than 20 g/day. Trans fat, processed food, cholesterol, and saturated fat should be avoided.
4. The amount of phenylalanine in the diet should be reduced.
5. Adequate nutrition having micronutrients (vitamins and minerals) should be consumed to minimize malnutrition.

13.4 Renal Diseases and Nutrition

Renal diseases occur due to glomerulonephritis, pyelonephritis, uremia, renal carcinoma, etc. Hematuria and proteinuria are the common symptoms of renal diseases. Renal hypertension occurs due to constriction of afferent arterioles either due to renal diseases or atherosclerosis. Expansion of the ECF and vasoconstriction due to increased RAAS cause renal hypertension. Hyperkalemia occurs due to decreased excretion. Renal diseases may lead to heart failure because of hyperkalemia and hypertension. Renal diseases cause metabolic bone disease due to lack of 1α-hydroxylase of the proximal renal tubules (Elia et al. 2013). Lack of erythropoietin secretion due to renal diseases causes anemia. Damage of nephrons in uremia causes oliguria or even anuria. Accumulation of toxic unwanted substances in the blood due to renal failure causes convulsion, and even death. Renal hypertension, hyperkalemia, metabolic bone disease, anemia, and uremia are the main features of renal diseases related to nutrition.

Constriction of the afferent arterioles before entering the glomeruli due to renal diseases or atherosclerosis increases the secretion of renin by the juxtaglomerular cells. Increased RAAS decreases sodium excretion and increases ADH secretion, resulting in water and sodium retention. Renal diseases also causes retention of potassium due to the failure of excretion. Angiotensin II is a potent vasoconstrictor. Expansion of the ECF and vasoconstriction results in renal hypertension. Renal diseases predispose to heart failure because of hypertension and expansion of the ECF. Deficiency of 1α-hydroxylase of the proximal tubules of kidneys prevents the synthesis of calcitriol. The patients may suffer from metabolic bone diseases due to decreased plasma levels of calcium and phosphate. Erythropoietin, probably secreted by interstitial cells of the peritubular capillaries in kidneys increases the number of committed stem cells (progenitor cells) in the bone marrow that are converted to RBC precursors and ultimately to mature RBCs.

Pluripotent stem cells → Progenitor cells → Normoblasts (erythroblasts) → Reticulocytes → RBCs. Thus, loss of erythropoietin due to renal failure causes severe anemia. Albuminuria occurs due to increased permeability of the glomerular capillaries especially in nephrosis. Loss of excessive amount of albumin overrides albumin synthesis in the liver. Edema results due to reduced oncotic pressure. Damage of nephrons causes uremia, leading to renal failure. Renal failure causes oliguria or even anuria. Uremia develops due to accumulation of urea, creatinine, phenols, and other toxic unwanted substances in the blood. Uremia causes abnormal taste, smell, appetite sensation and causes abnormal intestinal absorption, since uremia disrupts intestinal epithelia. Uremia also causes nausea, vomiting, convulsion, coma, and eventually death. Treatment of renal failure requires either hemodialysis/peritoneal dialysis or renal transplantation.

13.4.1 Nutritional Interventions (Anderson et al. 2016; Kalantar-Zadeh and Fouque 2017)

1. Renal hypertension: Excessive salt and fluid intake must be avoided. DASH diet should be followed due to hyperkalemia.
2. Metabolic bone disease: Administration of synthetic 1,25-dihydroxy vitamin D will be useful. Excessive intake of calcium phosphate is not recommended as hypercalcemia may cause calcification of tissues like kidneys as well as renal stone formation which may aggravate further renal diseases.
3. Loss of erythropoietin: Anemia can be corrected by injection of recombinant erythropoietin.
4. Uremia: Dietary protein should not exceed more than 40 g/day. An adequate intake of carbohydrate (about 250 g/day) and fat (about 50 g/day) is required in order to provide energy.

13.5 Gastrointestinal Diseases and Nutrition

Gastrointestinal carcinoma, untreated celiac disease, Crohn's disease, and cystic fibrosis can lead to severe malnutrition. Inflammation of the stomach due to Helicobacter pylori causes gastric carcinoma. Inflammation due to bacterial overgrowth, long-standing ulcerative colitis, and high consumption of non-vegetarian diet may cause colorectal cancer. Celiac disease due to flat mucosa of the small intestine prevents the absorption of nutrients from the intestine. Crohn's disease due to the inflammation of ileum and jejunum causes anorexia, steatorrhea, and malabsorption. Cystic fibrosis causes viscid secretion, which blocks pancreatic ducts, biliary ducts, bronchi, and bronchioles. As a result, pancreatitis, bronchiectasis, pulmonary fibrosis, and liver damage occurs. Diverticular disease is characterized by a sac formed at the weak point of the wall of the intestinal tract. It is common in the descending colon. Because of least resistance, a pouch is formed. Low consumption of dietary fibers may be responsible for diverticulosis. Constipation alternating with diarrhea are the common symptoms (Longo et al. 2011).

13.5.1 Gastric Carcinoma

Atrophic gastritis may cause susceptibility to *Helicobacter pylori* infection of the stomach. The cause of gastric cancer by infection with *H. pylori* is now well documented (Wroblewski et al. 2010). Anorexia, dysphagia, discomfort after meal, hematemesis, melena, and cachexia with loss of body weight are the common symptoms of gastric carcinoma.

13.5.2 Colorectal Carcinoma

Inflammation of the colon and rectum by bacterial overgrowth, especially *Fusobacterium nucleatum* promotes colorectal carcinoma (Liu et al. 2018). The patients suffering from ulcerative colitis for a long time are at a significantly higher risk for developing colonic cancer. Cachexia, borborygmia, rectal bleeding, increase in stool frequency, a sensation of incomplete emptying, and colicky pain are the common features.

13.5.3 Celiac Disease

Celiac disease is characterized by the presence of flat mucosa of the small intestine (normal appearance of the mucosa is finger-like villi). Reaction to gluten, a protein present in cereals, causes severe inflammation of the intestinal tract that hampers the absorption of nutrients from the intestine. Features of malnutrition are common. The patients suffer from diarrhea and abdominal distension.

13.5.4 Crohn's Disease

Long-standing inflammation of the alimentary tract, particularly ileum and jejunum leads to thickened wall of the intestine that can be palpable. Crohn's disease leads to malabsorption syndrome. Weight loss is due to malabsorption, leading to fat, protein, and vitamin deficiencies. Pain of the abdomen, nausea, vomiting, borborygmia, anorexia, and steatorrhea are the common features.

13.5.5 Cystic Fibrosis

Cystic fibrosis is an autosomal recessive disease due to mutation of a gene on the long arm of chromosome 7, which blocks pancreatic ducts, biliary ducts, bronchioles, etc. Thus, it causes pancreatitis, bronchiectasis, pulmonary fibrosis, and hepatic damage. It also causes male infertility due to blockage of vas deferens. Pancreatic insufficiency is caused by obstruction of pancreatic ductules, leading to destruction of acinar cells. Steatorrhea is common due to the loss of pancreatic lipase and proteolytic enzyme, leading to malabsorption syndrome. Fat and protein malnutrition occurs.

13.5.6 Diverticular Disease

Diverticular disease is a pouch formed at the weak point of the intestinal tract. Middle aged elderly person may suffer from diverticulosis.

13.5.7 Nutritional Therapies

1. Gastric carcinoma: The patients having atrophic gastritis or after partial gastrectomy should consume low sodium and high dietary fiber to reduce the occurrence of gastric carcinoma. Consumption of excessive amount of alcohol along with smoking must be avoided.
2. Colorectal cancer: The individuals must supplement the vegetarian diet with dietary fibers. Excessive intake of red meat and food containing toxins should be avoided.
3. Celiac disease: Features of malnutrition are common. Anemia occurs due to the deficiency of iron, vitamin B_{12}, and folate. Gluten-free diet with nutritional supplements and plenty of dietary fibers should be consumed in order to minimize the serious consequences of celiac disease.
4. Crohn's disease: Dietary intervention is required to rectify nutritional deficiencies. Macrocytic and megaloblastic anemia (pernicious anemia) occurs due to the involvement of terminal ileum. Due to loss of body weight, a high protein diet is necessary. A low-fat diet may improve steatorrhea. Lactose-free diet and soya milk should be taken to prevent lactose intolerance. Because of malabsorption, recommended intake of nutrients (macro- and micronutrients) must be supplemented.
5. Cystic fibrosis: Because of malabsorption, recommended daily intake of energy and nutrients is required. Supplement of fat-soluble vitamins is needed. One should be careful about food intolerance due to pancreatic insufficiency.
6. Diverticular disease: High fiber diet and plenty of fluid should be taken to prevent constipation.

13.6 AIDS and Nutrition

The acquired immunodeficiency syndrome (AIDS) is the immunodeficiency disease caused by the human immunodeficiency virus (HIV). AIDS causes a severely decreased number of CD4+ helper T cells, which leads to more or less complete loss of immune function. It is a global pandemic. It has wreaked havoc on mankind and killed more than a million people throughout the world. HIV-1 and HIV-2 are two retroviruses. The worldwide infection is mainly caused by HIV-1. Closely related HIV-2 is rare and less aggressive, and caused endemic in West Africa. HIV carries the genetic information in single stranded RNA retrovirus and docks with receptor of target host cell. The fusion of viral and cellular membrane allows viral core to enter inside the target cell. The RNA is then released from the viral core. In the next step, reverse transcriptase catalyzes the synthesis of double-stranded viral DNA from single stranded viral RNA in the cytoplasm. Finally integrase catalyzes the insertion of viral DNA into cellular DNA in the nucleus, causing the infection. Abnormalities of the immune system also include defects in lymphocytes, monocytes, macrophages, and natural killer cells.

HIV infection is acquired through unprotected sexual contact (heterosexual or homosexual), through transmission of contaminated blood/blood products, and through needles and syringes. Virus can be transmitted from the infected mother to fetus inside the uterus during pregnancy. Transmission can also occur during childbirth and during breast-feeding. Drug abusers are infected by sharing infected needles.

13.6.1 Stages of HIV Infection (Gibney et al. 2009)

Stage 1 (incubation period): After exposure to the HIV infection, there is an incubation period for about a month. During this stage, patients will not show any clinical features. ELISA test will be negative. The infected individual can transmit infection during the period.

Stage 2 (infective period): Clinical features that appear are fever, headache, weakness, enlarged lymph nodes, pharyngitis, pain in muscles, mucosal ulceration of the oral cavity, and rash of the body.

Stage 3 (asymptomatic): Persistent or severe symptoms may not surface for many years. This asymptomatic period is highly variable and may last for even weeks. Some patients may go rapidly into symptomatic clinical deterioration. This period depends on the nutritional status and drug therapy. Even in this period HIV is actively infecting and killing cells of the immune system. Viral replication takes place within lymphoid tissues. Weight loss, increased susceptibility to infection, unexplained diarrhea, deficiencies of vitamins, and lymphadenopathy (enlarged glands at the inguinal sites) may be the few signs and symptoms. During this stage, CD4+ cell counts decline from the normal value of 1200 cells/μL.

Stage 4 (symptomatic): CD4+ T cell counts fall further. Anorexia, marked loss of body weight, emaciation, fever, herpes zoster, pulmonary tuberculosis, chronic diarrhea, oral leukoplakia, oropharyngeal candidiasis, and numerous severe bacterial infections are the commonest symptoms.

Stage 5 (final stage): The patient becomes critically ill. CD4$^+$ T cell counts are extremely low and falls below a critical level (below 200 cells/µL). This stage is characterized by candidiasis (fungal infection) of the esophagus, airways or lungs, toxoplasmosis (parasitic disease caused by protozoa) of the brain causing dementia, Kaposi's sarcoma (nodules that may be red, purple, or black found on the skin and may spread to the gastrointestinal and respiratory tracts), carcinoma, viral infections, B cell lymphomas of bone marrow and gastrointestinal tract, and infections due to normally nonpathogenic bacteria. Any organ can be affected due to infections. Death will occur due to loss of immune function or due to malignant tumors.

13.6.2 Laboratory Diagnosis

1. Demonstration of HIV antibody by enzyme immunoassay—ELISA. ELISA test shows positive after 6–12 weeks.
2. Western blot is commonly used as a confirmatory test and detects antibodies to HIV antigens. Antibodies begin to appear within 2 weeks of infection.
3. CD4$^+$ T cell count is useful to evaluate and monitor HIV-infected individuals.

13.6.3 Management (Fauci et al. 2009)

A combination of several antiretroviral agents termed HAART has been effective in improving T cell counts. They act by

1. Inhibiting the viral reverse transcriptase enzyme. It falls into two categories.
   (a) Nucleoside analogues reverse transcriptase inhibitor (NRTI), for example, zidovudine (i.e., azidothymidine (AZT)).
   (b) Non-nucleoside reverse transcriptase inhibitors (NNRTI), for example, delavirdine.
2. By inhibiting HIV protease enzyme and thus preventing replication for example, ritonavir.
3. By inhibiting the binding of HIV to the receptor and thus inhibiting HIV entry, for example, maraviroc.
4. By inhibiting the viral integrase and interfering with the integration of viral DNA into the host cell genome, for example, raltegravir.

13.6.4 Role of Nutrition in HIV Infection

Anorexia and loss of body weight associated with undernutrition and malnutrition are the common features of patients infected with HIV and will be aggravated further if HIV patients are suffering from malnutrition. Furthermore, AIDS patients suffer from malabsorption syndrome, which may be due to inflammatory disease of the small intestine, resulting in maldigestion. Deficient nutrient absorption will cause malnutrition in HIV patients. Protein malnutrition causes suppression of the immune system. Individuals due to protein malnutrition are very susceptible to infections, resulting in diarrhea followed by dehydration, dermatitis, and other secondary infections. Thereby, protein malnutrition will aggravate further loss of immune function in patients infected with HIV. Malnutrition should be rectified by optimal diet having more proteins. Improved nutrition will slow the development and progression of HIV infection to AIDS. Furthermore, proper nutrition (both macro- and micronutrients) will minimize the side effects of antiretroviral drugs and will slow progression to the final stage.

13.6.5 Prevention

1. No vaccine against HIV is available. AIDS cannot be cured. There must be mass education (through advertisement or television) in order to prevent transmission of virus from person to person. It should be emphasized frequently from time to time.
2. Screening of all blood donors must be done.
3. Occupational risk of infection for health care workers and laboratory personnel who work with HIV-infected samples can be avoided by proper hygienic practices.
4. Disposable needles should be used. Needles or scalpels after handling with infected patient must be disposed of as per the guidelines.
5. Doctors and nurses after handling the patient should be careful about needle prick injury.
6. Sexual contact with people having many sex exposures must be avoided. Use of condoms for sex practices must be emphasized.
7. Infected mother should avoid breast-feeding.
8. Counselling is essential as the patient is shattered because of side effects of antiretroviral drugs. Note: AIDS and SCID (severe combined immunodeficiency) are compared in Table 13.1.

Table 13.1 Comparison between AIDS and SCID

AIDS	SCID
1. Acquired	1. Inherited disease
2. Common, global pandemic	2. Rare
3. Loss of CD4$^+$ cells	3. Loss of both B cells and T cells
4. Complete loss of immune function	4. Complete loss of immune function
5. Not curable	5. The patient's life can be prolonged by bone marrow transplant, which will provide stem cells to produce new B and T cells

13.7 Summary

Proper nutrition for cancer patients will reduce the side effects of chemotherapy and radiotherapy. Physical activity (brisk walking for about 1 h) distracts mind from negative thoughts, anxiety, and depression. Several viruses and bacterial flora in the gastrointestinal tract and various organs of the body are implicated for the development of cancer. Cancer cells cause inflammation by upregulation of proinflammatory genes. Chronic inflammation can also be responsible for development of cancer. Renal hypertension, metabolic bone disease, anemia, and uremia are nutrition-related renal diseases. Gastrointestinal cancer, celiac disease, Crohn's disease, and cystic fibrosis can lead to malnutrition. Atrophic gastritis, chronic gastritis, and chronic inflammation may cause gastrointestinal cancer. Malnutrition, anorexia, and cachexia along with loss of body weight are the main features of liver dysfunction. Complete dysfunction of the liver due to cirrhosis causes portal hypertension, ascites, hepatic encephalopathy, and increased morbidity and mortality. Malnutrition and atherosclerosis are risk factors for cardiovascular diseases. They must be prevented by nutritional interventions. AIDS is the immunodeficiency disease caused by HIV. It is a global pandemic leading to complete loss of immune function. Patients infected with HIV suffer from malabsorption syndrome. Nutritional therapy will slow progression to the final stage of AIDS.

References

Anderson CA et al (2016) Nutrition interventions in chronic kidney disease. Med Clin North Am 100(6):1265–1283

Chakrabarty AS, Chakrabarty K (2006) Fundamentals of respiratory physiology, 1st edn. I.K. International Publishing House Pvt Ltd, New Delhi

Colledge NR et al (eds) (2010) Davidson's principles and practice of medicine, 21st edn. Churchill Livingstone, London

Crosbie EJ et al (2013) Human papillomavirus and cervical cancer. Lancet 382(9895):889–899

Elia M et al (eds) (2013) Clinical nutrition, 2nd edn. Wiley-Blackwell, Wiley, Hoboken

Eliassen AH et al (2015) Plasma carotenoids and risk of breast cancer over 20 y of follow-up. Am J Clin Nutr 101:1197–1205

Farrell PJ (2019) Epstein-Barr virus and Cancer. Annu Rev Pathol 14:29–53

Fauci AS et al (eds) (2009) Harrison's manual of medicine. McGraw-Hill, New York

Ganong WF (2003) Review of medical physiology, 21st edn. McGraw-Hill, New York

Gibney MJ et al (eds) (2009) Introduction to human nutrition, 2nd edn. Wiley-Blackwell, Wiley, Nutrition Society, Hoboken

Grivennikov SI, Karin M (2011) Inflammatory cytokines in cancer: tumour necrosis factor and interleukin 6 take the stage. Ann Rheum Dis 70(Suppl 1):i104–i108

Kalantar-Zadeh K, Fouque D (2017) Nutritional management of chronic kidney disease. N Engl J Med 377(18):1765–1776

Liu L et al (2018) Diets that promote colon inflammation associate with risk of colorectal carcinomas that contain Fusobacterium nucleatum. Clin Gastroenterol Hepatol 16:1622–1631

Longo DL et al (eds) (2011) Harrison's principles of internal medicine, 18th edn. McGraw-Hill, New York

Mann J, Truswell AS (eds) (2007) Essentials of human nutrition, 3rd edn. Oxford University Press, Oxford

Marco S et al (2015) Nutrition in chronic liver disease. GE Port J Gastroenterol 22(6):268–276
WHO (2003) Diet nutrition and the prevention of chronic diseases. WHO technical report series
    916. World Health Organization, Geneva
Wroblewski LE et al (2010) *Helicobacter pylori* and gastric cancer: factors that modulate disease
    risk. Clin Microbiol Rev 23(4):713–739
Wu L et al (2018) Nasopharyngeal carcinoma: a review of current updates. Exp Ther Med
    15(4):3687–3692

Further Reading

Elia M et al (eds) (2013) Clinical nutrition, 2nd edn. Wiley-Blackwell, Wiley, Hoboken
Jochems SHJ et al (2018) Impact of dietary patterns and the main food groups on mortality and
    recurrence in cancer survivors: a systematic review of current epidemiological literature. BMJ
    Open 8(2):e014530
Mann J, Truswell AS (eds) (2007) Essentials of human nutrition, 3rd edn. Oxford University Press,
    Oxford
WHO (2003) Diet nutrition and the prevention of chronic diseases. WHO technical report series
    916. World Health Organization, Geneva

Miscellaneous Health Problems

14

Abstract

Fever is defined as the rise in body temperature due to the increase of set point of hypothalamic thermostat. The normal set point of core temperature is regulated within 37 ± 0.5 °C. Exogenous pyrogens (for example, bacterial toxins) act on macrophages and release endogenous pyrogens (cytokines). Cytokines act on hypothalamic thermostat and release prostaglandins, which raise the set point and body temperature. Fever may be caused by viruses, bacterial infections, protozoans, nematodes, endocrine disorders, pyrexia of unknown origin, and climatic factors. Proper diet plan should be constructed in order to prevent undernutrition/malnutrition or hypovolemia of the patient suffering from fever. Heat stroke is due to the failure of thermoregulatory mechanisms. Rehydration salt therapy and cold therapy is necessary to treat heat stroke without any delay. Heat exhaustion occurs in hot humid weather. Heat cramp, i.e., severe pain in the calf muscles, occurs due to exercise and excessive sweating of athletes. Pain due to heat cramp will disappear after immediate treatment of oral rehydration salt therapy. Prolonged cold exposure causes reduction of physiologic functions, leading to inhibition of respiration, low blood pressure, cardiac arrhythmia, and even ventricular fibrillation. Infants and elderly people, because of poor thermoregulation, are sensitive to hypothermia. Chronic intermittent hypoxia (CIH) is encountered when repeated adjourn to high altitude is followed by return to sea levels. Repeated cycles of hypoxia and reoxygenation produce many beneficial adaptive responses. CIH can optimize acclimatization and enhance athletic performance. Diseases which cause constipation should be rectified. Preventive measures will minimize or prevent constipation.

Keywords

Fever · Diet plan · Hypothalamic thermostat · Cytokines · Prostaglandins · Heat stroke · Heat exhaustion · Heat cramp · Hypothermia · CIH · Constipation

14.1   Normal Body Temperature

Morning oral temperature for adults is about 37 °C (98.6 °F). Evening oral temperature is slightly more than morning temperature due to daytime activities or due to circadian fluctuation and is about 37.5 °C (99.5 °F). Children have a temperature, i.e., 0.5 °C, above the adult temperature. The rectal temperature indicates the temperature of the core of body and is 0.5 °C higher than the oral temperature. Hypothalamic center, i.e., thermostat (preoptic nucleus of hypothalamus), regulates body temperature within a normal range under the fluctuations of environmental temperature (either cold or hot environment). Normal body temperature depends on the balance between heat production and heat loss due to thermoregulatory responses coordinated by the hypothalamus (Fig. 14.1).

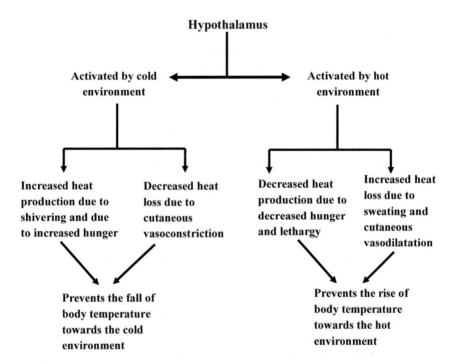

Fig. 14.1 Thermoregulatory reflex responses operated by the hypothalamus. Heat is retained inside the body due to cutaneous vasoconstriction, whereas due to cutaneous vasodilatation warm blood comes to the surface, i.e., skin, so that heat is transferred to the high environmental temperature. Shivering increases heat production due to involuntary muscular contraction. Due to sweating, evaporation of water removes heat from the body (evaporation of 1 g of water removes about 0.6 kcal of heat). Increased hunger because of specific dynamic action (SDA) of food increases heat production

14.2 Fever

Fever is defined as the rise of body temperature above the normal range due to the increase of set point of hypothalamic thermostat. The normal set point of core temperature is regulated within 37 ± 0.5 °C. The thermostat has been adjusted to a higher set point resulting in rise of body temperature. Exogenous pyrogens (for example, lipopolysaccharide endotoxin) act on the macrophages and release cytokines. Cytokines act on the preoptic nucleus (thermostat) and release prostaglandins, for example, prostaglandin E_2 (Ganong 2003), which raise the set point and body temperature (Figs. 14.2 and 14.3) (Guyton 1986).

14.3 Causes of Fever (Colledge et al. 2010; Longo et al. 2011)

A review of numerous causes of fever is beyond the scope of this nutrition-related book. Certain common causes are given below. Fever may be due to viruses, bacterial infections, protozoans, nematodes, endocrine factors or disorders, pyrexia of unknown origin, and climatic factors.

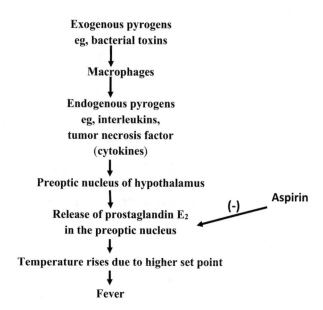

Fig. 14.2 Exogenous pyrogens, for example, bacterial toxins act on macrophages, releasing cytokines. Cytokines act on the preoptic nucleus and release prostaglandin E_2. Prostaglandin E_2 raises the set point of preoptic nucleus, causing fever. Aspirin inhibits prostaglandin synthesis and thus acts as an antipyretic agent

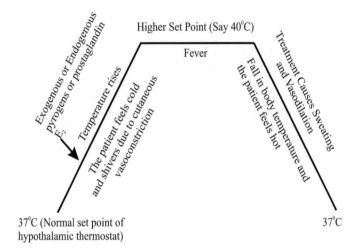

Fig. 14.3 Mechanisms of reset of hypothalamic thermostat to a higher set point. Figure also shows the mechanisms of fall of body temperature to the normal set point, i.e., 37 °C

14.3.1 Viruses

1. HIV (see Chap. 13).
2. Measles is caused by a paramyxovirus (incubation period of 6–19 days) and occurs in childhood with febrile onset accompanied by sneezing, watering of the eyes, photophobia, and Koplik's spot (red spot with bluish-white centers) on the mucus membrane of the mouth. Malnourished children suffering from measles may have serious complications such as encephalitis. All children aged 1 year should receive measles vaccination in combination with mumps and a further vaccination at the age of 4 years. Vitamin A therapy will be useful. Malnutrition should be prevented by recommended dose of macronutrients and micronutrients. Antibiotic therapy may be required if there is bacterial complication.
3. Mumps is caused by paramyxovirus. School children and young adults are mainly affected. Infection is transmitted by respiratory droplets. Fever with tender swelling of parotid glands is the characteristic feature. Sterility may occur due to orchitis (inflammation of the testis). Vaccination is described under measles. Analgesia may be given to relieve pain.
4. Dengue (breakbone fever) is caused by arboviruses and transmitted by *Aedes aegypti* mosquito. It causes high fever with severe backache and generalized pain of the body, severe pain in the joints, and an irritating rash. Dengue hemorrhagic fever is characterized by lowering of platelet count, bleeding from the gums and nose, and internal bleeding. Breeding places of *Aedes* mosquitoes must be destroyed by insecticides. Aspirin should be avoided due to the risk of hemorrhage. Platelet transfusion is needed due to thrombocytopenia.
5. Chikungunya virus (arbovirus) is transmitted in humans by mosquitoes of the genus *Aedes*. The symptoms are similar to dengue and causes fever, headache,

skin rash (maculopapular rash), joint and muscle pain. The incubation period is 2–12 days.

6. Zika virus is transmitted by *Aegypti* mosquitoes. They bite during the day. The virus causes fever, headache, muscle and joint pains, skin rashes, and redness of the eyes. It may cause Guillain–Barré syndrome (peripheral neuritis) and fetal malformation. It may be transmitted by blood transfusion, sexual transmission and from the mother to the child.

7. Hepatitis starts with jaundice accompanied by fever (see Chap. 16).

8. Viral encephalitis occurs due to various types of viruses. Inflammation can occur in various parts of the brain. It causes headache accompanied by fever, aphasia, hemiplegia, epilepsy, and unconsciousness.

9. Meningitis is due to the inflammation of meninges, i.e., pia and arachnoid membrane (CNS is enveloped by meninges). The causative agents are various types of viruses. Headache, irritability, neck stiffness, and high fever are the common symptoms.

10. Poliomyelitis starts with mild fever. Polio virus destroys the gray matter of anterior horn cells of the spinal cord, causing paresis or paralysis of muscles of the lower limbs. Prevention is by childhood immunization.

14.3.2 Bacterial Infections

1. Typhoid fever is caused by *Salmonella typhi*. Temperature rises in step-ladder fashion for 4–5 days followed by appearance of typical rash on upper abdomen and on the back. It may cause erosion of the intestinal wall, leading to hemorrhage.

2. Pulmonary tuberculosis due to *Mycobacterium tuberculosis* is characterized by evening pyrexia, anorexia, weight loss, cough, hemoptysis, and dyspnea.

3. Meningitis also occurs due to bacterial infection, for example, *Meningococcus*. It is characterized by fever, headache, drowsiness, and neck stiffness.

4. Pneumonia (inflammation of the lungs) is caused by *Streptococcus pneumoniae* (most common) and also by *Staphylococcus aureus*. Temperature rises in a few hours from 39 to 40 °C with rigor.

14.3.3 Protozoal Infections

Malaria is caused by *Plasmodium falciparum*, *P. vivax*, *P. ovale*, and *P. malariae* and is transmitted by female anopheles mosquitoes.

1. *P. vivax* and *P. ovale* cause fever with rigor. Temperature rises to about 40 °C. It lasts for several hours. Gradual fall of temperature with profuse sweating. After 48 h, the cycle is repeated.

2. *P. malariae* (quartan malaria) causes bouts of fever every third day.

3. *P. falciparum* (malignant malaria) is most dangerous due to complications like brain damage. It may lead to confusion and coma. Fever does not usually rise so high as found in other varieties.

14.3.4 Nematode Infections

Filariasis is caused by *Wuchereria bancrofti* and transmitted by infected female mosquitoes of a number of different species (most common is *Culex pipiens fatigans*). It starts with bouts of fever accompanied by pain along the inflamed lymphatic vessels. Eventually, blocking of lymph vessels causes marked swelling of the tissues (elephantiasis). The legs, scrotum, and breasts are commonly affected.

14.3.5 Endocrine Disorders

1. Hyperthyroidism causes rise of body temperature as thyroid hormones increase the metabolic rate (calorigenic action).
2. Epinephrine and norepinephrine secreted by adrenal medulla increase body temperature due to cutaneous vasoconstriction, which decreases heat loss.

14.4   Fever of Unknown Origin

The etiology of fever is uncertain. Extrapulmonary tuberculosis, neoplasms (lymphoma, malignancies, etc.), pulmonary embolism, endocarditis, thyroiditis, osteomyelitis, etc. should be ruled out.

14.5   Accompaniments of Fever

Tachycardia, hyperpnoea, concentrated urine, restlessness and delirium, anorexia, high BMR, headache, and dehydration due to water loss through the skin and lungs are the accompaniments of fever.

14.5.1 Diet Therapy and Diet Plan for Fever

14.5.1.1   Diet Therapy for Fever
The metabolic rate rises about 14% for each 1 °C of body temperature. Shivering occurs with the rise of body temperature and causes rise of the metabolic rate. Increased energy expenditure leads to glycogenolysis and lipolysis. Moreover, increased protein catabolism and increased tissue breakdown occur due to fever, especially fever of longer duration. Anorexia, nausea, and vomiting causes an aversion towards food intake. There will be marked loss of hunger and appetite. Loss of

body fluid and serum electrolytes occurs due to sweating during the falling phase of fever and due to vomiting. In view of the above, the following measures must be considered in order to prevent undernutrition/malnutrition and hypovolemia while constructing the diet charts.

1. Sufficient amount of easily digested and absorbed macronutrients is to be provided to compensate energy expenditure.
2. Well-cooked starch and simple sugars like glucose are to be selected as they are easily assimilated.
3. Proteins of high biologic value like boiled egg, fish, and milk are to be provided.
4. Emulsified form of fats like egg yolk, whole butter, and cheese should be preferred.
5. Essential fatty acids are to be provided.
6. Excessive amount of macronutrients should be restricted as they may initiate nausea and vomiting.
7. Adequate micronutrient must be provided.
8. Soluble dietary fibers present in vegetables and green leafy vegetables must be included in the diet.
9. Avoid trans fat.
10. Oral rehydration therapy to correct dehydration and electrolyte deficiencies is essential. Plenty of fluids like soups and juices should be taken.

14.5.1.2   Diet Plan
Diets must be selected according to the taste and liking of the patient. Forced feeding may initiate nausea and vomiting.

- Morning: Tea with one or two biscuits.
- Breakfast: Cheese sandwiches, one boiled egg, one cup of whole milk, and one glass of fruit juice.
- Lunch: Mixed pulses and rice or curd rice, soup of mixed vegetables with green leafy vegetables especially spinach, fish or tender chicken, curd, stewed apple or banana and one glass of orange or mixed fruit juice.
- Evening: Tea with biscuits and cheese sandwich.
- Dinner: Mixed vegetables or tomato soup, rice, mashed potato, mixed pulses without spices, and papaya.
- Bed time: One cup or glass of whole milk.

14.6   Health Problems Due to Climatic Factors (Colledge et al. 2010; Longo et al. 2011)

1. Tropical anhidrosis: Diminished sweating due to decreased number of sweat glands increases body temperature when the individual is exposed to high environmental humid temperature (for example, 40 °C). Due to loss of evaporation of water, body temperature approaches towards the environmental temperature.

2. Heat exhaustion occurs due to excessive exhaustion in hot and humid weather. Progressive rise of core body temperature occurs as water is not evaporated because of humid weather. Evaporation of water is not able to remove heat from the body. The individual must be removed from hot and humid weather with oral rehydration salt therapy and cold therapy. Otherwise the individual may progress to heat stroke.

3. Heat stroke: Thermoregulatory mechanisms may not operate due to the following conditions.
   (a) The individual with high fever is exposed to high humid temperature.
   (b) Due to congenital absence of sweat glands.
   (c) Due to overexertion in high humid temperature. Due to the failure of thermo-regulatory mechanisms, body temperature may be more than 45 °C. The individual loses consciousness and dies.

4. Heat cramps: Pain in the muscles of the legs occurs due to exercise associated with excessive sweating in hot weather. Core temperature remains more or less normal. Heat cramps occur due to sweating-induced sodium depletion. Painful muscular contraction disappears rapidly after oral rehydration salt or intravenous saline.

5. Effects of CIH at high altitude: CIH is often encountered when repeated adjourn to high altitude is followed by return to sea level. Due to CIH, an individual is exposed to repeated cycles of hypoxia and reoxygenation. CIH stimulates expression of transcription factor HIF-1 that has many beneficial adaptive responses. CIH can optimize acclimatization and enhance athletic performances. It has been reported that the attenuated maximal contraction and sensitivity of tracheal smooth muscle to ACh may be one of the mechanisms reducing the airway hyperresponsiveness of asthmatic patients after CIH exposure to high altitude (Chakrabarty and Fahim 2005).

6. Hypothermia: Body temperature may fall below 24 °C (75 °F) when an individual is exposed to cold environment (cold air or cold water) for a long time, for example, long distance swimming in cold water. Prolonged cold exposure causes reduction of physiologic functions, leading to inhibition of respiration, reduction of heart rate, and lowering of blood pressure. Cardiac arrhythmia and ventricular fibrillation may occur. ECG finding due to severe hypothermia may show J wave at the junction of QRS complex and ST segment. The individual is cold and pale. Consciousness is lost on further prolonged exposure to cold temperature. The individual can survive after rapid rewarming with heat. Infants and elderly people, because of poor thermoregulation, are sensitive to hypothermia.

7. Frostbite: Cold environment below 0 °C, especially at high altitude causes frostbite characterized by severe pain of exposed parts, i.e., fingers and feet. Lack of blood supply and oxygen due to severe vasoconstriction may cause necrosis and even gangrene of fingers and feet. High altitude, because of reduction of partial pressure of oxygen, will aggravate frost bite.

14.7 Constipation and Preventive Measures (Forootan et al. 2018)

Constipation may be defined as a decrease in frequency of motion or infrequent passage of hard stool or difficult/painful defecation. A sensation of incomplete evacuation is felt after defecation.

14.7.1 Symptoms

Slight anorexia and mild abdominal discomfort with bloating are the common symptoms.

14.7.2 Causes

1. Inadequate intake of dietary fibers.
2. Inactivity and lack of exercise.
3. Dehydration.
4. Drugs such as analgesics, antacids (calcium- or aluminum-based antacid), anticholinergics (acetylcholine increases the motility of intestine), antidepressants, oral intake of iron salt and calcium channel blocker.
5. Anorectal pain (fissures, abscesses, and piles).
6. Organic disease, for example, colorectal tumors.
7. Drug abuse, for example, opiates.
8. Hirschsprung's disease is a congenital disorder characterized by megacolon (colonic dilatation) due to the absence of ganglion cells. Contents of the bowel distend the upper colon, whereas the rectum is without the bowel contents.
9. Chagas' disease (caused by the protozoan parasite *Trypanosoma cruzi*). Dilatation of the various parts of the bowel causes constipation.
10. Hypothyroidism.
11. Cushing's syndrome.
12. Depression.
13. Irritable bowel syndrome associated with recurrent abdominal pain, distension of the abdomen, and feeling of incomplete defecation.
14. Diverticular disease.
15. Anorectal fissures or hemorrhoids.
16. Crohn's disease.

14.7.3 Preventive Measures

1. Adequate intake of dietary fibers.
2. Plenty of warm water in the morning.
3. Morning exercise. A brisk walk for half an hour to 1 h daily.

4. Sufficient time for defecation.
5. Avoid drugs which will cause constipation.
6. Diseases which cause constipation must be treated. Colorectal tumor should be removed by surgery.
7. Prescribed medication, for example, lactulose made from lactose is osmotic laxative and may be used in treatment of chronic constipation. But it may cause increased gas formation, bloating, and even diarrhea in some individuals.

14.8   Summary

Hypothalamic center, i.e., thermostat, regulates body temperature within a normal range under the fluctuation of environmental temperature. The normal set point of core temperature is regulated within 37 ± 0.5 °C. Fever is defined as the rise of body temperature above the normal range due to the increase of set point of hypothalamic thermostat. The thermostat is adjusted to a higher set point resulting in the rise of body temperature. Exogenous pyrogens act on the macrophages and release cytokines. Cytokines act on the thermostat and release prostaglandin E_2, which raise the set point and thus body temperature. Health problems due to climatic factors have been reviewed and highlighted. Various factors in the pathogenesis of constipation along with preventive measures are discussed.

References

Chakrabarty K, Fahim M (2005) Mechanism of the contractile responses of Guinea pig isolated tracheal rings after chronic intermittent hypobaric hypoxia with and without cold exposure. J Appl Physiol 99:1006–1011
Colledge NR et al (eds) (2010) Davidson's principles and practice of medicine, 21st edn. Churchill Livingstone, London
Forootan M et al (2018) Chronic constipation: a review of literature. Medicine (Baltimore) 97(20):e10631
Ganong WF (2003) Review of medical physiology, 21st edn. McGraw-Hill, New York
Guyton AC (1986) Textbook of medical physiology, 8th edn. Saunders, Philadelphia
Longo DL et al (eds) (2011) Harrison's principles of internal medicine, 18th edn. McGraw-Hill, New York

Further Reading

Colledge NR et al (eds) (2010) Davidson's principles and practice of medicine, 21st edn. Churchill Livingstone, London
Longo DL et al (eds) (2011) Harrison's principles of internal medicine, 18th edn. McGraw-Hill, New York

An Integrated View of Cognition, Oxidative Stress, Brain Functions, and Nutritional Interventions in Aging

15

Abstract

Aging is manifested within an individual due to cognitive dysfunction, oxidative stress, decline in the volumes and functions of certain areas of brain, central neurotransmitter deficits, and undernutrition/malnutrition. Progressive dysfunction of the body's homeostasis and increased susceptibility to stress and diseases are manifested with advancing age. Aging is heritable either due to allelic variation or due to polymorphism of genes. Aging alters the function of cells and various organs of the body. Despite numerous theories (cellular senescence theory, molecular inflammatory theory, mitochondrial aging theory, glycation theory, shorter telomeres theory, etc.) have been proposed, the mechanism of decline of progressive physiological functions with advancing age is poorly understood. Nutritional interventions will help to age successfully with higher functional ability in all aspects (physical, mental, and social), even at old age.

Keywords

Aging · Cognition · Brain dysfunction · Oxidative stress · Nutritional interventions

15.1 Cognitive Functions

Limbic system and certain areas of cerebral cortex contribute to the genesis of cognition. Cognition is the "study of any kind of mental operation by which knowledge is acquired" (Hedden and Gabriel 2004). It includes learning, memory, judgement, language, awareness of perception, planning and execution of tasks, all other intelligent functions and acquisition of information. Age-related decline of cognitive function is the characteristic feature in old age. Decline of all types of memory (except autobiographical memory and semantic memory regarding specific

© Springer Nature Singapore Pte Ltd. 2019
K. Chakrabarty, A. S. Chakrabarty, *Textbook of Nutrition in Health and Disease*,
https://doi.org/10.1007/978-981-15-0962-9_15

experience) was observed with advancing age. The inability to execute important information and also to initiate appropriate action are the features of getting older.

15.2   Age-Related Decline of Physiological Functions

Several changes occur in physiological functions due to aging:

1. With advancing age, downregulation of immunity occurs due to age-related atrophy of the thymus, resulting in decreased activity of cellular immunity in response to antigen. Dysfunction of B lymphocyte from the liver and bone marrow causes decreased function of plasma cells, resulting in decreased formation of antibodies in response to antigens. Thereby, elderly persons are more susceptible to various infections, especially lung infection.
2. Decreased motility of gastrointestinal tract causes constipation. Maldigestion and malabsorption will lead to nutritional deficiencies.
3. Impaired glucose tolerance occurs due to degeneration of beta cells of the pancreas. Hypothyroidism may occur due to decreased secretion of thyroid hormones. The ovaries decrease in size and become insensitive to gonadotropins.
4. Progressive decline of cardiac muscle strength and cardiac output may lead to postural hypotension. Hypercholesterolemia associated with advancing age occurs due to high LDL and low HDL. Atherosclerosis due to hypercholesterolemia, loss of elasticity and elastin of the blood vessel with advancing age increases blood pressure as well as increases the incidence of CIHD and cerebral stroke.
5. Increased loss of bone density due to aging increases the incidence of bone fracture. Fall is common owing to loss of skeletal muscle mass and strength. Fall may be due to postural hypotension, osteoarthritis, joint pain, Parkinson's disease (due to rigidity of the limbs and "festinating" gait), dizziness and vertigo (due to vestibular dysfunction), ataxia (due to cerebellar dysfunction), and poor vision. Baroreceptors are stimulated by external pressure (for example, tight collar) and may cause bradycardia and vasodilatation due to stimulation of vagal discharge, and inhibition of sympathetic discharge. This "carotid sinus syndrome" may also cause fall in the elderly people (Colledge et al. 2010).
6. Wrinkles of the skin occur due to decrease in the number of fibroblasts.
7. Aging leads to loss of nephrons with decreased glomerular filtration and tubular functions. Urinary incontinence of aged persons may be common either due to prostate enlargement in males or due to loss of muscle strength of pelvic floor in female.
8. Reduced vital capacity and increased residual volume occur due to the weakness of respiratory muscles. Loss of function of ciliary escalator of the airways increases risk of lung infection. Increased work of breathing with decline of lung elasticity causes difficulty in breathing.
9. Visual sensation is impaired due to age-related degeneration of visual receptors and macular lutea. Age-related glycosylation of lens protein causes senile cataract. Hearing sensation is impaired due to the degeneration of organ of corti.

Taste perception is impaired due to degeneration of taste buds. Impairment of olfactory functioning is described below.

15.3   Age-Related Changes of Brain Functions

Degeneration of the entorhinal cortex leads to a decrease in the perception of odor strength. The elderly people are not able to name odors due to impaired olfactory functioning. Memory functioning is impaired due to decline in the volume of PFC and hippocampus. PFC is concerned with working memory and short-term memory, whereas hippocampus is concerned with retention of explicit memory (declarative memory). Episodic memory indicates new memories of episodes and is a component of explicit memory. Alzheimer's disease occurs in middle age people with the loss of short-term and long-term memory. Amyloid peptide plaques and neurofibrillary tangles are present in the temporal lobe of patients suffering from Alzheimer's disease. Elderly people suffer from senile dementia with the loss of cognitive function. At later stage, apraxia, aphasia, and visuospatial impairment occur. Postmortem studies indicate atrophy of cerebral cortex and hippocampus. Postmortem studies also indicate that elderly people tend to have lower volumes of gray matter compared with younger adults (Carroll 2018). Impairment of recent memory correlates with the lesions of mamillary bodies and anterior nucleus of thalamus. Hippocampus projects to the mamillary bodies via fornix. Mamillary bodies project to the anterior nucleus of thalamus via mamillo-thalamic tract. Thalamus projects to cingulate gyrus, which is connected to hippocampus. Anterior nucleus of thalamus concerned with recent memory projects to PFC (Fig. 15.1).

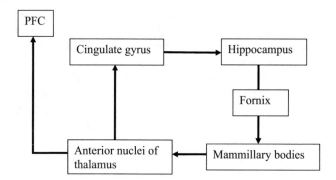

Fig. 15.1 Papez circuit

15.4   Neurotransmitter Deficits of the Brain

Low serotonin level due to decline in the function of PFC may cause depression in the elderly individuals. Age-related loss of the dopaminergic deficit of the striatum of basal ganglia leads to Parkinson's disease. Cholinergic hypothesis of aging indicates that cognitive decline in aging may be due to deterioration of cholinergic receptors of frontal and temporal lobe. Damage of cholinergic pathways is associated with Alzheimer's disease. Acetylcholinesterase inhibitors such as tacrine, donepezil, and galantamine may be useful for patients suffering from Alzheimer's disease.

15.5   Aging and Oxidative Stress (Kregel and Zhang 2007)

ROS are produced due to oxygen utilization. Highly toxic ROS causes oxidative stress and damage by modulating various enzymes (kinase 1 and 2) and transcription factors. Transcription factors activated in response to ROS or oxidative damage enter from the cytoplasm to the nucleus of the cell and bind to promoter region of genes altering gene expression. Gene expression will determine the fate of a cell (cell survival, cell senescence, or even cell death). Oxidative stress is determined by the balance between oxidative damage and the body's ability to detoxify ROS (Fig. 15.2). The cellular antioxidant defenses will try to prevent cellular and tissue damage.

Various theories have been suggested as shown below. However, aging cannot be accounted for by a single theory.

1. Cellular senescence theory: ROS have been found to modulate various signals leading to accelerated mitogenesis and premature cellular senescence. ROS are one of the main culprits for premature aging (Fig. 15.3).
2. Molecular inflammatory theory: Age-related oxidative damage due to ROS causes upregulation of proinflammatory molecules (cytokines and chemokines). The inflammatory cascade is exaggerated during aging and causes chronic inflammation leading to cellular and tissue damage, loss of cognitive functions (for example, poor performance of memory), and progression to neurodegenerative diseases like Alzheimer's and Parkinson's disease.
3. Mitochondrial aging theory: Aging-related mitochondrial dysfunction will cause leakage of electrons from the electron transport chain (ETC), which will interact with oxygen to produce superoxide or hydroxyl radicals. Production of ROS causes damage of mitochondrial DNA (mtDNA). High levels of ROS can also inflict direct damage to macronutrients such as lipids and proteins. DNA bases are very susceptible to ROS. mtDNA is prone to oxidative damage due to proximity to a primary source of ROS and its deficient repair capacity compared with nuclear DNA. Oxidative damage to both mtDNA and nuclear DNA has detrimental effect, leading to uncontrolled cell proliferation or accelerated cell death. Oxidative damage is highly correlated with biological aging (Fig. 15.4).

Note: "Glycation theory" suggests that post-translation protein modification by the linking of glucose to lysine results in decreased function of protein. Protein is resistant to breakdown. The decreased function of protein will cause anorexia and ultimately malnutrition.

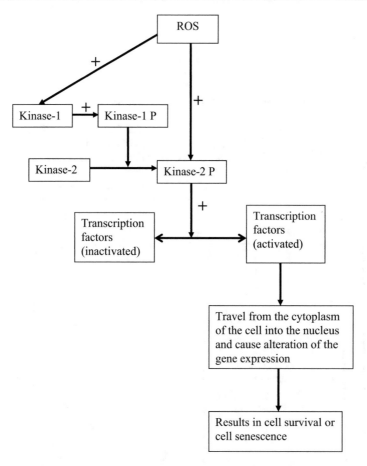

Fig. 15.2 ROS causes oxidative stress by modulating various enzymes and transcription factors. Transcription factors alter the gene expression, which will determine the fate of cell

15.6   Aging and Telomeres

Telomeres are the stretches of DNA at the ends of chromosomes. Due to cell division each time the telomeres become shorter and shorter. When they are too short, the cells cannot divide further and become inactive or "senescent." Progressive shortening of telomeres causes progressive decline in physiological and cognitive functions (Shay 2018). Normally telomeres prevent uncontrolled cell division and cancer. Shorter telomeres lead to shorter lives. Coronary disease, cancer, and infectious diseases are common in people over 60 having shorter telomeres. Cancer and infectious diseases may lead to increased incidence of malnutrition or undernutrition. Werner's syndrome is characterized by accelerated aging even at an early age due to mutation of DNA repairing enzymes. It is associated with neurodegenerative diseases and malignancy. It has been suggested that shorter telomeres may lead to premature aging in Werner's syndrome.

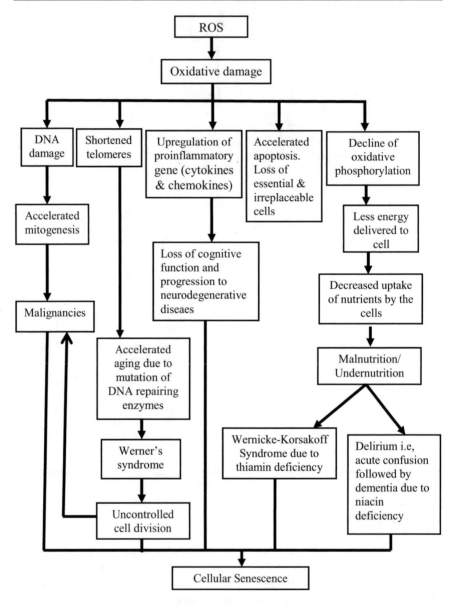

Fig. 15.3 Multifactorial pathways induced by ROS lead to cellular senescence

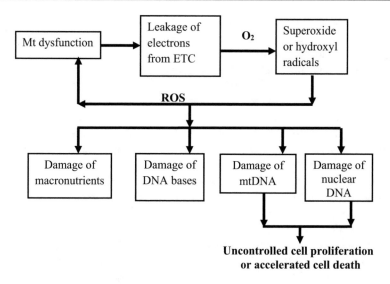

Fig. 15.4 Effects of oxidative damage

15.7   Nutritional Interventions and Additional Measures for Successful Aging

The following measures will help to age successfully in all aspects (physical, mental, and social).

1. Diets that are rich in antioxidants, i.e., vitamin E, vitamin C, β-carotene, and selenium, will minimize oxidation of cell membrane, cellular and tissue damage by ROS. Antioxidants may minimize loss of cognitive functions and may also slow the development of neurodegenerative diseases (Alzheimer's and Parkinson's diseases).
2. It has been demonstrated that regular exercise increases BDNF. BDNF supports the survival of existing neurons and stimulates the growth of new neurons and synapses. BDNF is especially active in the hippocampus (prevents decline of memory), cerebral cortex (improves intellectual and mental functions), and basal ganglion (prevents progression to Parkinson's disease). It has also been demonstrated that exercise decreases decline in age-related loss of volume of cerebral cortex (frontal, parietal and temporal lobes, and prevents the development of Alzheimer's disease). Exercise benefits hippocampal-dependent memory and prefrontal cortex-mediated executive function and may maintain gray and white matter over the lifespan (Vivar and van Praag 2017). Exercise has "global effects on the brain." Aged people must avoid sedentary life. One can prevent getting older by exercise as well as by active daily living (both indoor and outdoor activities). Brisk walking for 1 h should be advised. The individual must take rest if there is dyspnea. Exercise increases muscle mass and strength and thus improves balance.

3. Aging results in a decreased resistance to various forms of stress and distress. Elderly people are not able to withstand minor stresses (frailty). Increased glucocorticoid levels for a prolonged period reduce glucose metabolism in the hippocampus and thus may damage hippocampal neurons. At an old age, one should not live in social isolation. Isolation syndrome is a disease and aggravates multiple forms of stress. Chronic stress increases susceptibility to numerous diseases. Social interaction and especially meditation will minimize stress. Age-related depression is a common feature. Depression of elderly individuals is resistant to antidepressant drugs. Depression can be minimized by social interaction and meditation.

4. Intellectual and enriched environments stimulate neurogenesis. Older people should attend religious discussions, seminars, and conferences.

5. Aged people lose body weight due to less food intake. Anorexia may lead to malnutrition. Positive nitrogen balance should be maintained. Extra vitamin D and calcium are needed to prevent osteoporosis and fractures. There is tendency for the old people to stay inside the house. They should be exposed to sunlight for sometime for the synthesis of vitamin D. Bedridden elderly people suffer from bedsore/decubitus ulcer and even thromboembolism due to reduced blood supply of the buttocks, heels, elbows, etc. The position should be changed frequently to prevent local gangrene.

6. Supplementation of vitamins, especially thiamin, cyanocobalamin and folate, and microminerals such as zinc, selenium and chromium improves cerebral and cognitive functions in the elderly person. The beneficial effect of dietary folate intake on oxidative stress in obese women has been observed (Ribeiro et al. 2018).

7. Alcohol consumption must be avoided. Loss of brain volume of cerebral cortex is evident in chronic alcoholics, causing marked loss of cognitive function.

8. Tobacco smoking must be forbidden as tobacco smoke produces ROS, which augment age-related cellular and tissue damage.

9. Daytime sleep is to be avoided. Minimum 4–5 h night sleep is essential to maintain chronobiological cycle. Stimulants like coffee and tea should be avoided to prevent insomnia. Tobacco smoking increases metabolic rate and causes insomnia.

10. Saturated fat and trans fat must be restricted to prevent atherosclerosis. Foods rich in omega-3 fatty acids should be preferred. It has been reported that loss of DHA from the neural tissues leads to dysfunction of CNS, which causes anxiety, irritability, susceptibility to stress, dyslexia, and cognitive decline during aging (Cardosa et al. 2016). Significant association between higher saturated fat intake and risk of Alzheimer's disease/dementia was reported (Ruan et al. 2018). Sweets should be restricted.

11. Green leafy vegetables containing soluble dietary fibers and plenty of fluids will prevent constipation because of decreased motility in aged people.

12. It has been suggested that omega-3 fatty acids reduce or slow the development of depression, Alzheimer's disease as well dementia in the elderly people. Deficiency of dietary omega-3 fatty acids may even accelerate cerebral aging.

13. Dietary interventions as mentioned above can prevent cognitive dysfunction and neurological disorders such as Alzheimer's, multiple sclerosis, and epilepsy (Francis and Stevenson 2018).

15.8 Summary

Progressive decline of cognitive functions occurs due to aging. Decline of all types of memory (except autobiographical memory and semantic memory) is the feature of getting older. Hippocampus, prefrontal cortex, and mamillary bodies mediated memory functioning is impaired. Degeneration of entorhinal cortex impairs olfactory functioning. Deterioration of dopaminergic neurons of the striatum of basal ganglia and cholinergic receptors of cerebral cortex is manifested with advancing age. ROS causes oxidative stress and damage. ROS is the main cause for premature aging. Nutritional interventions and exercise will slow the development of premature aging and improve physiological functions at old age. Recommended macronutrients and micronutrients must be supplemented to prevent malnutrition.

References

Cardosa C et al (2016) Dietary DHA and health: cognitive function ageing. Nutr Res Rev 29:281–294

Carroll MA (2018) Cognitive aging and changes in brain morphology: a narrative review. Top Geriatr Rehabil 34(1):1–7

Colledge NR et al (eds) (2010) Davidson's principles and practice of medicine, 21st edn. Churchill Livingstone, London

Francis HM, Stevenson RJ (2018) Potential for diet to prevent and remediate cognitive deficits in neurological disorders. Nutr Rev 76:204–217

Hedden T, Gabriel JDE (2004) Insights into the ageing mind: a view from cognitive neuroscience. Nat Rev Neurosci 5:87–96

Kregel KC, Zhang HJ (2007) An integrated view of oxidative stress in aging: basic mechanism, functional effects, and pathological considerations. Am J Physiol Regul Integr Comp Physiol 292:R18–R36

Ribeiro MR et al (2018) Influence of the C677T polymorphism of the MTHFR gene on oxidative stress in woman with overweight or obesity: response to a dietary folate intervention. J Am Coll Nutr 27:1–8

Ruan Y et al (2018) Dietary fat intake and risk of Alzheimer's disease and dementia: a meta-analysis of cohort studies. Curr Alzheimer Res 15(9):869–876

Shay JW (2018) Telomeres and aging. Curr Opin Cell Biol 52:1–7

Vivar C, van Praag H (2017) Running changes the brain: the long and short of it. Physiology 32:410–424

Further Reading

Carroll MA (2018) Cognitive aging and changes in brain morphology: a narrative review. Top Geriatr Rehabil 34(1):1–7

Kregel KC, Zhang HJ (2007) An integrated view of oxidative stress in aging: basic mechanism, functional effects, and pathological considerations. Am J Physiol Regul Integr Comp Physiol 292:R18–R36

Poscia A et al (2018) Effectiveness of nutritional interventions addressed to elderly persons: umbrella systematic review with meta-analysis. Eur J Pub Health 28(2):275–283

Poor Maintenance of Food Hygiene and Food Safety

16

Abstract

Nutrition is adversely affected due to food spoilage, food adulteration, and food-borne/water-borne diseases. Food safety is essential for neonates, infants, children, and elderly people as their immune system is downregulated. Excessive consumption of toxins liberated due to food spoilage and food adulteration can cause malnutrition. Chronic food-borne and water-borne diseases can also lead to malnutrition.

Keywords

Food spoilage · Food adulteration · Food-borne/water-borne diseases · Food safety · Malnutrition

16.1 Food Spoilage

Food spoilage is due to the deterioration of food from the normal fresh conditions. Discoloration, bad odor, and unpleasant smell, and dysgeusia indicate food spoilage. Humidity, moisture, high environmental temperature (thermophilic), oxidizing enzymes, anaerobic condition, accumulation of toxic chemicals, food processing, and insect infestation are mainly responsible for causing food spoilage.

Causes

1. Toxic glycoalkaloids solanine and chacomine cause browning of the apples and potatoes. Due to consumption of these alkaloids, one can suffer from diarrhea, vomiting, and abdominal pain. Excessive accumulation of toxins such as linamarin in the cassava (fleshy tuberous edible roots) can cause food spoilage. Toxic substances present in soil and in the food, for example, cadmium, can cause food spoilage and can cause renal damage. Toxins are present in many foods, for example, amygdalin in the almonds and dhurrin in the sorghum.

© Springer Nature Singapore Pte Ltd. 2019
K. Chakrabarty, A. S. Chakrabarty, *Textbook of Nutrition in Health and Disease*,
https://doi.org/10.1007/978-981-15-0962-9_16

2. The bacteria *Clostridium botulinum* cause a severe form of food poisoning called botulism. It is due to the ingestion of toxins produced by this anaerobic bacteria present in preserved food (in canned vegetables, fish, and meat). This bacteria thrives in anaerobic conditions. The cans are inflated and show bubbles on opening, exhibiting sign of food spoilage. Symptoms are vomiting, diarrhea, dryness of mouth, difficulty in swallowing and breathing, and blurring of vision. Even death may occur due to the paralysis of respiratory muscles, leading to respiratory failure.

3. Food exposed to the outside environment favors the growth of bacteria. Bacteria can cause food spoilage characterized by bad odor and unpleasant smell.

4. Psychrophilic bacteria are capable of growth and reproduction in cold temperature ranging from 15 to 10 °C and can cause food spoilage.

5. Storage of food grains (rice, wheat, etc.) in humid and moist condition causes food spoilage. Water within or around the stored food is a good medium for the growth of bacteria and fungus. Growth of fungus *Aspergillus flavus* and *A. parasiticus* produces aflatoxin, which is carcinogenic and hepatotoxic. Aspergillus producing mycotoxin is a risk factor for cancer. Production of the toxic ergotoxin due to the fungal growth causes constriction of peripheral blood vessels resulting in painful sensation of the legs. Post-harvest spoilage of stored food occurs due to ochratoxin A, a fungal toxin. Fungi such as *Aspergillus* grow in stored cereals, beans, pulses, etc. and produce nephrotoxic ochratoxin A. Post-harvest spoilage of stored food in humid conditions occurs due to microbial toxins.

6. Many weeds, for example, crotalaria, grow along with the cereals and the millets. Weed seeds contain toxic pyrrolizidine alkaloids and may be contaminated with the food grains. As a result, human suffers from pain in the epigastrium.

7. Food processing and packaging techniques may cause the formation of toxic acrylamide, which causes food spoilage.

8. Due to the storage of fruits and vegetables in lead-soldered cans or due to burning of discarded battery cases near the agriculture lands, lead may get deposited on the food and can cause food spoilage. Children may be affected if they chew and eat lead-based pencil and paints. Ingestion of lead-contaminated food will cause anemia due to the inhibition of δ-aminolevulinic acid dehydratase (ALAD) and basophilic stippling of RBC.

   Clinical features are pain in muscle and joints, a bad taste, renal tubular damage, and even encephalopathy, particularly in the children. Blue line along the gums margin adjacent to the teeth may be present.

9. Polycyclic aromatic hydrocarbon (PAH) formed during barbequing or grilling meat is carcinogenic. Meat after barbequing or grilling should not be stored.

10. Methyl mercury compounds present in fish in contaminated water (for example, due to burning of fossil fuels near the river) cause Minamata disease. Ingestion of mercury-contaminated fish may cause mercury poisoning characterized by dysarthria, paresthesia, vomiting, circulatory collapse, impairment of hearing, ataxia, and even encephalopathy and renal failure. It caused many deaths in the Japanese coastal town of Minamata.

11. Arsenic poisoning can occur due to food cooked only with ground water over a long period. Arsenic inhibits enzyme action by acting on –SH groups and causes uncoupling of phosphorylation due to the inactivation of pyruvate dehydrogenase of the citric acid cycle. When ingested in large amounts, arsenic causes the symptoms of nausea, vomiting, skin cancer, cancer of the bladder and kidneys, diarrhea, convulsion, and even coma.

12. Ground water contains high levels of heavy metals and microminerals, containing toxins, which is beyond permissible limits. Food cooked with ground water over a long period causes the following hazards on health.

   (a) Nitrate toxicity occurs due to methemoglobinemia. Normally some oxidation of Fe^{2+} of hemoglobin to Fe^{3+} forms small amount of methemoglobin. But, NADH-methemoglobin reductase converts methemoglobin back to hemoglobin. After ingestion, nitrate gets converted to nitrite by gastrointestinal microflora, leading to methemoglobinemia. Methemoglobinemia causes anemic hypoxia (lack of O_2 at the tissue level). Thus, amount of O_2 carried to the tissue is less than normal. Color of methemoglobin is brownish purple (coffee/slate gray), resembling cyanosis (bluish coloration of the skin and mucous membrane induced by hypoxic hypoxia and stagnant hypoxia).

   (b) Iron toxicity (see iron metabolism—Chap. 4).

   (c) Arsenic toxicity is described before.

   (d) Lead toxicity is described before.

   (e) Salinity may lead to hypertension and gastric cancer.

   (f) Manganese toxicity (see Chap. 4).

   (g) Copper toxicity (see Chap. 4).

   (h) Fluoride toxicity (see Chap. 4).

   (i) Cadmium toxicity causes renal damage, bronchitis, and emphysema.

   (j) Chromium toxicity causes cancer of the lung.

16.2   Food Adulteration

Food is adulterated if it meets any of the following criteria:

1. It contains any deleterious substance, which may be detrimental to health.
2. It contains adulterant added to foods to increase quantity and to reduce manufacturing cost. Cheaper substance is added wholly or in part ("economic adulteration").
3. It is obtained from diseased animals.
4. It is packed after cooking under unhygienic condition.
5. It contains prohibited excessive coloring substance or excessive preservative.

Examples of Food Adulteration

1. Argemone oil is added to mustard oil. Argemone seeds (seeds of the poppy weed) derived from a wild plant are similar to mustard seeds and contain the

alkaloid sanguinarine. Argemone seeds can be identified from mustard seeds. Argemone seeds on pressing are white inside, whereas mustard seeds on pressing are yellow inside. Sanguinarine prevents the oxidation of pyruvic acid. Accumulation of pyruvic acid causes dilatation of the capillaries. Fluid comes out of the capillaries, causing edema (epidemic dropsy). Vomiting, diarrhea, and fever are other clinical features.

2. *Lathyrus sativus* seeds are added to pulses which causes lathyrism. Neurotoxin present in the *Lathyrus sativus* causes paralysis of the lower limbs and thus causes serious health problem. Neurotoxin inactivates lysyl oxidase and inhibits cross-linkage of collagen fibers, resulting in the abnormalities of bone, joints, and large blood vessels.

3. Additive/preservative/packaged food in excess is equivalent to food adulteration. Pulses, fruits, and vegetables are adulterated with the toxic coloring agents to make them look fresh. An individual suffering from hypertension should not consume packaged frozen meat or fish containing high sodium chloride for prolonged periods. Prolonged use of monosodium glutamate, a food additive, especially used in the Chinese food causes symptom like palpitation. Boric acid or sulfites may be used as an additive. Prolonged use of boric acid may cause kidney damage. Sulfite may aggravate asthma.

4. Oleomargarine or lard is added to butter.

5. Alum is added to low quality flour, making expensive flour.

6. Melamine was added to milk in China causing contamination (Chinese milk scandal). As a result of the contamination, many children died. Melamine is a white crystalline compound made by heating cynamide and used in making plastics.

7. Mogdad is a leafy tropical weedy shrub, whose seeds have been used as an adulterant for coffee. Similarly, coffee may be adulterated with chicory powder, having a large amount of caramel.

8. Brick powder is used as an adulterant for chili powder (Identification: When put in water, brick powder will settle down faster than chili powder).

9. Metanil yellow powder is used as an adulterant for turmeric powder or pulses (Identification: Instant appearance of violet color after adding drops of HCl in a test tube indicates metanil).

10. Urea or polluted water is added to milk.

11. Vanaspati is added to pure ghee or butter.

12. Powdered beechnut husk aromatized with cinnamic aldehyde is sold as a powdered cinnamon.

13. Tinopal (bleaching agent) may be added to the rice noodles to make the noodles whiter.

14. Methanol is added to alcohol. Methanol is an organic solvent used in paints. Methanol is very toxic and may cause blindness and even death.

15. Calcium carbide is used as a source of acetylene gas for ripening mangoes although it has been banned for ripening fruits by the prevention of Food

Table 16.1   Effects of food spoilage and food adulteration on health

Source	Hazard to health
1. Solanine and chacomine cause browning of the apples and potatoes	Diarrhea, vomiting, and abdominal pain
2. Toxins linamarin present in cassava, amygdalin present in almonds, and dhurrin present in sorghum	Diarrhea, vomiting, renal and hepatic damage
3. Toxin produced by anaerobic bacteria *Clostridium botulinum* (botulism) Present in preserved food (canned vegetables, fish, and meat)	Vomiting, diarrhea, difficulty in swallowing and breathing. Death may be due to the paralysis of respiratory muscles
4. Aflatoxin produced by fungus *Aspergillus flavus* and *A.parasiticus*	Carcinogenic and hepatotoxic
5. Ergotoxin, ochratoxin A due to fungal growth	Nephrotoxic
6. Lead poisoning due to using underground water or due to the storage of food in lead-soldered cans	Anemia, renal damage, and encephalopathy
7. Arsenic poisoning due to food cooked with ground water	Nausea, vomiting, diarrhea, convulsion, and even coma
8. Epidemic dropsy due to adulteration of mustard oil with argemone seeds, containing the alkaloid sanguinarine	Edema, vomiting, diarrhea, and fever
9. Lathyrism due to adulteration of pulses with seeds of *Lathyrus sativus*, containing neurotoxin	Abnormalities of bone, joints, and large vessels
10. Alcohol may be adulterated with methanol	Loss of vision and even death

Adulteration Act (under section 44A). Acetylene gas can damage the liver and kidneys.

16. Tartrazine (toxic coloring agent known to cause asthma attacks and brain damage) is used as an adulterant for yellow lentils.

Note: Wheat or rice lying on the floor of the factories may be contaminated by rat's urine, which may contain leptospira bacteria. This may lead to fatal disease known as leptospirosis (Weil's disease) (See Chap. 17).

Note: Effects of food spoilage and food adulteration on health is shown in Table 16.1.

16.3   Food-Borne and Water-Borne Diseases

Food-borne diseases (Table 16.2) (Gibney et al. 2009; Chatterjee and Chatterjee 2009)

1. *Clostridium botulinum* (described under food spoilage).
2. *Bacillus* cereus: Soil and dust are the sources of bacteria. Spores are heat resistant. Consumption of the contaminated milk, cereals, vegetables, meat, and poultry produces enterotoxin in the small intestine and causes diarrhea and abdominal pain.

Table 16.2 Food-borne diseases

Food-borne viruses	Food-borne bacteria	Food-borne parasites

1. Hepatitis A virus
   Hepatitis E virus
	1. *Clostridium botulinum*	1. Protozoa
		   (a) *Giardia intestinalis*
		   (b) *Entamoeba histolytica*
		   (c) *Toxoplasma gondii*

2. Noroviruses	2. *Bacillus cereus*	2. Trematodes
		   (a) *Clonorchis sinensis* (liver fluke)
		   (b) *Fasciolopsis buski* (intestinal fluke)
		   (c) *Paragonimus westermani* (lung fluke)

3. Astroviruses	3. *Staphylococcus aureus*	3. Nematodes
		   (a) *Ascaris lumbricoides* (roundworm)
		   (b) *Enterobius vermicularis* (thread worm)

4. Rotaviruses	4. *Clostridium perfringens*	4. Cestodes
		   (a) *Taenia solium* (pork tapeworm)
		   (b) *Taenia saginata* (beef tapeworm)
		   (c) *Echinococcus granulosus* (dog tapeworm)

5. Avian flu influenza virus (H5N1)	5. *Escherichia coli*	
6. Swine flu influenza virus (H1N1)	6. *Campylobacter jejuni*	
7. Nipah virus	7. *Listeria monocytogenes*	
	8. Salmonella species (e.g., *S. typhimurium*)	
	9. *Vibrio cholerae*	
	10. *Vibrio parahemolyticus*	
	11. *Yersinia enterocolitica*	
	12. *Shigella dysenteriae*	

3. *Staphylococcus aureus*: Skin wound infection of the hand, nose, and throat infections may be infiltrated by *S. aureus*. Poor handling practices and exposed nose infection can transfer the bacteria to the food. Consumption of the contaminated food produces heat stable toxin, which produces enterocolitis characterized by nausea, vomiting, diarrhea, and abdominal colic pain. Even certain percentage of healthy individuals carries *S. aureus* in the nose and throat and may transmit the bacteria, resulting in enterocolitis.

4. *Escherichia coli*: Human and animals (cattle, sheep, and pigs) feces are the sources of *E.coli*. Consumption of the contaminated food (undercooked beef, salad, and raw milk) causes intermittent gastroenteritis (diarrhea and abdominal pain), hemorrhagic colitis (presence of blood in the stool), and urinary tract infection with fever. *E. coli* is killed by cooking. According to Central Pollution Control Board, the number of fecal bacteria is very high in the Yamuna river. Foods obtained by farming along the Yamuna banks contain excessive amount of fecal bacteria, including *E. coli*.

5. *Clostridium perfringens*: Soil and animal feces are the sources of *C. perfringens*. Consumption of the contaminated food (meat, precooked food, and poultry) produces enterotoxin in the large intestine, which causes abdominal pain and diarrhea.

6. *Campylobacter jejuni*: Domestic animals and birds shed *C. jejuni* in their feces. Contaminated undercooked food of animal origin, unpasteurized milk and water cause fever, abdominal pain, and watery foul smelling diarrhea (sometimes with blood and mucus).

7. *Listeria monocytogenes*: It occurs as a saprophyte in soil, water, and sewage. Human infection results from contact with the infected animals (mammals, ticks, fish, etc.), inhalation of contaminated dust, and consumption of the contaminated food (milk, cheese, meat, and vegetables). Immuno-compromised individuals (neonates, pregnant women, and elderly persons) are highly vulnerable to infection. Fever, watery diarrhea, and pain in joints and muscles are the common symptoms. Infection in pregnant woman may lead to abortion or stillbirth. It may cause meningitis and septicemia.

8. *Salmonella* species (e.g., *S. typhimurium and S.enteritidis*): Droppings of rats, lizards, poultry, and other animals cause food contamination. Human carriers also do occur. Sources of salmonella food poisoning are undercooked poultry, milk, egg, meat, raw fruit, and raw salad. Proper cooking destroys the bacteria. Fever, diarrhea, nausea, vomiting, and abdominal pain are the common symptoms.

9. *Vibrio cholera*: Described in detail under water-borne diseases. Contamination of food by flies, unclean hands and consumption of shellfishes cause cholera.

10. *Vibrio parahemolyticus*: It inhabits costal and estuarine waters. It is found in fishes, shrimps, crabs, and oysters. Consumption of the infected fishes and shellfishes causes gastroenteritis.

11. *Yersinia enterocolitica*: It has been isolated from a wide range of domestic animals such as cats, dogs, and pigs. Consumption of undercooked pork and raw milk causes gastroenteritis with fever and ileitis, which may mimic appendicitis.

12. *Shigella dysenteriae* (Bacillary dysentery): It is a worldwide endemic. Outbreaks occur in crowded population with poor sanitation due to natural catastrophes. This disease occurs by ingestion of infected food (contact through unwashed hands after defecation) and is spread by the flies. The bacteria cause inflammation of the large intestine. Fever, diarrhea, colicky abdominal pain, tenesmus (continuous and frequent desire to defecate with little fecal matter), purulent exudate and blood in stools with little fecal material are the common symptoms.

13. *Giardia intestinalis* (*Giardia lamblia*): Geographical distribution is worldwide, but more common in the tropics. Mode of infection in man is by ingesting cysts present in the contaminated food or water. After ingestion, the cyst hatches out trophozoites, which multiply in enormous numbers in the duodenum and jejunum. They often localize in the biliary tract or gall bladder to avoid the acidity of duodenum. Enteritis, diarrhea, steatorrhea due to malabsorption of fat and

inflammation of the gall bladder, flatulence, and abdominal distension are the common symptoms.

14. *Entamoeba histolytica*: Geographical distribution is worldwide, but more common in the tropics than in the temperate zones. The mode of infection is fecal contamination of cysts in the vegetables, food, and drinking water, which causes amoebiasis. Eating of uncooked vegetables and fruits causes amoebiasis. Clinical features are: (1) Amoebic dysentery characterized by the intermittent passage of blood and mucus in the stool. (2) Thickening of the bowel wall, rendering it palpable due to ulceration. (3) The trophozoites of *E. histolytica* are carried to the liver by the portal vein from the large intestine causing hepatitis and even liver abscess characterized by pain of the upper part of right abdomen and fever.

15. *Toxoplasma* gondii: Worldwide distribution. Mode of infection is: (1) Congenital toxoplasmosis occurs when the fetus is infected in the uterus through the placental route of the infected mother. Infection may lead to abortion or stillborn. If the fetus survives, there may be hydrocephalus (enlargement of the head due to increased cerebrospinal fluid within ventricles of the brain) or microcephaly (smallness of the head due to the underdeveloped brain) or blindness (2) Acquired toxoplasmosis due to ingestion of raw inadequately cooked infected meat, cow's milk, and egg. Humans may be affected by ingestion of cysts discharged in the feces of infected cat. Clinical features are fever, enlarged lymph nodes, rashes of skin, hepatitis, retinitis leading to blindness, pneumonia, and neurological disorders such as encephalomyelitis (demyelination of ascending or descending tracts of the spinal cord and brain).

Note: *Trypanosoma cruzi* causes Chagas' disease. Although Chagas' disease is not a food-borne disease, it is transmitted to humans by the protozoan parasite (Edwards and Bouchier 1991). Protozoan parasite *T.cruzi* is transmitted to the humans by the trypanosomes present in the feces of nocturnal blood sucking reduviid bugs. Infected feces are rubbed in through the mucosa of mouth or nose or wound on the skin. Young children are commonly affected. Acute infection is characterized by fever and enlargement of the liver and spleen. After many years chronic infection appears characterized by dilatation of the colon, esophagus, bile ducts, and bronchi. The presence of parasite in the cardiac muscle results in inflammation, which can be fatal.

16. *Clonorchis sinensis* (liver fluke): It is a parasite of the Far East. The endemic areas include China, Japan, Korea, Taiwan, and Vietnam. Mode of infection is eating inadequately cooked or pickled fish. Habitat is fish-eating mammals, for example, cat. The adult worm lives in the biliary tract of the liver. It infects the bile duct and gall bladder. Anorexia, abdominal pain, diarrhea, recurrent jaundice (as it infects the bile ducts of man), and hepatomegaly are the common symptoms.

17. *Fasciolopsis buski* (giant intestinal fluke): It is an Asiatic trematode found in China, Malaysia, India, and other oriental regions. Principal reservoirs of infection for man are pigs. Mode of infection is eating infected (encysted cercariae) water plants as raw foodstuff. Site of localization is the small intestine.

Clinical features are asthenia (weakness or loss of strength), anasarca (edema of legs and trunks), anemia, fever, abdominal pain, and intestinal obstruction.

18. *Paragonimus westermani* (lung Fluke): Geographical distribution—Japan, Korea, India (Bengal, Assam, and South India), and Nepal. Principal reservoirs are domestic and wild carnivora that feed on crustaceans. Adult worms live in the lungs of humans. Mode of infection is eating raw or improperly cooked flesh of an infected crab or crayfish. "Drunken crab" (raw crab meat soaked in alcohol) is a popular delicacy in China. Cough with recurring attacks of hemoptysis (coughing up of blood) is the characteristic symptom.

19. *Ascaris lumbricoides* (roundworm): Worldwide distribution but more prevalent in the tropics such as India, China, and South-East Asia. The adult worm lives in lumen of the small intestine of humans and is the largest intestinal nematode. Eating raw vegetables cultivated on soil containing contaminated human excreta (*Ascaris* eggs) can cause infection. Infection may also occur by drinking contaminated water. Eggs may be transmitted to the mouth by dirty fingers. Clinical features are: (1) Intestinal obstruction, especially in children due to large number of adult roundworms in the intestine. (2) The body fluids of roundworm when absorbed may cause allergic manifestations such as urticaria, conjunctivitis, and irritation of the respiratory tract. (3) It may cause protein malnutrition by preventing the absorption of amino acids from the intestine. (4) Larvae penetrate the intestinal mucosa, pass through portal circulation to the liver and enter inside the lungs via the right heart. It is characterized by pneumonia, fever, cough, and difficulty in breathing. It is caused by the presence of larvae inside the lungs ("Loeffler's syndrome").

20. *Enterobius vermicularis* (thread worm): Worldwide distribution. Adult worms (gravid females) live in the cecum and appendix. Mode of infection is by ingestion of ova present in contaminated food and water. The movement of worms and laying eggs around the anus causes intense itching. Eggs present in the fingers due to scratching of the anus are transmitted in the food or water or directly into the mouth (especially, in the children). Presence of ova causes pruritus and eczematous condition around the anus. Bedwetting due to nocturnal enuresis (involuntary micturation) and even appendicitis are the common symptoms.

21. *Taenia solium* (pork tapeworm): Worldwide distribution. It causes cysticercosis. Adult worm lives in the small intestine. The larvae penetrate the intestinal mucosa and are carried to many parts of the body forming cysticerci. Mode of infection is by eating insufficiently cooked pork, containing cysticercus cellulosae. Vague abdominal discomfort, indigestion, and diarrhea are the common symptoms. Palpable nodules in the subcutaneous tissue and muscles are due to the presence of cysticerci. Presence of cysticerci in the brain leads to epileptic attack.

22. *Taenia saginata* (beef tapeworm): All the features are same as *Taenia solium*. Only difference is in the mode of infection, i.e., infection is due to eating of insufficiently cooked beef, containing cysticercus bovis.

23. *Echinococcus granulosus/Taenia echinococcus* (dog tapeworm, the hydatid worm): Worldwide distribution. Humans harbor the larval form in the small intestine. Mode of infection is due to ingestion of eggs present in the dog's feces by humans. It occurs by (1) allowing the dog to take food from the same plate, (2) eating uncooked food contaminated with infected dog feces, and (3) handling and fondling of the dog. Disease (echinococcosis or hydatid disease) remains silent for many years. After many years, disease is detected due to pressure effect of hydatid cyst or due to the rupture of cyst. The larval forms via the small intestine enter inside various organs such as the liver, lungs, and brain, which, in turn, form hydatid cyst. The pressure symptoms will vary according to the site of cyst. Rupture of cyst causes severe allergic reactions, causing fever and urticaria or even anaphylactic reaction as the fluid of the hydatid cyst is highly toxic and contains antigen.

24. Hepatitis A Virus (HAV): Severe food-borne/water-borne disease is caused by Hepatitis A virus (belong to picornavirus group of enteroviruses). It is transmitted by oral-fecal route. Infected person excretes virus in the feces for up to 14 days before the onset of illness. The virus may be shed in the feces for 7–14 days after the onset of symptoms. The food or water becomes contaminated with the feces of infected individuals. Infected shellfishes, oysters, and raw mussels can cause outbreaks. Infection spreads via the intestine to the liver. Anorexia, fever, nausea, vomiting, abdominal pain (right upper abdomen due to stretching of peritoneum over the enlarged liver), tender liver, and jaundice are the common symptoms. Diagnosis is based on the detection of IgM antibody.

25. Hepatitis E virus (HEV): Severe food-borne and water-borne disease is caused by HEV (belongs to the genus Hepevirus). It is transmitted by fecal-oral route. Parenteral transmission may be possible. Acute liver failure accompanied by anorexia, nausea, vomiting, tender liver, right abdominal pain over the enlarged liver, and jaundice are the common features.

26. Norovirus (Calicivirus): Food or water becomes contaminated with feces of the infected patients (virus may be shed for up to 7 days after the illness). Shellfish eaters are common sufferers as shellfish beds are contaminated with the human feces from sewage. It is characterized by gastroenteritis (nausea, vomiting, and diarrhea).

27. Astrovirus: The infants less than 1 year are most susceptible. Contaminated food may cause outbreaks in creches and schools. Diarrhea is the common symptom.

28. Rotavirus: Food and water are contaminated and causes infantile gastroenteritis characterized by fever, vomiting, and diarrhea.

29. Avian flu influenza virus (H5NI): Bird flu can infect the humans who have no immunity against it. Apart from people working with poultry, consumption of the undercooked infected poultry meat or eggs can cause infection. Avian flu epidemic first occurred in Hong Kong. Afterwards, it has been reported in Asia, Africa, Europe, Indonesia, and Vietnam. Bird flu may pose a pandemic threat. The birds who have recovered from the infection may continue to shed the virus in the feces for about 10 days. Diarrhea, cough, dyspnea, sore throat, high fever,

and headache are the common symptoms. Death may occur due to pneumonia, sepsis, and organ failure.

30. Swine flu influenza A virus (H1N1): In 2009, swine flu outbreaks occurred in the USA. Consumption of undercooked infected pork can cause infection. It is a contagious respiratory disease caused by the virus, which also enters the body through the inhalation of droplets from the mouth or the nose. Immuno-compromised individuals, cancer patients, diabetic patients, and AIDS patients are more at risk. Fever, chills, sore throat, cough, body ache, headache, fatigue, nausea, vomiting, and diarrhea are the common symptoms. Death may occur due to electrolyte imbalance as a result of dehydration, pneumonia, and respiratory failure.

31. Nipah virus was first recognized in 1999 among pig farmers in Malaysia. Apart from pigs, outbreaks of the Nipah virus were also reported in other domestic animals such as horses, goat, sheep, cats, and dogs. Fruit bats are the natural host for Nipah virus. Nipah virus is a zoonotic virus, which is transmitted from animals to humans. The virus can also be transmitted through contaminated food (either by bat or by pig) or by direct contact with infected people. People should avoid being in contact with infected pigs. Fruits with sign of bat bites should be discarded. Fruit products such as raw dates contaminated with saliva from infected fruit bats must be washed thoroughly and peeled before consumption. Infected people initially suffer from fever, headache, myalgia (muscle pain), vomiting, and acute respiratory illness. Later on infected people suffer from dizziness and drowsiness followed by pneumonia, acute respiratory distress, seizures, encephalitis, and unconsciousness. Death occurs due to fatal encephalitis.

Water-Borne Diseases (Gibney et al. 2009; Chatterjee and Chatterjee 2009)
1. *Vibrio cholera*: Infected drinking water, ingestion of shellfish, and contamination of the food by flies or by hands cause cholera. Epidemics occur following large religious festivals. Bacteria pass out in the stool or vomit of the patient. Severe "rice-water" type of diarrhea, vomiting, and abdominal pain are the common symptoms. The excessive loss of fluid and electrolytes leads to intense dehydration. The blood pressure falls, the pulse becomes rapid and feeble, and the urine output decreases. The skin becomes cold and the eyes are sunken. Death occurs due to circulatory failure unless fluid and electrolytes are replaced.
2. *Campylobacter jejuni.*
3. *Giardia intestinalis.*
4. *Entamoeba histolytica.*
5. *Ascaris lumbricoides.*
6. *Enterobius vermicularis.*
7. Hepatitis A virus.
8. Noroviruses.

Water-borne disease from 2 to 8 are described under food-borne diseases.

Table 16.3 Effects of food-borne and water-borne diseases on health

Source	Hazard to health
1. *Bacillus cereus*, producing enterotoxin in small intestine	Diarrhea and abdominal pain
2. *Staphylococcus aureus*, producing enterocolitis	Nausea, vomiting, diarrhea, and abdominal pain
3. *Escherichia coli*, causing gastroenteritis	Diarrhea, abdominal pain, hemorrhagic colitis, and urinary tract infection with fever
4. *Salmonella* species (*S.typhimurium* and *S. enteritidis*)	Fever, nausea, vomiting, diarrhea, and abdominal pain
5. *Vibrio cholerae*	Rice-water type of diarrhea, vomiting, intense dehydration, and circulatory failure
6. *Shigella dysenteriae* (bacillary dysentery) causing inflammation of the large intestine	Fever, diarrhea, colicky abdominal pain, tenesmus, and purulent exudates with blood in stool
7. *Giardia intestinalis* trophozoites present in the duodenum and jejunum, biliary tract, and gall bladder	Steatorrhea, inflammation of the gall bladder, and abdominal distension
8. *Entamoeba histolytica* (amoebiasis)	Blood and mucus in the stool, palpable bowel wall, hepatitis, and even liver abscess
9. *Toxoplasma gondii*	Hydrocephalus, microcephaly, and blindness. Humans may suffer from fever, hepatitis, blindness, pneumonia, and encephalomyelitis
10. *Ascaris lumbricoides* (roundworm) causes Loeffler's syndrome	Intestinal obstruction, allergic manifestation, pneumonia, and dyspnea

Note: New Delhi metallo-beta-lactamase-1 (NDM-1) refers to gene's protein product that some bacteria (for example, *E. coli*) synthesize. Superbug NDM-1 makes bacteria highly resistant to antibiotics. The bacteria can lead to severe dysentery, septicemia, and urinary tract infection. It has been found in the drinking water of Asian continent. It has been also reported in developed countries.

Note: Effects of food and water borne diseases on health are shown in Table 16.3.

16.3.1 Vulnerable Individuals to Food-Borne and Water-Borne Diseases

Chronic food-borne and water-borne diseases can cause malnutrition. Immuno-compromised individuals are more susceptible to severe food-borne or water-borne infections. Immune system of neonates, infants, and children are not well developed. Pregnant women because of downregulation of immunity suffer more from severe illness. Elderly people are more prone to infection as immune function deteriorates with aging process. Children with protein malnutrition are adversely affected. Transplant patients having immunosuppressant drugs and AIDS patients are also adversely affected due to the downregulation of immunity.

16.3.2 Prevention Strategies for Food-Borne and Water-Borne Diseases

Whenever there is outbreak of infection, one must take the following preventive measures:

1. Hand washing with soap after the use of lavatory. Standard of personal hygiene must be improved by education. Good hygiene will prevent the spread of infection.
2. Vaccination for a specific disease, if available, must be done where there is endemic, epidemic, or pandemic infection.
3. Food and water are to be protected from contamination by flies, cockroaches, rats, and lizards.
4. One should observe personal cleanliness and elementary hygiene while taking meals.
5. Water supplies should be protected from fecal pollution.
6. Salads and fruits must be washed thoroughly.
7. One should not take any undercooked food. Food must be cooked properly.
8. Purified water must be taken (potable water).
9. Carrier, if possible, should be detected and isolated.
10. Habitat of the vector should be destroyed.
11. Poultry infected with avian influenza virus must be destroyed in order to prevent pandemic threat.
12. It will be safe to bury or burn infective discharge, soiled clothing, and stools of infected patients at a designated disposal point.

16.4 Nausea and Vomiting

Nausea refers to the sensation or desire to vomit. Irritation of the mucosa of bowel causes activation of vomiting center and area postrema in the medulla, which via visceral reflexes causes vomiting. Impulses are projected to the vomiting center and area postrema via sympathetic nerves and vagi. The area postrema situated on the lateral wall of the fourth ventricle is stimulated by circulating chemical agents. Because of reverse peristalsis, contents are emptied into the stomach. Vomiting ultimately refers to the forceful ejection of the gastric contents through the mouth due to increased intra-abdominal and intra-thoracic pressure. Simultaneous closure of the glottis prevents aspiration of the contents into the trachea. Simultaneous relaxation of the lower esophageal sphincter and the esophagus helps in ejection of the gastric contents (Ganong 2003).

Causes of nausea and vomiting
1. Enteritis (either due to bacterial infection, especially *Staphylococcus aureus, Salmonella* species and *Vibrio cholerae* or due to viral infection, especially hepatitis A virus, norovirus and rotavirus).
2. Inflammatory diseases, for example, hepatitis.

3. Psychological causes, for example, anorexia nervosa and bulimia nervosa.
4. Uremia due to high concentration of unwanted toxic nitrogenous compounds in the blood.
5. Pregnancy (during the first trimester, pre-eclampsia and eclampsia).
6. Toxins liberated due to hepatic failure.
7. Excessive consumption of alcohol.

Complications of vomiting: Dehydration, malnutrition, metabolic alkalosis (Fig. 16.7), hypokalemia, hypochloremia, hyperaldosteronism, and hematemesis due to mucosal tear of the stomach.

16.5 Diarrhea

Diarrhea is liquid or loose or watery fecal output with increased frequency of bowel movement. Fecal output may contain blood, mucus, and purulent exudate.

Diseases causing diarrhea are given below:
1. Due to parasites: (1) *Giardia intestinalis,* (2) *Entamoeba histolytica,* (3) *Clonorchis sinensis* (liver fluke), (4) *Taenia solium* (pork tapeworm), (5) *Taenia saginata* (beef tapeworm). Giardia causes diarrhea with steatorrhea. Stools are greasy and yellowish color due to malabsorption of fat. *Entamoeba histolytica* causes intermittent blood and mucus diarrhea.
2. Due to bacterial infections: (1) *Clostridium botulinum,* (2) *Clostridium perfringens,* (3) *Staphylococcus aureus,* (4) *Escherichia coli,* (5) *Campylobacter jejuni,* (6) *Listeria monocytogenes,* (7) *Salmonella enteritidis,* (8) *Salmonella typhimurium,* (9) *Vibrio cholerae,* (10) *Vibrio parahemolyticus,* (11) *Yersinia enterocolitica,* (12) *Shigella dysenteriae,* (13) *Bacillus cereus.*

Characteristic features of bacterial infections: (1) *E. coli* causes intermittent blood (due to hemorrhagic colitis) and mucus diarrhea. (2) Purulent exudate and blood in the stools with little fecal material occurs due to *Shigella dysenteriae.* Other symptoms are colicky abdominal pain, fever, and tenesmus. (3) Typical rice-water diarrhea occurs due to *Vibrio cholerae.* (4) Watery foul smelling diarrhea with blood and mucus occurs due to *Campylobacter jejuni.*
3. Due to viruses: (1) Hepatitis A virus, (2) Noroviruses, (3) Astrovirus, (4) Rotavirus, (5) Avian flu influenza virus (H5N1).

(See more details under food-borne and water-borne diseases.)
4. Miscellaneous causes: (1) Lactose intolerance is due to the deficiency of enzyme lactase. Lactose cannot be hydrolyzed to glucose and galactose. As a result, non-absorbed lactose increases intraluminal osmotic pressure, which holds water and causes diarrhea (osmotic diarrhea). Lactose passes into the colon, where it is fermented by bacteria, causing flatulence and discomfort (see more details in Chap. 2). (2) Celiac disease leads to severe diarrhea. (3) Endocrine disorders, causing abnormal intestinal motility, for example, diabetic diarrhea, hyperthyroidism, and adrenal insufficiency. (4) Inadequate absorptive surface due to surgical removal of the intestine (due to malignancy). The defect of digestion and

absorption of macro- and micronutrients will cause diarrhea with bulky fecal material and malnutrition. (5) Tropical sprue occurs mainly in the West Indies and Asia. It is characterized by malabsorption and villous atrophy (more or less similar to celiac disease) of the small intestine. Common feature are steatorrhea, abdominal distension, and edema (due to hypoalbuminemia).

Prevention of diarrheal diseases has been discussed under food-borne and water-borne diseases.

16.5.1   Management and Nutritional Therapy of Diarrheal Diseases (Elia et al. 2013)

A fluid or semifluid diet without any dietary fiber should be given. Dehydration and deficiency of electrolytes must be corrected by either oral administration or intravenous administration of electrolytes (sodium, potassium, and chloride) and dextrose or glucose, depending on the severity of diarrhea. Rehydration therapy contains both salt and glucose as glucose augments the absorption of salts both in the kidneys and intestine due to glucose-Na^+cotransport (secondary active transport). If there is associated vomiting, intravenous therapy must be started to prevent hypovolemic shock. Diarrhea should be controlled by the administration of specific antibiotics against bacterial infections or specific drugs for protozoal infections and parasitic diseases as prescribed by the physician. Oral rehydration therapy is recommended to correct dehydration. In addition to oral rehydration therapy, coconut water, buttermilk, barley water, pulse water, cereal water, etc. may be given depending on liking of the patient.

16.6   Control of Osmolality and Water Balance of the Body

Thirst is a subjective sensory perception that stimulates a desire to drink water. The intake of water is regulated by balancing the loss of water from the body. Increased motivation to drink water occurs in response to increased osmolality of the ECF and decreased volume of the ECF. The following factors stimulate drinking: (1) Osmolality, (2) Angiotensin II, (3) Oropharyngeal receptors, (4) Inhibition of volume receptors in the low pressure system of the atria as well as in ventricles of the heart (Elia et al. 2013). Dehydration occurs due to vomiting, diarrhea, sweating, physical exercise, starvation, major abdominal surgery, especially complicated by infection, etc. Dehydration also occurs in diabetes insipidus and diabetes mellitus, which is secondary to polyurea. Fluid can be shifted to the interstitial compartment during starvation or surgical stress and can lead to edema. Atrial natriuretic peptide (ANP) is produced by the atria of the heart in response to a rise in atrial pressure. Brain natriuretic peptide (BNP) is produced in the brain and ventricles of the heart. Expansion of the ECF volume increases the secretion of ANP and BNP due to stretching of the atrial and ventricular receptors of the heart and causes natriuresis and diuresis. In congestive heart failure, stretching of the ventricular receptors increases the secretion of BNP and causes natriuresis and diuresis in response to high sodium concentration.

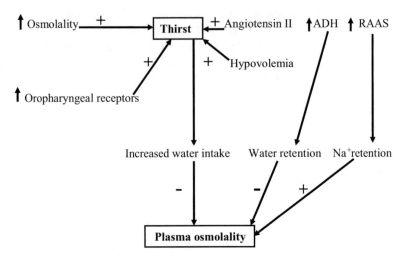

Fig. 16.1 Control of osmolality and water balance of the body. All the pathways shown above are additive and integrated. Summation of these pathways controls osmolality and water balance of the body

ANP and BNP cause decreased secretion of renin, aldosterone, and ADH. Increased renal sympathetic nerve via baroreceptor decreases sodium excretion and increases RAAS. Increased renin-angiotensin decreases sodium excretion and increases aldosterone and ADH secretion. Angiotensin II increases the secretion of aldosterone by acting directly on the adrenal cortex. It cannot cross the blood–brain barrier and acts on the circumventricular organs [subfornical organ (SFO) and organum vasculosum of the lamina terminalis (OVLT)], which project to the neural areas concerned with thirst/osmoreceptors of the anterior hypothalamus and stimulate thirst. Osmoreceptors are present in preoptic nucleus of the anterior hypothalamus and sense the osmolar concentration of the surrounding ECF. Shrinkage of osmoreceptors by surrounding hypertonicity causes stimulation. Reverse happens due to hypotonicity. Hypovolemia produced by hemorrhage causes increased drinking despite no change in the osmolality of plasma. The combination of thirst and simultaneous ADH release due to blood loss during hemorrhage will try to minimize hypovolemia. Furthermore, the vagi carry information from the oropharyngeal receptors due to dehydration-induced dryness to the neural areas/osmoreceptors of the anterior hypothalamus and induce drinking (Figs. 16.1 and 16.2).

16.7 Regulation of Acid-Base Balance of the Body

The anion composition of ECF depends upon the acid-base balance of the body. Acidosis is present when the arterial pH is below 7.4, whereas alkalosis is present when it is above 7.4.

Henderson–Hasselbalch equation

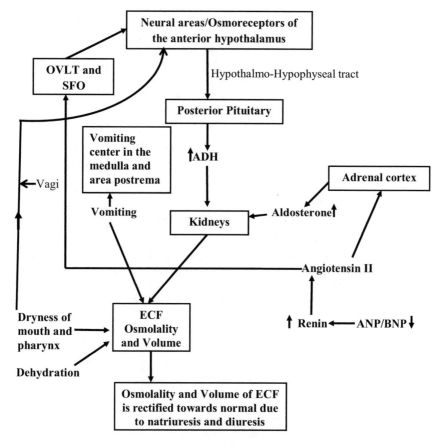

Fig. 16.2 Control of ECF volume and tonicity (see details in the text)

$$pH = pK_1 + \log \frac{HCO_3^-}{H_2CO_3}$$

It is clear from the equation that any condition which increases the amount of HCO_3^- or decreases the amount of H_2CO_3 will increase the pH. On the other hand, any condition which increases the amount of H_2CO_3 or decreases the amount of HCO_3^- will decrease the pH. Fundamentally, the kidneys and the respiratory system regulate the acid-base balance (Guyton 1986). The kidneys regulate the acidosis by increasing reabsorption of HCO_3^- into the ECF, which is secondary to the secretion of H^+ in the tubular fluid (Fig. 16.3). This reabsorption of HCO_3^- will correct acidosis, taking the example of strong acid, i.e., HCl.

$$NaHCO_3 + HCl = NaCl + H_2O = NaCl + H_2O + CO_2$$

As H_2CO_3 is a weak acid compared to HCl, the pH of the ECF will be higher. Moreover, H_2CO_3 will be converted into CO_2, which will be exhaled through the

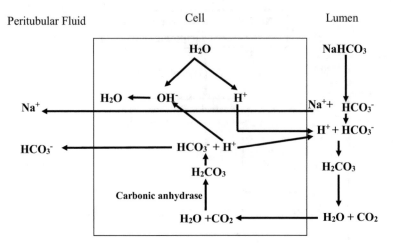

Fig. 16.3 CO_2 and H_2O is hydrated by carbonic anhydrase into H_2CO_3, which is converted into H^+ and HCO_3^-. H^+ enters the tubular lumen and combines with HCO_3^- to form H_2CO_3, which is ultimately converted into CO_2 and H_2O. CO_2 enters into the tubular cell for further secretion of H^+. With the secretion of H^+, Na^+ from the tubular lumen is reabsorbed into the peritubular fluid

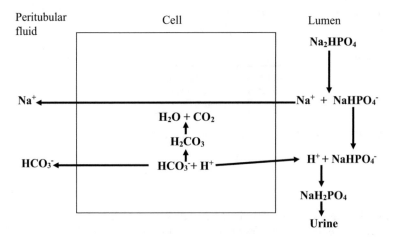

Fig. 16.4 The pH of tubular lumen should not reach 4.5 (limiting pH). H^+ secreted is removed by $NaHPO_4^-$

lungs. H^+ is secreted by the renal tubules (about 80% of H^+ is secreted by the proximal tubules, about 19% is secreted by the distal tubules, and a small amount is secreted by the collecting tubules).

Secretion of H^+ will stop when the pH of tubular lumen reaches 4.5 (limiting pH). Secreted H^+ is removed continuously by dibasic phosphate (Na_2HPO_4) and glutamine/deamination of other amino acids (Figs. 16.4 and 16.5).

Cell

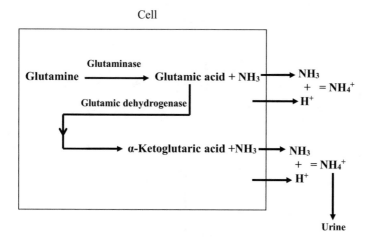

Fig. 16.5 Secreted H^+ is removed by deamination of amino acids, for example, glutamine

Renal disease or decreased renal mass results in decreased reabsorption of bicarbonate. The amount of endogenous acids (for example, sulfuric acid or phosphoric acid derived from dietary proteins) becomes more, compared with bicarbonate and thus acidosis develops.

Respiratory acidosis occurs due to hypoventilation

Hypoventilation $\rightarrow \uparrow CO_2 \rightarrow \uparrow H_2CO_3 \rightarrow \uparrow H^+$ secretion $\rightarrow \uparrow HCO_3^-$ reabsorption will try to rectify acidosis.

Respiratory alkalosis occurs due to hyperventilation (Seifter and Chang 2017).

Hyperventilation $\rightarrow \downarrow CO_2 \rightarrow \downarrow H_2CO_3 \rightarrow \downarrow H^+$ secretion $\rightarrow \downarrow HCO_3^-$ reabsorption will try to rectify alkalosis.

CNS interstitial fluid may acidify due to intracranial hemorrhage. Acidosis will be minimized or rectified towards normal by respiratory alkalosis due to hyperventilation (Seifter and Chang 2017). Metabolic acidosis occurs due to the increased amount of noncarbonic acids (for example, increased lactic acid due to strenuous exercise and increased beta-hydroxybutyric acid and acetoacetic acid in diabetic ketosis will increase acid load). Metabolic acidosis is mainly regulated by the lungs (Fig. 16.6).

$\uparrow H^+ \rightarrow$ Hyperventilation $\rightarrow CO_2$ excreted out $\rightarrow \downarrow pCO_2$ and $\downarrow H^+$ (Chakrabarty and Chakrabarty 2006).

Increased excretion of H^+ will try to rectify acidosis. Metabolic alkalosis occurs due to vomiting (Fig. 16.7). Metabolic alkalosis is mainly regulated by lungs. Hypoventilation causes a decrease in pCO_2 and simultaneously also decreases the renal ability to reabsorb HCO_3^- and will try to rectify alkalosis.

Factors affecting hydrogen ion secretion and bicarbonate reabsorption
1. Substances inhibiting carbonic anhydrase (for example, acetazolamide) inhibit hydrogen ion secretion.

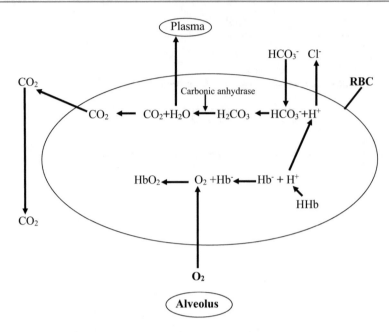

Fig. 16.6 Transport of CO_2 at the level of alveolus. After chloride shift, HCO_3^- enters inside the RBC for exhalation of CO_2

2. Hyperkalemia inhibits hydrogen ion secretion as potassium and hydrogen compete with each other for secretion into the tubules.
3. Administration of aldosterone increases hydrogen ion concentration as hydrogen is secreted in exchange with sodium reabsorption.
4. Parathormone decreases bicarbonate reabsorption and hydrogen secretion.
5. Hypercalcemia increases hydrogen ion secretion either through the inhibition of parathormone or through the activation of carbonic anhydrase.
6. Decrease in the ECF volume increases acid secretion by the following mechanism. Decrease in ECF volume causes renal vasoconstriction, which subsequently decreases peritubular pressure and activates the RAS. As a result, there is increased Na^+ reabsorption and increased H^+ secretion from the renal tubules.
7. Fruits contain Na^+ and K^+ salts of weak organic acids, which are metabolized to $NaHCO_3$ and $KHCO_3$ in the body. This indicates that fruits are the dietary sources of alkali. The slight alkalinity is necessary for the function of nerves, muscles, and heart. In contrast of the dietary source of alkali in fruits, some fruits contain acid. For examples, tomatoes contain citric acid which is oxidized to CO_2 and H_2O, and apples contain malic acid which is again oxidized to CO_2 and H_2O. Acidic fruits will not create problem as CO_2 is exhaled out. Furthermore, acid present in some other fruits will not be absorbed. Therefore, the alkalinity of fruits is maintained (Mann and Truswell 2007).

Note: Contrast between respiratory acidosis and metabolic acidosis is shown in Table 16.4.

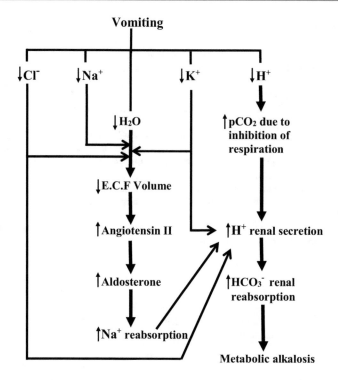

Fig. 16.7 Causes of metabolic alkalosis due to vomiting

Table 16.4 Contrast between respiratory acidosis and metabolic acidosis

Respiratory acidosis	Metabolic acidosis
1. Hypoventilation	1. Hyperventilation
2. Increase in pCO_2	2. Decrease in pCO_2
3. Increased bicarbonate reabsorption by the kidneys	3. Decreased bicarbonate secretion by the kidneys
4. Acidosis is mainly corrected by the kidneys	4. Acidosis is mainly corrected by the lungs

Reverse occurs due to respiratory alkalosis and metabolic alkalosis

16.8 Summary

Food spoilage, food adulteration, food-borne and water-borne diseases will have adverse effects on a healthy life. Deterioration of food from a fresh condition may cause cancer, renal damage, hepatic damage, encephalopathy, etc. and can even lead to death. Prolonged intake of toxins will lead to malnutrition or undernutrition. Immuno-compromised individuals are vulnerable to the above diseases. Apart from causing malnutrition or undernutrition, food-borne and water-borne diseases are detrimental for health as they may cause meningitis, septicemia, hepatitis, pneumonia, and even organ failure. Preventive strategies for food spoilage, food adulteration, food-borne and water-borne diseases must be followed as prevention is better than cure.

References

Chakrabarty AS, Chakrabarty K (2006) Fundamentals of respiratory physiology, 1st edn. I.K. International Publishing House Pvt Ltd, New Delhi

Chatterjee KD, Chatterjee D (2009) Parasitology (prootozoology and helminthology), 13th edn. CBS Publishers and Distributers Pvt Ltd, New Delhi

Edwards CRW, Bouchier IAD (1991) Davidson's principles and practice of medicine, 16th edn. Churchill Livingstone, London

Elia M et al (eds) (2013) Clinical nutrition, 2nd edn. Wiley-Blackwell, Wiley, Hoboken

Ganong WF (2003) Review of medical physiology, 21st edn. McGraw-Hill, New York

Gibney MJ et al (eds) (2009) Introduction to human nutrition, 2nd edn. Wiley-Blackwell, Wiley, Nutrition Society, Hoboken

Guyton AC (1986) Textbook of medical physiology, 8th edn. Saunders, Philadelphia

Mann J, Truswell AS (eds) (2007) Essentials of human nutrition, 3rd edn. Oxford University Press, Oxford

Seifter JL, Chang HY (2017) Extracellular acid-balance and ion transport between body fluid compartments. Physiology 32:367–369

Websites

https://www.who.int/news-room/fact-sheets/detail/nipah-virus
http://www.encyclopedia.com/topic/food_adulteration.aspx

Further Reading

Chatterjee KD, Chatterjee D (2009) Parasitology (prootozoology and helminthology), 13th edn. CBS Publishers and Distributers Pvt Ltd, New Delhi

Cissé G (2019) Food-borne and water-borne diseases under climate change in low- and middle-income countries: further efforts needed for reducing environmental health exposure risks. Acta Trop 194:181–188

Gibney MJ et al (eds) (2009) Introduction to human nutrition, 2nd edn. Wiley-Blackwell, Wiley, Nutrition Society, Hoboken

Oliver SP (ed) (2019) Foodborne pathogens and disease, vol 16. Mary Ann Liebert, Inc, New Rochelle

Potable Water

17

Abstract

Water from rivers, ground water, streams, rain water collection, and lakes is not potable as these contain parasites, bacteria, and viruses, leading to water-borne diseases. Ground water contains excessive unwanted minerals such as nitrates, lead, arsenic, chromium, and fluoride. Toxic minerals may lead to serious health problems. Water can be purified by various methods to make it potable. Boiling water, chemical disinfection, filtration, UV purification, solar water disinfection, and reverse osmosis (RO) are various water safety measures for potable water. Water should not be contaminated by rat's urine. Rat's urine may contain leptospira bacteria, which may lead to fatal disease known as Weil's disease.

Keywords

Untreated water · Boiling water · Chemical disinfection · UV purification · Solar water disinfection · RO · Weil's disease

17.1 Sources of Untreated Water

Potable water is high quality drinking water, which should not lead to any water-borne disease. It should not produce any immediate or long-term hazard or harm to the human body (Ford et al. 2017). High quality drinking water remains a critical issue in underdeveloped countries. According to WHO, nearly 10% of population in the world does not have access to potable water resources. Exposure to chemicals in drinking water may lead to carcinoma. Safe drinking water is a global challenge, especially in rural areas. Sources of untreated water are from rivers, ground water through tube wells or dug wells, streams, rain water collections, and lakes. Such water is not potable as it may contain parasites, bacteria, viruses, high level of suspended solids, etc. It may cause water-borne diseases. Rivers may be polluted with sewage effluent, industrial pollutants, or surface runoff. The presence of dead

K. Chakrabarty, A. S. Chakrabarty, *Textbook of Nutrition in Health and Disease*,
https://doi.org/10.1007/978-981-15-0962-9_17

animals in untreated sources of water is common. Open pit latrine and bucket latrine cause diseases. Ground water may contain excessive unwanted minerals such as nitrates, lead, arsenic, chromium, and excessive fluoride dissolved from the granite rocks. The presence of coliform bacteria in water indicates fecal contamination by sewage. Presence of nitrates and arsenic causes serious impact on health.

17.2   Purification of Drinking Water (Treacy 2019)

Purification of water (water safety measures for potable water)

1. Boiling water will kill bacteria and many parasites like *Giardia intestinalis* and *Cryptosporidium parvum*, which are commonly found in rivers and lakes. Water temperature above 70 °C will kill water-borne pathogens within 30 min. However, it cannot kill *Clostridium botulinum*. This method is not possible for the poor people as it requires abundant fuel. Moreover, boiling cannot remove heavy metals.
2. Chemical disinfection by addition of chlorine/iodine/sodium hypochlorite (bleach) to natural sources of water. After the treatment, water is piped to homes through a tap. However, these additions to water give unpleasant taste. Excessive ingestion of iodine may lead to adverse health effect (iodine toxicity).
3. Filtration: Granular active carbon filtering adsorbs many toxic compounds and removes chlorine from treated water. Filter of this kind can remove most bacteria and protozoa, but not viruses. It can remove chlorine from treated water. However, it cannot remove thread-like leptospira bacteria. Water contaminated by infected rat's urine contains *L. icterohemorrhagiae*. This will lead to fatal disease known as leptospirosis (Weil's disease) characterized by hepatitis, renal tubular necrosis, hemorrhagic conjunctivitis, and even meningitis.
4. Ultraviolet purification (Johnson et al. 2010): UV light deactivates microbes by forming covalent linkages on DNA and thus prevents microbes from reproducing.
5. Solar water disinfection: Microbes are destroyed by temperature and UV radiation provided by the sun. It is a low-cost method of purifying water.
6. Reverse osmosis (Ahuchaogu et al. 2018): Normally due to osmosis water moves from low solute concentration through a membrane to high solute concentration until there is equilibrium of solute concentration on both sides of the membrane. By RO water having minerals (underground water obtained from wells or tube wells and sea water) is allowed to pass through a thin membrane by applying pressure leaving behind the larger molecules of solutes. Thus, water is forced to pass through a membrane by applying pressure from high solute concentration. This osmosis is termed RO. Larger particles cannot pass through the pores of the membrane as pore size varies from 0.1 nm depending on the type of filter. RO system also includes carbon filter to trap chlorine and organic substances and ultraviolet lamp for disinfecting microbes.

17.3 Summary

Safe drinking water is a global challenge, especially in rural areas. Water from rivers, ground water, rain water collection, and lakes is not potable as it contains parasites, bacteria, viruses, etc. Ground water contains excessive unwanted minerals, which causes serious impact on health. Potable water is high quality drinking water, which should not cause any water-borne disease. Potable water should not cause any immediate or long-term hazard to the human body. Exposure to chemicals in drinking water may lead to cancer, gastrointestinal diseases, renal diseases, etc. High quality drinking water is a critical issue in underdeveloped countries. Water can be purified by various methods to make it potable. Boiling, chemical disinfection, filtration, UV purification, solar water disinfection, and RO are various water safety measures.

References

Ahuchaogu AA et al (2018) Reverse osmosis technology, its applications and nano-enabled membrane. Int J Adv Res Chem Sci 5(2):20–26

Ford L et al (2017) Human health risk assessment applied to rural populations dependent on unregulated drinking water sources: a scoping review. Int J Environ Res Public Health 14(8):846

Johnson KM et al (2010) Ultraviolet radiation and its germicidal effect in drinking water purification. J Phytology 2(5):12–19

Treacy J (2019) Drinking water treatment and challenges in developing countries. In: The relevance of hygiene to health in developing countries. IntechOpen. https://doi.org/10.5772/intechopen.80780

Websites

http://www.who.int/water_sanitation_health/dwq/GDWQ2004web.pdf
https://www.who.int/news-room/fact-sheets/detail/drinking-water

Further Reading

Cotruvo J (2018) Drinking water quality and contaminants guidebook, 1st edn. CRC Press, Taylor and Francis, Boca Raton

Levallois P, Belmonte CV (eds) (2019) Drinking water quality and human health. Int J Environ Health Public Res 16:631. https://doi.org/10.3390/books978-3-03897-727-8

Spellman FR (2017) The drinking water handbook, 3rd edn. CRC Press, Taylor and Francis, Boca Raton

Appendix A

Normal values [Serum (S); Plasma (P); Blood (B)]

Albumin (S) 3.5–5.5 g/dL (35–55 g/L)

Ascorbic acid (P) 0.4–1.5 mg/dL (23–85 µmol/L)

Angiotensin-II (P) 1–3 ng/mL

ADH (P) 1–13 pg/mL

Bilirubin (S) Total 0.2–1.2 mg/dL; Direct (conjugated) 0.1–0.4 mg/dL; Indirect 0.2–0.7 mg/dL

Calcium (S) 8.5–10.5 mg/dL (2.1–2.6 mmol/L); Ionized calcium (S) 4.24–5.24 mg/dL

β-carotene (S) 50–300 µg/dL

Ceruloplasmin (S) 25–45 mg/dL

Chloride (S/P) 96–106 meq/L (96–106 mmol/L)

Cholesterol (S/P) 150–200 mg/dL (4–6 mmol/L); Cholesterol (HDL) 35–60 mg/dL; Cholesterol (LDL) 80–180 mg/dL

Copper (P) 100–200 µg/dL (16–30 µmol/L)

Creatinine (S/P) 0.7–1.5 mg/dL (60–130 µmol/L)

Cyanocobalamin (S) 200 pg/mL (148 pmol/L)

Ferritin (S) Adult women 20–120 ng/mL; Adult men 30–300 ng/mL

Folic acid (S) 5–20 ng/mL (10–30 nmol/L)

Glucose (S/P) (Fasting) 70–110 mg/dL (4–5.5 mmol/L)

Hemoglobin (B) Male 14–16 g/dL (15.5 g%); Female 13–15 g/dL (14 g%)

HbA$_{1C}$ (S) Normal 4–5.6%; Diabetes 6.5% or higher

IgG (S) 800–1200 mg/dL

IgM (S) 50–200 mg/dL

IgA (S) 150–300 mg/dL

IgE (S) 1.5–4.5 µg/dL

Insulin (S) 4–25 µU/mL

Iodine (S) 5–10 µg/dL

Iron (S) 60–150 µg/dL

Iron binding capacity (S) 250–400 µg/dL; Percent saturation 20–55%

Magnesium (S/P) 2–3 mg/dL (1–1.5 mmol/L)

pH (B) 7.35–7.45 (H$^+$ 44.7–45.5 nmol/L)

Phosphate (S) 3–4 mg/dL (1–1.5 mmol/L)

© Springer Nature Singapore Pte Ltd. 2019
K. Chakrabarty, A. S. Chakrabarty, *Textbook of Nutrition in Health and Disease*,
https://doi.org/10.1007/978-981-15-0962-9

Potassium (S/P) 3.5–5 meq/L (3.5–5 mmol/L)
Sodium (S/P) 136–145 meq/L (136–145 mmol/L)
Transferrin (S) 200–300 mg/dL (23–35 μmol/L)
Triglyceride (S) 76–150 mg/dL (0.9–1.8 mmol/L)
Urea (S) 20–40 mg/dL (2.4–4.8 mmol/L)
Vitamin A (S) 15–50 μg/dL (0.5–2 μmol/L)
Vitamin B_{12} (S) >200 pg/mL (>148 pmol/L)
Vitamin D (calcitriol) (S) 25–60 pg/mL (60–100 pmol/L)
Zinc (S) 50–100 μg/dL (8–16 μmol/L)

Appendix B

Greek alphabets	Units of length	Units of mass
Alpha α	1 megameter (Mm) = 10^6 m	1 megagram (Mg) = 10^6 g
Beta β	1 kilometer (km) = 10^3 m	1 kilogram (kg) = 10^3 g
Gamma ϒ	1 meter = m	1 gram = g
Delta δ	1 centimeter (cm) = 10^{-2} m	1 centigram (cg) = 10^{-2} g
Epsilon ε	1 millimeter (mm) = 10^{-3} m	1 milligram (mg) = 10^{-3} g
Zeta ζ	1 micrometer (μm) = 10^{-6} m	1 microgram (μg) = 10^{-6} g
Eta η	1 nanometer (nm) = 10^{-9} m	1 nanogram (ng) = 10^{-9} g
Theta θ	1 angstrom (Å) = 10^{-10} m	1 picogram (pg) = 10^{-12} g
Iota ι	1 picometer (pm) = 10^{-12} m	1 femtogram (fg) = 10^{-15} g
Kappa κ	1 femtometer (fm) = 10^{-15} m	
Lambda λ		
Mu μ		
Nu ν		
Xi ξ		
Omicron o		
Pi π		
Rho ρ		
Sigma σ,ς		
Tau τ		
Upsilon υ		
Phi φ		
Chi χ		
Psi ψ		
Omega ω		

© Springer Nature Singapore Pte Ltd. 2019
K. Chakrabarty, A. S. Chakrabarty, *Textbook of Nutrition in Health and Disease*,
https://doi.org/10.1007/978-981-15-0962-9

Appendix C

RDA/Adequate daily dietary intakes of lipid-soluble and water-soluble vitamins at different age groups and during pregnancy and lactation [Adapted from National Institute of Nutrition (Indian Council of Medical Research)]

	Infant	Schoolchildren	Adult	Elderly	Pregnancy	Lactation
Vitamin A (µg)	350–400	600	600	600	800[a]	1000
Vitamin D (µg)	10	10	10	10	10[b]	10[b]
Vitamin E[c] (µg)	4–6	4–6	10	10	10	12
Vitamin K (µg)	15–20	15–20	60	60	70–80	70–80
Vitamin B_1 (mg)	0.3	0.3	1.5	1.5	1.5	1.5
Vitamin B_2 (mg)	0.3	1.4–1.6	1.4–1.6	1.4–1.6	1.4–1.6	1.4–1.6
Vitamin B_3 (mg)	0.5	15	15	15	>15	>15
Vitamin B_5 (mg)	0.4	4–6	4–6	4–6	4–6	4–6
Vitamin B_6 (mg)	0.3–0.4	2–2.5	2–2.5	2–2.5	2–2.5	2–2.5
Biotin (µg)	100–200	100–200	100–200	100–200	100–200	100–200
Vitamin B_{12} (µg)	0.2	1.5	1.5	1.5	1.5	1.5
Folate (µg)	25–30	100	100	100	300–400	300–400
Vitamin C (mg)	25	50	50	50	50	>50

[a]Excessive amount to be avoided to prevent developmental abnormalities of the fetus
[b]Pregnant women and elderly person require additional amount due to inadequate exposure to sunlight
[c]Requirement depends on the intake of PUFA

© Springer Nature Singapore Pte Ltd. 2019
K. Chakrabarty, A. S. Chakrabarty, *Textbook of Nutrition in Health and Disease*,
https://doi.org/10.1007/978-981-15-0962-9

Appendix D

RDA/Adequate daily dietary intakes of minerals at different age groups and during pregnancy and lactation [Adapted from National Institute of Nutrition (Indian Council of Medical Research)]

	Infant	Schoolchildren	Adult	Elderly	Pregnancy	Lactation
Calcium/ phosphorus (mg)	400–600	400–600	800–1000	800–1000	1200	1200
Magnesium (mg)	40–60	200	400	400	400	400
Iron (mg)	8	20	20[a]	>20	30	30
Copper (µg)	>200–400	>200–400	1000	1000	1000	1000
Iodine (µg)	40	40	150	150	>150[b]	150
Zinc (mg)	4	4	15	15	15	15
Manganese (mg)	0.5–1.5	0.5–1.5	2.5	2.5	2.5	2.5
Selenium (µg)	15–20	15–20	60	60	60	60

[a]Adult females require >20 to compensate loss of blood during menstruation
[b]To prevent congenital hypothyroidism and cretinism

Appendix E

Types of vegetarian diets (all vegetarian diets include cereal and millets, grain legumes, vegetables including green leafy vegetables, fruits, roots, tubers, nuts, and seeds)

Sub-classification of vegetarian diet:

1. Vegans: Excludes all animal flesh or any animal by-product, ingredients or additives, for example, honey and gelatine. Vegans do not consume eggs and dairy products. They avoid sugars that are processed using animal bone char.
2. Lactovegetarians: Do not consume any animal flesh or eggs. They consume dairy products. However, cheese is made with animal-derived rennet.
3. Ovovegetarians: Do not consume any animal flesh, animal by-product, and dairy products. They consume eggs.
4. Lacto-ovovegetarians: Similar to lacto-/ovovegetarian, but they consume dairy products and eggs. Most common type of vegetarian.
5. Piscatarian: Consumes fish, seafood, and dairy products, but excludes red meat and poultry.
6. Fruitarian: A type of vegan, which consume mainly all types of fruits.
7. Pollotarian: Consumes poultry or fowl, but excludes red meat, fish, and seafood.
8. Flexitarian: High consumption of plant-based diet with occasional consumption of meat.
9. Semi-vegetarian: Include pollotarian and flexitarian. They usually consume fish or poultry, but exclude red meat.

Note: Non-vegetarian refers to a person who is not a vegetarian. Consumes animal flesh and includes all sub-classification of vegetarian diet from 1 to 9

Website

https://vegetarian-nation.com/resources/common-questions/types-levels-vegetarian/

© Springer Nature Singapore Pte Ltd. 2019
K. Chakrabarty, A. S. Chakrabarty, *Textbook of Nutrition in Health and Disease*,
https://doi.org/10.1007/978-981-15-0962-9

Glossary

Acid value　It is the mass of potassium hydroxide in milligrams to neutralize 1 g of oil or fat.

Acrocyanosis　Bluish-purple discoloration of the digits due to slow circulation of the blood through capillaries.

Alactasis　Lactose is not digested due to the deficiency of lactase.

Allergy　Hypersensitive individuals to a particular antigen, leading to rashes, dermatitis, asthma attack, and other symptoms depending on a particular allergen (e.g., food allergy)

Alopecia　Loss of hair.

Alzheimer's disease　Dementia of middle age or later age characterized by loss or impairment of memory, language problem, and apraxia (loss of skilled movement) due to the development of neurofibrillary tangle of temporal lobes.

Amenorrhea　Absence of the menstrual periods.

Amnesia　Total or partial loss of memory.

Amphipathic lipid　Part of fatty acid is hydrophobic and part is hydrophilic.

Amphoteric property　Molecule or ion that can react as an acid as well as a base.

Anaphylaxis　Severe immediate allergic response to a sensitized antigen causing flushing, itching, vomiting, severe hypotension, asphyxia due to airway obstruction, etc. and even death.

Anasarca　Edema of legs and trunk.

Anencephaly　Absence of bones of the skull.

Angular stomatitis　Inflammation at the corners of mouth.

Anorexia　Lack or loss of appetite leading to weight loss.

Aphonia　Loss of voice.

Apoenzyme　Protein portion of holoenzyme.

Apoptosis　Programmed cell death.

Appetite　Desire to eat food associated with pleasant sensation.

Ascites　Accumulation of fluid in the peritoneal cavity.

Asphyxia　Combined effects of acute severe hypoxemia and hypercapnia.

Asthenia　Weakness or loss of strength.

Asthma　Narrowing of the lumen of the airways (especially bronchi) causes difficulty in breathing mainly during expiration, cough, and wheezing.

Ataxia　Defect about the direction of movement.

© Springer Nature Singapore Pte Ltd. 2019
K. Chakrabarty, A. S. Chakrabarty, *Textbook of Nutrition in Health and Disease*,
https://doi.org/10.1007/978-981-15-0962-9

Atelectasis   Collapse of the alveoli of lungs.

Bartter syndrome   A rare condition of hypovolemia due to loss of sodium in the urine and hyperkalemia.

Bitot's spots   Cheesy foamy grayish-white triangular spots on the surface of conjunctiva.

BMR   It is the minimum energy expenditure necessary to carry out the basic physiologic functions of the body and the vital life processes of the body when a person is at rest and awake.

Calorie (equal to 4.184 J)   It is the amount of heat that raises the temperature of 1 g of water from 14.5 to 15.5 °C.

Catalyst   It is a substance that increases the rate of a chemical reaction and remains unchanged after the reaction.

Celiac disease   Abnormal mucosal cell defect (finger-like or leaf-like villi) of the small intestine induced by immunological responses to the gluten protein of wheat.

Chagas' disease   *Trypanosoma cruzi* (protozoan parasite) causes dilatation of the various parts of the bowel.

Cheilosis   Fissures at the corner of the mouth and cracked swollen red lips.

Chemokines   Protective substances that attract neutrophils to inflammation areas.

Cirrhosis   Strands of fibrous tissue and nodules of the liver, e.g., alcoholic cirrhosis.

Coarctation of the aorta   Congenital narrowing of a segment of aorta causes hypertension in the arm and hypotension in the legs.

Coenzymes   Heat stable low molecular weight organic molecules that are essential for the activity of enzymes.

Cofactor   A non-protein ion associated with an enzyme that is essential for the activity of the enzyme.

Cognition   Study of any kind of mental operation by which knowledge is acquired. It includes learning, memory, language, perception, reasoning, acts of creativity, problem-solving, planning and execution of tasks, and so on.

Colostrum   Yellowish fluid secreted from the mammary gland for about 2–3 days after the delivery.

Conn's syndrome   Tumor of adrenal cortex causes excessive secretion of aldosterone.

Convalescence   Gradual recovery of strength of body after a disease.

Cor pulmonale   Right ventricular hypertrophy.

Crohn's disease   Edematous inflammation of the bowel leading to ulceration.

Cushing's syndrome   Tumor of glucocorticoid-secreting adrenal cortex causes excessive secretion of glucocorticoids.

Cytokines   Hormone like molecules that regulate immune functions.

Deamination   Removal of an amino group to form a ketoacid and NH_3.

Delirium   Mental disturbances, disorientation, hallucination, and extreme excitement.

Delirium tremens   Acute state characterized by agitation, bad dreams, anxiety, tremor, visual and sensory hallucination of animals and insects. It is generally due to withdrawal symptom in chronic alcoholics.

Delusion   A false belief that cannot be altered by rational argument and is outside the person's socio-cultural and educational background.

Dementia   Deterioration of higher intellectual functions, poor memory, deterioration of personal care, changes in personality, impaired reasoning ability, and disorientation.

Dermatitis   Inflammation of skin caused by a particular agent.

Dysphagia   Difficulty in swallowing.

Dyspnea   Difficulty in breathing.

ECF   The fluid that surrounds the cells of the body.

Eczema   Itchy red skin leading to vesicle formation.

Edema   Accumulation of fluid in the tissues (outside the vascular system).

Elephantiasis   Marked swelling of the legs, scrotum, and breast due to the obstruction of lymph vessels.

Emphysema   Loss of elastic tissue of the lungs.

Enantiomers   Mirror images of the pairs of structures.

Epimers   Different configuration of two monosaccharides around one specific carbon.

External respiration   Exchange of oxygen and carbon dioxide occurring between alveolar air and pulmonary capillaries.

Fanconi syndrome   A disorder of proximal convoluted renal tubule leading to deficient reabsorption of phosphate, amino acids and glucose, and large amount of urinary excretion.

Gangrene   Cessation of blood flow of peripheral arteries (e.g., leg) causes decay or death of a part of leg.

Glossitis   Inflammation of the tongue. Tongue appears purplish known as "magenta tongue."

Gluconeogenesis   The conversion of nonglucose molecule to glucose.

Glycogenesis   The process of glycogen formation.

Glycogenolysis   Glycogen breakdown.

Glycolysis   The breakdown of glucose to pyruvate or lactate.

Hallucination   A false sensation that occurs in the absence of a stimulus, e.g., visual, auditory, and tactile hallucination.

Hartnup disease   A hereditary defect that causes defective intestinal absorption and renal tubular reabsorption of tryptophan and thus causes deficiency of niacin.

Hematemesis   Blood vomiting.

Hemiplegia   Paralysis of one side of the body.

Hemoptysis   Coughing up of blood.

Hemosiderosis   Excessive deposition of hemosiderin in the tissues.

Hepatomegaly   Enlargement of liver.

Holoenzyme   The combination of apoenzyme and coenzyme.

Hunger   An intrinsic instinct of sensation to eat food.

Insulinoma   An insulin producing benign tumor of the beta cells of the pancreas.

Internal respiration Oxygen consumption and carbon dioxide production due to mitochondrial respiration.

Iodine value Iodine value of a fat indicates the number of grams of iodine used by 100 g of fat.

Isoenzymes Isoenzymes are different molecular forms of the same enzyme.

Isomers Compounds having the same chemical formula.

Keratomalacia Corneal opacities, necrosis, and ulceration.

Keshan's disease Cardiomyopathy, heart enlargement, and heart failure occur in certain areas of China due to deficiency of selenium in the soil.

Knock-knee (genu valgum) Abnormal curving of the legs results in gap between the ankles when the knees are in contact.

Laurence–Moon–Biedl syndrome Characterized by obesity, short stature, mental retardation, and hypogonadism.

Lipogenesis The synthesis of lipids from amino acids or glucose.

Lipolysis The breakdown of triglycerides into glycerol and fatty acids.

Lipotropic A substance (e.g., methionine or choline) transports fatty acids from the liver to the tissues for the utilization.

Menkes syndrome A genetic disease which is fatal in infants. Signs and symptoms are mental retardation, fragile kinky hair, and convulsions.

Metaplasia Abnormal cell growth.

Micelles The molecular aggregates of hydrophilic and hydrophobic groups in which polar regions line the surface and nonpolar regions line the center. Fatty acids, monoglycerides, and bile salts form the micelles.

Neuritis Inflammation of peripheral nerves.

Nystagmus Involuntary oscillation of eyeball.

Oliguria Abnormally small amount of urine.

Ophthalmoplegia Paralysis of the eye muscle.

Oxidation and reduction Oxidation is the combination of a substance with oxygen or loss of hydrogen or loss of electron. Reduction is the reverse of oxidation.

PAL The ratio of total energy expenditure and BMR.

Papilledema Swelling of the optic disc.

Paranoid personality disorder Suspiciousness, hostility, delusion of sexual infidelity and self-importance.

Paraplegia Paralysis of both legs.

Parkinson's disease Disease of basal ganglia due to the deficiency of dopamine characterized by hypokinesia (for example, mask-like face), static or resting tremor (pill rolling movement of the fingers), and rigidity (contraction of both extensor and flexor muscles).

Pendred's syndrome Goiter occurs due to deficiency of peroxidase that is essential for the utilization of iodine to form thyroid hormones.

Pheochromocytomas Tumor of adrenal medulla causes increased secretion of epinephrine and norepinephrine.

Photophobia Discomfort of the eye due to exposure of light.

Pickwickian syndrome Characterized by obesity and sleep apnea.

Polymer A substance formed by linkages of a large number of molecules.

Prader–Willi Syndrome An autosomal dominant trait due to an abnormality of chromosome 15 characterized by obesity, hypogonadism, and mental retardation.

RDA RDA is the average daily amount of a nutrient required by the body and indicates a margin of safety of a nutrient requirement.

Retinopathy Hypertension may damage or rupture the blood vessels of the retina and cause impairment or loss of vision (for example, diabetic retinopathy).

RQ The ratio of volume of CO_2 production to the volume of O_2 consumption per unit time.

Saponification The hydrolysis of triglycerides by alkali.

Saponification value The number of milligrams of potassium hydroxide required to saponify 1 g of fat.

SDA SDA of food is the energy expenditure due to digestion and absorption.

Steatorrhea Abnormal altered stool which is pale, bulky, greasy, and foul smelling.

Stomatitis Inflammation of the mucous membrane of the mouth.

Strategic memory Ability to work with memories.

Stroke Flow of blood to a part of brain prevented by thrombosis or embolism or rupture of an artery may lead to hemiplegia, coma, and death.

Subacute combined degeneration of the spinal cord Paresthesia (pricking and burning sensation), muscular weakness and impairment of sense of position and ataxia. It occurs due to complication of vitamin B_{12} deficiency.

Surfactant A mixture of various lipids present in the alveoli of lungs.

Telomeres Stretches of DNA at the ends of chromosomes.

Transamination The conversion of one amino acid to a keto acid with simultaneous conversion of another keto acid to an amino acid.

Ventricular fibrillation Asynchronous rapid irregular ventricular contraction which ultimately ceases to pump blood.

Weaning The transition period of liquid and semisolid diet other than breast milk from 6 months to a year.

Wernicke's encephalopathy Characterized by nystagmus, ataxia, neuritis, and loss of memory due to chronic alcoholism and thiamin deficiency.

Wilson's disease A genetic disease which causes low plasma ceruloplasmin, leading to deposition of free copper in various organs, e.g., liver and brain. It causes jaundice, cirrhosis, and symptoms similar to Parkinsonism.

Xerophthalmia Dry, thickened and wrinkled cornea.

Index

© Springer Nature Singapore Pte Ltd. 2019
K. Chakrabarty, A. S. Chakrabarty, *Textbook of Nutrition in Health and Disease*,
https://doi.org/10.1007/978-981-15-0962-9

Printed in the United States
By Bookmasters